The Post-Recombination Universe

NATO ASI Series

Advanced Science Institutes Series

A Series presenting the results of activities sponsored by the NATO Science Committee, which aims at the dissemination of advanced scientific and technological knowledge, with a view to strengthening links between scientific communities.

The Series is published by an international board of publishers in conjunction with the NATO Scientific Affairs Division

A Life Sciences
B Physics

Plenum Publishing Corporation
London and New York

C Mathematical
   and Physical Sciences
D Behavioural and Social Sciences
E Applied Sciences

Kluwer Academic Publishers
Dordrecht, Boston and London

F Computer and Systems Sciences
G Ecological Sciences
H Cell Biology

Springer-Verlag
Berlin, Heidelberg, New York, London,
Paris and Tokyo

Series C: Mathematical and Physical Sciences - Vol. 240

The Post-Recombination Universe

edited by

N. Kaiser

Institute of Astronomy,
Cambridge, U.K.

and

A. N. Lasenby

Mullard Radio Astronomy Observatory,
Cavendish Laboratory,
Cambridge, U.K.

Kluwer Academic Publishers

Dordrecht / Boston / London

Published in cooperation with NATO Scientific Affairs Division

Proceedings of the NATO Advanced Study Institute on
The Post-Recombination Universe
Cambridge, U.K.
July 27 – August 7, 1987

Library of Congress Cataloging in Publication Data

NATO Advanced Study Institute (1987 : Cambridge, Cambridgeshire)
   The post-recombination universe : proceedings of the NATO Advanced
Study Institute held at Cambridge, U.K., July 27-August 7, 1987 /
edited by N. Kaiser and A.N. Lasenby.
      p.   cm. -- (NATO ASI series. Series C, Mathematical and
physical sciences ; vol. 240)
   ISBN-13: 978-94-010-7864-1   e-ISBN-13: 978-94-009-3035-3
   DOI: 10.1007/ 978-94-009-3035-3
   1. Cosmology--Congresses. 2. Radiation, Background--Congresses.
3. Red shift--Congresses. 4. Galaxies--Congresses. 5. Quasars-
-Congresses.   I. Kaiser, N. (Nick), 1954-   . II. Lasenby, A. N.
(Anthony N.), 1954-   . III. Title. IV. Series: NATO ASI series.
Series C, Mathematical and physical sciences ; no. 240.
QB980.N37 1987
523.1--dc19
                                                      88-17583
                                                          CIP

ISBN 978-94-010-7864-1

Published by Kluwer Academic Publishers,
P.O. Box 17, 3300 AA Dordrecht, The Netherlands.

Kluwer Academic Publishers incorporates the publishing programmes of
D. Reidel, Martinus Nijhoff, Dr W. Junk, and MTP Press.

Sold and distributed in the U.S.A. and Canada
by Kluwer Academic Publishers,
101 Philip Drive, Norwell, MA 02061, U.S.A.

In all other countries, sold and distributed
by Kluwer Academic Publishers Group,
P.O. Box 322, 3300 AH Dordrecht, The Netherlands.

TABLE OF CONTENTS

III ABSORPTION LINE SYSTEMS

IV GALAXY CLUSTERING AND LARGE-SCALE STRUCTURE

PREFACE

This volume consists of invited talks and contributed papers presented at the NATO Advanced Study Institute "The Post Recombination Universe" which was held in Cambridge in the summer of 1987.

There have, in recent years, been numerous meetings devoted to problems in observational cosmology. The attention given reflects the exciting rate of development of the subject, and a survey of the proceedings from these symposia reveals that a great deal of emphasis has been given to consideration of the very early universe on the one hand, and to large scale structure in the universe at the present epoch on the other. The theme of this meeting was chosen to complement these efforts by focussing on the state of the universe at quite early times, but at those epochs which are still accessible to direct observations. The meeting provided a broad coverage of the post recombination universe by drawing on experts from a wide variety of fields covering theory, background radiation fields and discrete sources at high redshift. Events in the moderately early universe will have left their mark in a great range of wavebands, from X-rays to the microwave region, and the evolution of the universe can be revealed by studies of the intergalactic medium, gravitational lensing and the abundance and clustering of high redshift sources. All of these subjects received much attention at the meeting, and the papers demonstrate the rich interplay between these areas in the rapidly expanding world of observational cosmology.

The invited talks were intended to give a survey of the state of the art of the subject, whilst retaining a pedagogical flavour. We feel that the contributions collected here have amply succeeded in attaining that goal and that this volume will provide an indispensible resource both for specialists and students.

The contributions have been grouped into six broad subject areas, although the interconnections revealed by many of the papers made the grouping somewhat arbitrary. Within each section we have first presented the invited talks, and have followed these with the contributed papers.

We are grateful to all those who helped with the practical organization of the meeting, and in particular express our thanks to Dr. Michael Ingham (Secretary of the Institute of Astronomy) and Mrs. Alice Julier. The original idea for the meeting came from George Efstathiou and Bob Carswell, and we thank them for launching the enterprise so successfully. The major part of the financial support was provided by NATO, and we would like to express our gratitude to Dr. Craig Sinclair and the NATO Science Committee for all the assistance provided by them. Finally, many thanks to those whose efforts have made this volume possible, above all the invited speakers and contributors whose papers appear below.

Nick Kaiser and Anthony Lasenby

Joan Najita M F Ingham R Flores B Grieger Nck Anderson E Martinez-Gonzalez J Cholaniewski M Banek B Rudak M Rees C Frenk M Davis S Bonametto Shobo Veeraraghavan Albert Stebbins José P S Lemos J L Sanz S J Maddox
X Barcons R Carswell D Alexander M Portilla P C van der Kruit B J Carr D Bennett A Dekel B Espey M Shull I Korner R Blandford T Schramm D Zaritsky S T Myers F Sanchez
R Valdarnini H M P Couchman O Lahav B Jones E Linder L Sparke H Sireussazio R Scaramella A Iovino Y Rephaeli C Hogan R Daly X Zhu R Durrer Y Chu D Crampton P Shaver G T Rixon C Moss
W Saslaw P Coles E Bertschinger W Sutherland A Wolfe J Primack S Refsdal V Antonuccio L Danese N Kaiser S Bridgeman G Efstathiou A Juher A N Lasenby S D M White M Donahue J McDowell D L Clemens P B Lilje

D. Alexander	U. of Glasgow, UK
N. Anderson	Princeton, USA
V. Antonuccio	ISAS Trieste, Italy
J.E. Baldwin	MRAO, Cambridge
X. Barcons	U. of Cantabria, Santander, Spain
D. Bennett	U. of Chicago, USA
J. Bergeron	Institut d'Astrophysique, Paris
E. Bertschinger	MIT, USA
R. Blandford	CALTECH, Pasadena, USA
D. Bond	CITA Toronto, Canada
S. Bonometto	U. of Padua, Italy
B. Boyle	U. of Edinburgh, UK
R. Brandenberger	DAMTP, Cambridge
B. Carr	Queen Mary College, London
R. Carswell	Institute of Astronomy, Cambridge
J. Choloniewski	U. of Warsaw, Poland
Y. Chu	U. of Bonn, FRG
D. Clements	Imperial College, London
P. Coles	U. of Sussex, UK
S. Collin	Institut d'Astrophysique, Paris
P. Corbyn	Queen Mary College, London
H. Couchman	CITA Toronto, Canada
L. Cowie	Institute for Astronomy, Hawaii
D. Crampton	Dominion Astrophysical Observatory, Canada
R. Daly	U. of Boston, USA
L. Danese	U. of Padua, Italy
R. Davies	Jodrell Bank, UK
M. Davis	U. of California at Berkeley, USA
A. Dekel	U. of Jerusalem, Israel
M. Donahue	U. of Colorado, JILA, USA
R. Durrer	U. of Zurich, Switzerland
G. Efstathiou	Institute of Astronomy, Cambridge
A. Evrard	Institute of Astronomy, Cambridge
A. Fabian	Institute of Astronomy, Cambridge
R. Flores	CERN, Switzerland
C. Frenk	U. of Durham, UK
O. Gerhard	Institute of Astronomy, Cambridge
W. Green	Tallahassee, USA
B. Grieger	Hamburger Sternwarte, Hamburg, FRG
R. Griffiths	Space Telescope Science Inst., Baltimore, USA
A. Heavens	Royal Observatory Edinburgh, UK
P. Hickson	U. of British Columbia, Canada
C. Hogan	Steward Observatory, Tucson, USA
A. Iovino	ESO, Munich, FRG
B. Jones	Nordita, Denmark
N. Kaiser	Institute of Astronomy, Cambridge
I. Kovner	Weizmann Institute, Israel
P. van der Kruit	Kapteyn Astronomical Inst., Groningen, Netherlands

H. van der Laan	U. of Leiden, Netherlands
O. Lahav	Institute of Astronomy, Cambridge
A. Lasenby	MRAO, Cambridge
C. Lawrence	CALTECH, Pasadena, USA
P. Lilje	Institute of Astronomy, Cambridge
E. Linder	Stanford University, USA
E. Martinez-Gonzalez	U. de Cantabria, Santander, Spain
J. McDowell	Jodrell Bank, UK
R. McMahon	Institute of Astronomy, Cambridge
F. Melchiorri	Physics Dept., U. of Rome, Italy
S. Morris	Steward Observatory, Tucson, USA
S. Myers	CALTECH, Pasadena, USA
J. Najita	U. of California at Berkeley, USA
J. Ostriker	Princeton, USA
M. Panek	U. of Warsaw, Poland
M. Portilla	Physics Dept., Valencia, Spain
J. Primack	U. of California at Santa Cruz, USA
A. Readhead	CALTECH, Pasadena, USA
M. Rees	Institute of Astronomy, Cambridge
S. Refsdal	U. of Hamburg, FRG
Y. Rephaeli	Tel Aviv University, Israel
G. Rixon	Royal Greenwich Observatory, Sussex, UK
M. Rowan-Robinson	Queen Mary College, London
B. Rudak	Copernicus Astronomical Center, Warsaw, Poland
P. Salucci	ISAS, Trieste, Italy
F. Sanchez	Inst. de Astrofisica de Canarias, Spain
J.L. Sanz	U. of Cantabria, Santander, Spain
W. Sargent	CALTECH, Pasadena, USA
R. Saunders	MRAO, Cambridge
R. Scaramella	ISAS, Trieste, Italy
T. Schramm	U. of Hamburg, FRG
T. Shanks	U. of Durham, UK
P. Shapiro	U. of Texas at Austin, USA
P. Shaver	ESO, München, FRG
M. Shull	U. of Colorado, JILA, USA
H. Sirousse Zia	Inst. Henri Poicare, Paris
L. Sparke	Kapteyn Laboratory, Groningen, Netherlands
A. Stebbins	Fermilab, Chicago, USA
W. Sutherland	Institute of Astronomy, Cambridge
R. Terlevich	Royal Greenwich Observatory, Sussex, UK
V. Trimble	U. of California at Irvine, USA
J. Trümper	Max Planck Institute, Garching, FRG
N. Turok	Imperial College, London
R. Valdarnini	U. of Sussex, UK
H. Vedel	Nordita, Denmark
S. Veeraraghavan	U. of California at Berkeley, USA
S. White	Steward Observatory, Tucson, USA
A. Wolfe	U. of Pittsburgh, USA
D. Zaritsky	Steward Observatory, Tucson, USA
X. Zhu	U. of Bonn, FRG

PROTOGALAXIES

Lennox L. Cowie
Institute for Astronomy
University of Hawaii
2680 Woodlawn Drive
Honolulu, HI 96822

ABSTRACT. We argue that the observed sample of flat-spectrum galaxies seen in recent deep surveys must contain both early disk systems and early spheroid systems in order to match observed number counts if $q_0 = 0.5$. The low average density of neutral hydrogen in damped L_α systems at $z = 2 - 3$ separates the disk formation at $z \lesssim 2$ from spheroid formation at $z \gtrsim 3$. Based on color arguments, we assign the period of spheroid formation to $z \sim 4$. The forming spheroids may ionize the IGM at $z = 2.5 - 3.5$ and contribute the thermal portion of the diffuse X-ray background. If we are indeed seeing the protospheroid population it must form slowly over the local Hubble time. Furthermore, if $q_0 = 0.5$, there must be some subsequent merging to form L_* ellipticals. Searches for L_α emission and absorption together with U band number counts should decide these questions.

1. INTRODUCTION

In this paper we shall argue that we may now have seen the protogalaxy population, that spheroid formation may have taken place at around $z = 4$ and disk formation from $z = 2$ to the present. This hypothesis leads to a large number of testable predictions. We will first outline the arguments in the introduction and then demonstrate them in more detail in the following sections.

In order to distinguish the two categories of galaxy formation which can produce comparable amounts of light and metals, we call the early stages of disk star formation protodisks and the progenitors of the ellipticals and bulges protospheroids.

The traditional model protogalaxy is a protospheroid corresponding to a giant elliptical galaxy collapsing on short ($10^7 - 10^8$ yr) timescales (*e.g.* Meier 1976, Davis 1980). Such an object would be extremely luminous. However, even if we do not assume such short collapse timescales but allow for slower dissipational collapse taking place over the Hubble time at the epoch of formation (*e.g.* Baron and White 1987), a protogalaxy would have an I band magnitude ≤ 27 at $z \leq 7$. (For a closed universe, $I \lesssim 24$.) Subject to the usual caveats of hierarchical merging, low surface-brightness, extinction and of course higher-z formation, these objects can be observed in existing deep red number counts (*cf.* Hall and Mackay 1984, Tyson 1987, Lilly, Cowie and Gardner 1987 [LCG]). In this case the problem is to identify what may be a minority population. The problem is further confused by the fact

1

N. Kaiser and A. N. Lasenby (eds.), The Post-Recombination Universe, 1–18.
© *1988 by Kluwer Academic Publishers.*

that the protodisks may look very like protospheroids unless extensive multicolor data is available.

It is widely accepted that the faint blue galaxies at $B > 22$ are too numerous to be explained by a non-evolving galaxy population and require a major contribution from a cosmologically distant actively star-forming population (e.g. Tinsley 1980, Ellis 1983, King and Ellis 1985). Kron (1980) found that at $B > 23$ there is a significant population of blue sources, while Tyson (1987) has recently demonstrated that very blue sources constitute a substantial fraction of the fainter number counts between J (roughly B) = 24 and 27. There is little choice but to identify these flat-spectrum objects of this latter population as actively star-forming galaxies at moderately high z ($\gtrsim 0.8$) on the basis of color alone. (That is, even an irregular galaxy would appear too red at lower redshift (e.g. Koo 1986a).) They cannot have $z\gtrsim3.5$ because at this redshift the Lyman continuum break passes through the B band. As we shall demonstrate in section 2 using simple and very model-independent arguments about surface brightness and metal production, this blue population alone could easily produce much of the presently-observed metals. A simple corollary of the argument is that disk formation and some spheroid formation may have occurred at $z \sim 3.5$. It is pointed out that protodisks and $z\lesssim3.5$ protospheroids may be hard to distinguish on the basis of magnitudes or blue colors and that ultraviolet colors are required for this purpose. Even more recently, LCG have shown that at $I\gtrsim24$ there exists a large population of 'flat red' spatially extended sources with flat power-law spectra from V to I but with a very steep drop between B and V. The most reasonable explanation for these systems is that they are actively star-forming galaxies lying between $z = 3.5$ and 5 with the Lyman continuum break redshifted to lie between the B and V bands. LCG argue that this is a major part of the protospheroid population. This population may correspond to the turnup in R and I counts versus the turndown in J counts at $J \approx 25$ which is marginally seen in Tyson's (1987) data. LCG also show that there are relatively few extended ultrared sources seen at faint I magnitudes, suggesting that either there is little galaxy formation beyond $z \sim 5$ or that the sources are too faint to be seen in their survey which extends to $I \sim 26$.

As pointed out by Rees during this talk, the suggestion that protospheroid formation takes place at $z \sim 4$ is supported by the recent realisation that significant ionizing sources over and above the observed quasars are required from $z = 2.5$ to $z = 3.5$, both to satisfy the high-z Gunn-Peterson test and to ionize the L_α forest clouds (e.g. Shapiro and Giroux 1987, Bartjlik, Duncan and Ostriker 1987). A natural mechanism for producing the ionization is active star formation in the protospheroids (Bechtold et al. 1987) if the forming spheroids are fairly dust-free and also transparent to the ionizing flux.

In section 3 we turn to considerations of the gas clouds from which the galaxies are forming. It is argued that until galaxy formation is very nearly complete these clouds will be nearly neutral and will have a high cross-section for producing damped L_α absorption against background quasars. We then apply a simple modification of the Gunn-Peterson formalism to Wolfe et al.'s (1986) observations of such damped L_α systems to show that from $z = 1.7$ to $z = 3.2$ (their survey redshift limit) there is too little change in the gas density to allow the protogalactic gas-to-star conversion which would be required to form the blue counts. This is our second conclusion: that the blue count population forms either at $3 < z < 4$ or at $z < 2$. In the former case they would be identified with spheroid formation and in the latter case with disk production. As we have suggested already, both populations may contribute at some level.

We next turn in section 4 to the problem of matching the number count-magnitude relation. Tyson's flat blue spectrum galaxy population is well fitted by a Schechter function with $R_* = 23.9$, as would be expected if galaxies form with a fixed timescale in a relatively narrow redshift interval. We next show that we can understand the luminosity function of the blue counts if they correspond either to spheroid formation alone or to disk formation alone only if the star-formation process is slow (extending over the entire redshift interval) and the universe is extremely open ($q_0 \lesssim 0.1$). In the case of the spheroids, there would also need to be some amount of subsequent merging at the bright end of the luminosity function to form a present-day L_* elliptical galaxy. One method of allowing a higher q_0 is to assign about half the blue population to protodisks and half to protospheroids. The two populations would then have to be distinguished on the basis of ultraviolet color or (more problematically) morphology.

In section 5 we argue that existing $L\alpha$ observations already imply that dust destruction must reduce $L\alpha$ emission from the protospheroids. Weak $L\alpha$ emission should be present in at least some cases but may be difficult to detect making spectroscopic confirmation hard. In section 6 we discuss observations of known extended objects at $z > 2$. Recent observations of faint radio galaxies by Lilly (1987) confirm that (at least for these objects) spheroid formation is more or less complete by $z = 2$ to 2.5. Very high spatial resolution observations of a sample of $z \sim 3$ quasars by Cowie and Hu (1987) are used to measure limits on underlying galaxies and gas clouds and to show that these objects have consumed most of their local gas at this point. Finally the $L\alpha$ companion to the quasar PKS 1614 (Djorgovski et al. 1985, 1987; Hu and Cowie 1987) at $z = 3.2$ is used to argue that the mass units are already considerably too large at this period to require very extensive subsequent merging.

Section 7 is devoted to tests of these suggestions. The simplest prediction is that faint number counts in the near UV should be lower than the B-band counts, a conclusion which follows immediately from the epoch range assigned to the formation of the spheroids. However, there should be comparable numbers of U-faint and U-bright galaxies in the faint blue population with the U-brights corresponding to the disks and the U-faints being the protospheroids. Searches for damped $L\alpha$ absorption in ultrahigh-z quasars ($z \sim 4$) appear to be an excellent route for tracing the evolution of spheroid formation. The protospheroids may also be the dominant contribution to the X-ray background (Bookbinder et al. 1980) and this is also discussed.

2. MODEL-INDEPENDENT RESULTS FROM NUMBER COUNTS

Before proceeding, we need a model-invariant description of the distant active star-forming population. Such a description is quite straightforward. Below a rest wavelength of about 3500 Å the light from a galaxy with significant ongoing star formation is totally dominated by the contributions from short-lived massive stars which are transient on galactic evolution timescales (e.g. Bruzual 1983). The measured B or J magnitude of any distant ($z \gg 0.3$) galaxy population is then a measure of the star formation rate $\dot{M}$ (e.g. Baron and White 1987). Conversion from absolute magnitude to $\dot{M}$ requires knowledge of the quite uncertain initial mass function (IMF) of the stars. However, because the massive stars which dominate the far UV flux are also those which return metals to the galaxy, the conversion from absolute magnitude to the metal mass formation rate $(\dot{M}Z)$ is relatively invariant to the details of the assumed IMF. The conversions may be obtained straightforwardly from spectral synthesis and yield calculations, or from previous calculations such as those of Meier (1976) which are based on the Larson (1974) galactic evolution models. The latter procedure gives a spectrum

$$F_\nu = 2 \times 10^{29} \left\{ (\dot{M}Z)/1 \ \mathrm{M_\odot \ yr^{-1}} \right\} \quad \mathrm{ergs/s \ Hz.} \tag{1}$$

The spectrum is flat because it is dominated by the earliest type stars whose blackbody peak lies just beyond the continuum break.

Equation (1) is very robust, as can be seen by deriving it instead from nucleosynthesis arguments (e.g. Partridge and Peebles 1967, Davis 1980). For, consider that 0.7% of the rest-mass energy of the material that is formed into metals is converted to bolometric luminosity and the resultant spectrum is flat for $0 < \nu < \nu_L$. Then, for 1 $\mathrm{M_\odot \ yr^{-1}}$ of produced metals,

$$F_\nu = 1.2 \times 10^{29} \quad \mathrm{ergs/s \ Hz.} \tag{2}$$

Only about 67% of produced metals are returned and the rest are retained in the stellar remnants, and we then obtain

$$F_\nu = 1.8 \times 10^{29} \left\{ (\dot{M}Z)/1 \ \mathrm{M_\odot \ yr^{-1}} \right\} \quad \mathrm{ergs/s \ Hz.} \tag{3}$$

We adopt equation (3) as our basic equation.

For some purposes, it may be more natural to relate F_ν to $\dot{M}$, the star formation rate, which may be done by assuming a mean yield of returned metals. For a Salpeter IMF (with a spectral index of -1.35) stretching from $0.02 \rightarrow 50 \ \mathrm{M_\odot}$ and using the results of Talbot and Arnett (1973) we obtain a mean yield of 1.4%. Thus

$$F_\nu = 2.5 \times 10^{29} \quad \mathrm{ergs/s \ Hz} \ (\dot{M}/100 \ \mathrm{M_\odot \ yr^{-1}}). \tag{4}$$

However, it must be recognised that equation (4) is quite uncertain because of the assumptions concerning the IMF.

Now, because of the flat-spectrum property, the equivalent visual (AB) magnitude of such a galaxy is wavelength-independent and given by

$$m_{AB} = 2.5 \log \left(F_\nu (1+z)/4\pi \, d_\ell^2 \, f_{\nu 0} \right), \qquad \nu < \nu_L (1+z)^{-1} \tag{5}$$

where $f_{\nu 0} = 3.63 \times 10^{-20}$ ergs/cm^2 s Hz determines the zero-point in the AB magnitude system (Oke 1974) and d_ℓ is the luminosity distance. In Figure 1 we plot this magnitude as a function of redshift for a galaxy with a star formation rate of $100 \ h^{-2} \ \mathrm{M_\odot \ yr^{-1}}$ ($h \equiv H_0/50 \ \mathrm{km \ s^{-1} \ Mpc^{-1}}$) and for one in which $10^{11} \ h^{-2} \ \mathrm{M_\odot}$ is formed in the time $t(z) \, h^{-1}$. This figure makes three points. Firstly, if galaxy formation occurs at $z \lesssim 7$, the active objects must appear in I-band counts extending to 27th magnitude or so, irrespective of cosmology or star formation history. (We shall argue later that intrinsic extinction and hierarchical merging cannot affect this conclusion.) Secondly, there is a very significant q_0 dependence. High q_0 guarantees much brighter objects, both because of the smaller luminosity distances

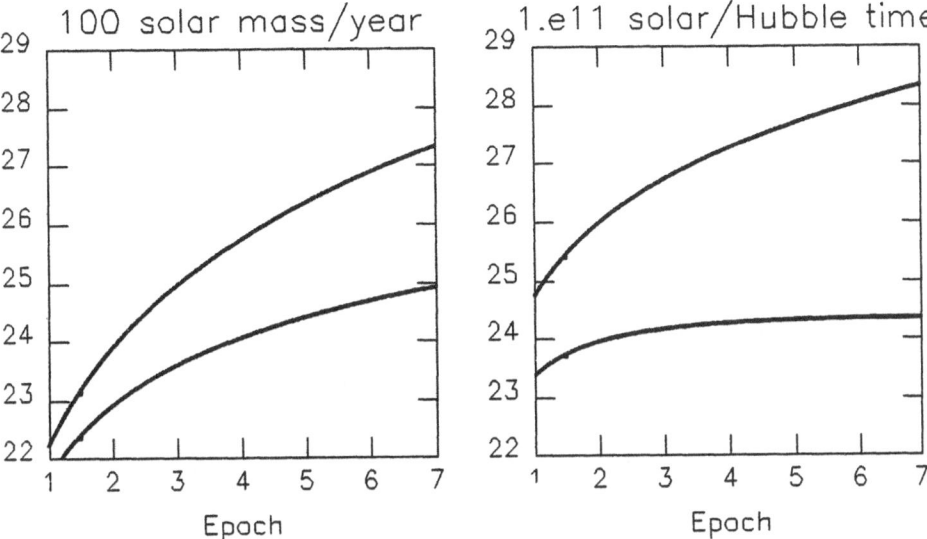

Figure 1 Equivalent visual (AB) magnitude of a galaxy forming 100 $M_\odot$ yr^{-1} (left panel) or 10^{11} $M_\odot$ in a Hubble time at epoch z (right panel). In each case $H_0 = 50$ km s^{-1} Mpc^{-1} and $q_0 = 0$ (upper curve) and $q_0 = 0.5$ (lower curve). The curves scale as $\Delta m = -5 \log h$.

and because of the shorter available timescales. Thirdly, because magnitudes vary extremely slowly with z, a forming population at high z may be indistinguishable from one at low z in colors where the Lyman limit has not passed through the band. This is particularly true for $q_0 = 0.5$.

Turning back to the more model-independent description, in terms of metal formation we can rewrite equation (3) as

$$\varepsilon_\nu = 2840 \quad \text{ergs g}^{-1} \text{ Hz}^{-1} \tag{6}$$

where ε_ν is the energy per unit frequency per unit mass of released metals. The surface brightness of the objects can now be expressed in terms of the present density of the metals they released (ρZ) as (*e.g.* Lilly and Cowie 1987)

$$\begin{aligned} S_\nu &= \frac{1}{4\pi}\, \varepsilon_\nu (\rho Z)\, c \\ &= 2.1 \times 10^{-25} \left\{ \frac{\rho Z}{10^{-34} \text{ g cm}^{-3}} \right\} \quad \text{ergs/cm}^2 \text{ s Hz deg}^2, \quad \nu < \nu_L (1+z)^{-1}. \end{aligned} \tag{7}$$

Equation (7) is extremely powerful because it is independent of cosmology, epoch of formation, details of star formation and galaxy evolution or the form of the IMF. If we can identify a population of distant actively star-forming objects, we may use equation (7) to obtain its contribution to the current metal density. Conversely, for known galactic classes we can use (7) to obtain their surface brightness during any cosmologically distant metal-forming phase.

Table 1 summarizes plausible (though quite uncertain) estimates of the relative light, mass, density and metal density in various galactic components at the present

TABLE 1. ADOPTED LIGHT DENSITIES, MASS DENSITIES AND METAL MASSES

	Fraction Light	Bulge Light Fraction	Blue Light Density (hL/Mpc^3)	Disk $\langle\rho\rangle$ (h^2g/cm^3)	Bulge $\langle\rho\rangle$ (h^2g/cm^3)	Disk $\langle\rho_z\rangle$ (h^2g/cm^3)	Bulge $\langle\rho_z\rangle$ (h^2g/cm^3)
Sab/Sbc	47%	10%	4×10^7	8×10^{-33}	2×10^{-33}	...	...
Scd/Scm	26%	2%	2×10^7	2×10^{-33}	5×10^{-34}	...	...
E/S0	28%	28%	2×10^7	...	7×10^{-33}	...	...
TOTAL	100%	40%	9×10^7	1.1×10^{-32}	10^{-32}	2.2×10^{-34}	1.4×10^{-34}

NOTES: Bulge and light fractions are from the summary of King and Ellis (1985). Total blue light density is from Kirshner, Oemler and Schechter (1979). Mass to light ratios derived from Kent (1987) converted assuming $(M/L_B) = 3h$ for disk and $4h$ for bulge. Metal masses have been calculated using $z = 0.02$ for both the disk and for the massive ellipticals which dominate the bulge light.

time, and the predicted surface brightnesses based on these numbers and equation (7) are placed in the context of recent measurements of the extragalactic background light in Figure 2. The most uncertain value is perhaps the assumed (MZ/L_B) for the spheroids. It is therefore worth noting that this quantity may also be derived from observations of the gas in rich clusters of galaxies since the gas contains much of the metals in these systems. Adopting a metallicity of one-half solar for the gas in Coma and $(M_{gas}/L_B) = 24\ h^{-0.5}$ at $2h^{-1}$ Mpc (Cowie, Henriksen and Mushotzky 1987) gives $(MZ/L_B) = 0.24\ h^{-0.5}$ compared with the value $(MZ/L_B) = 0.08\ h$ adopted in Table 1.

On this diagram we have also placed the surface brightness of number counts in approximately fixed flux ranges of $B = 22 - 27$, $R = 21.5 - 26.5$ and $I = 21.5 - 26.5$ using the data of Tyson (1987). (This data is slightly extrapolated in the I band using the power-law fit to Tyson's counts at brighter magnitudes.) The extremely blue $B - R$ color of the average light in this flux range should be noted. It is compared in Figure 2 with an irregular galaxy (from Coleman, Wu and Weedman 1980) shifted to $z = 1.5$, which would provide a reasonable match to the observed colors. It can be seen directly from this data that a considerable fraction of the light at these magnitudes cannot arise in any local population since the average $B - R$ colors are too blue (Tyson 1987) and much of the blue light must arise in a relatively distant ($z \gtrsim 0.8$) actively star-forming population as we discussed in the introduction. The total J flux of 1.4×10^{-24} ergs/cm^2 s Hz deg^2 would correspond to a present metal density of 7×10^{-34} g/cm^3, if it were fully formed in such a population, but of course there is a major local contribution here which should not properly be included. If we restrict ourselves to very blue ($J - R < 0.4$) objects we can find the fractional population of the total number counts which may be directly allocated to distant star-forming galaxies. This is summarized in Figure 3, based on data from Koo (1986a) and Tyson (1987). The blue population emerges from being a trace population at $J = 22$ to being a substantial fraction of the counts at $J = 24$. As can be seen in Figure 3, this blue population is quite well fitted by a Schechter function with $\alpha = -1.25$ and $J_* = 24.25$ (or $R_* = 23.9$). These very blue galaxies constitute a large fraction of the excess number counts over a non-evolving galaxy population.

Restricting ourselves to this very blue population, we find that it contains $\sim 30\%$ of the blue light from $J = 22$ to $J = 27$ or about 4×10^{-25} ergs/cm^2 s Hz deg^2.

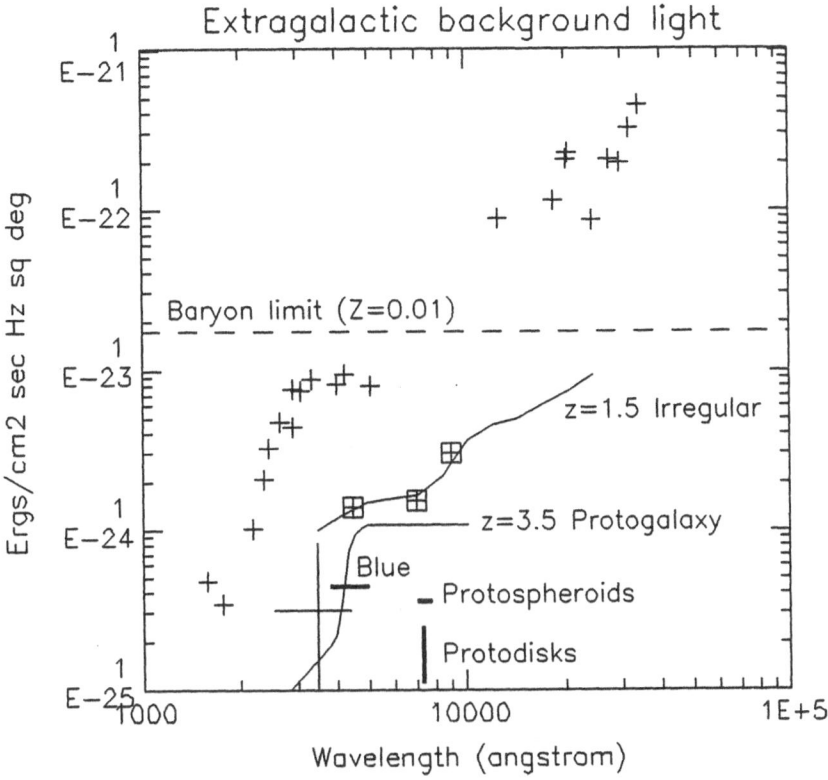

Figure 2 Predicted background light from $z > 0.8$ disks (solid line shows range) and from protospheroids (both scaling as h^2) is compared with upper limits on the extragalactic background light (Hurwitz *et al.* 1986, Feldman *et al.* 1981, Maucherat-Joubert *et al.* 1980, Paresce and Jacobsen 1980, Dube *et al.* 1977, Spinrad and Stone, 1979, Masumoto *et al.* 1987) shown as crosses and with the light of counts between $J = 22$ and 27, $R = 21.5$ and 26.5 and $I = 21.5$ and 26.5 (Tyson 1987) (open boxes). The heavy solid line (Blue) shows the contribution of the blue sources in the B band. Also shown are the continuum shapes (arbitrary normalisation) of a $z = 3.5$ protospheroid (Meier 1976) and a $z = 1.5$ irregular (Coleman, Wu and Weedman 1980). The large cross is Bartjlik *et al.*'s (1987) estimate of the Lyman edge flux required to ionize the L_α forest. The dashed line shows the upper bound on the surface brightness, based on the Big-Bang nucelosynthesis limit on baryons, assuming these have an average metallicity of $Z = 0.01$.

From equation (7) this corresponds to a current metal density of 2×10^{-34} g cm^{-3}. As can be seen from Table 1, both disk and spheroid populations contain enough metals to account for this blue population within the uncertainty in the metal mass to light ratios and the Hubble constant. However, it should be noted that we have overestimated the available disk metals for blue count formation in Table 1 since, from the age-metallicity relation in our own disk, only 25% to 60% of the metals formed before $z = 0.8$ (Twarog 1980, Carlberg *et al.* 1985). Including this correction we show in Figure 2 the plausible contributions of protospheroids and

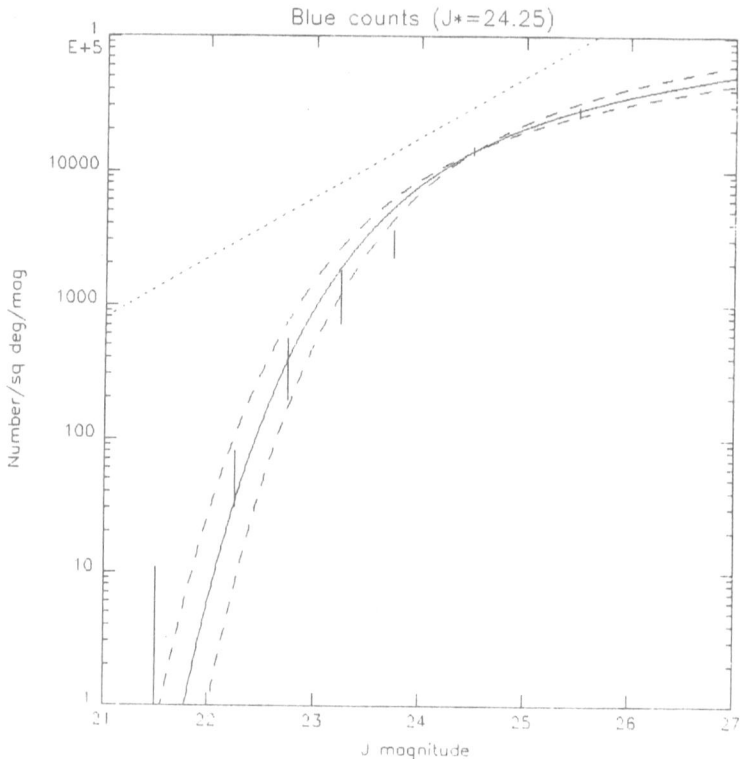

Figure 3 Blue counts ($J - R < 0.6$) shown as error bars versus J magnitude, while dotted line shows total counts. Solid line shows Schechter function fit for $J_* = 24.25$, and dashed lines are for $J_* = 24.0$ and $J_* = 24.5$.

protodisks to the blue count surface-brightness. Within reasonable uncertainty in the metal density and the Hubble constant, the blue counts may be formed either by a protodisk population (disks at $z \gtrsim 0.8$) or by a protospheroid population or by both types of sources.

While the blue population may be produced either by modest-z (~ 1) protodisks or higher-z protospheroids, LCG have recently identified another population which can only be assigned to high-z star-forming galaxies. LCG used BVI colors of a red-selected population ($I \lesssim 26$) to show that at $I \gtrsim 24$ there exists a flat-spectrum $V - I$ population which is very faint in B. This population may produce the excess R and I counts in Tyson's (1987) sample and would correspond to $z = 3.5 - 5$ protospheroids with a surface brightness of $\sim 2 \times 10^{-25}$ ergs/cm^2 s Hz deg^2. This population only begins to be seen at $I > 24$ and we may be missing many of the fainter objects. Thus the surface-brightness above may underestimate the contribution of this population. However, it could be that as much as half of the protospheroid population is directly seen here. If a comparable amount of light is contributed to the flat blue spectrum sources by a $z \lesssim 3.5$ protospheroid population we could easily account for all the observed spheroid metal production but, as we shall discuss below, it seems probable that the total protospheroid light is significantly higher (by a factor of two or more) and dominated by faint-end sources.

LCG identify relatively few extended and very red sources (very large $V - I$) so that any protospheroids at $z > 5$ would have to be very faint. It is clear from the discussion so far that U-band colors constitute the crucial diagnostic between a $z \lesssim 3.5$ protospheroid component and a lower-z protodisk population. Unfortunately, because of previous instrumental constraints, there is relatively little U-band data. The most useful data is the four-color work of Koo (1986a) which, however, extends only to $J = 23.5$ and suffers serious selection effects on color-selected samples at the fainter end. There are 31 objects in the SA68 population with $22.5 < J < 23$ and $J - N < 0.4$, of which 15 have $U - F < 0.4$ and 16 have $U - F > 0.4$. In comparison, Koo's no-evolution simulations show 12 objects with $U - F < 0.4$ and 5 with $U - F > 0.4$. There are too few objects for this result to be compelling and the data is subject to selection effects against U-faint objects, but the excess of U-faint objects over the simulated results allows the possibility of a small population of U-faint blue galaxies which might be associated with a protospheroid population at $z = 3 - 3.5$. It is clearly essential to resolve this issue with deep U-band studies.

Finally, the requirement of providing an ionizing source for the IGM at $z = 2.5 - 3.5$ allows us to make an alternative estimate of the $z \sim 4$ protospheroid population. A model protospheroid transparent to ionizing radiation at $z = 3.5$ is shown in Figure 2 (Meier 1976) and compared with the ionizing flux required at the Lyman edge (Bartjlik et al. 1987). We can see at once from the figure that the required I band surface brightness is in the range $0.5 - 4 \times 10^{-24}$ ergs/cm^2 s Hz deg^2. This would suggest that we are counting only a fraction of the protospheroids in the flat red sources and the fraction of the flat blue population assigned to protospheroids rather than protodisks corresponding to the bright end population. This is reasonable if I_*(protospheroid) lies around 24.

A corollary of this argument is that the galaxies must be reasonably transparent and dust-free. Furthermore, we require a spheroid (MZ/L_B) of $0.07 - 0.6\ h^{-1}$ to produce the required surface brightness compared with the value of $0.08\ h$ estimated in Table 1. Again these values (at least at the lower end) are reasonable within the uncertainties in metal density and Hubble constant.

3. DAMPED LYMAN ALPHA ABSORPTION

As Wolfe et al. (1986) have recently shown, early disks or protodisks should be nearly neutral and should easily be seen as damped L_α absorbers in the spectra of background quasars. This is also true of a protospheroid gas cloud, provided star formation does not take place extremely rapidly. To show this, let us suppose for the moment that a large fraction of the blue and red flat spectrum objects do correspond to slowly-forming spheroids at $z \sim 4$. If we adopt $(MZ/L_B) = 0.08\ h^{-1}$ then a current L_* galaxy with $L_{*\ Blue} = 3.9 \times 10^{10}\ h^{-2}$ L$_\odot$ would have produced $3 \times 10^9\ h^{-3}$ M$_\odot$ of metals in the formation of the blue counts, and would have initially contained $\sim 2 \times 10^{11}\ h^{-3}$ M$_\odot$ of baryons. If now the galaxy formed this gas into stars over a cosmological timescale (say from $z = 4$ to $z = 3$) rather than as an impulsive burst (when the gas would be neutral prior to onset and abruptly consumed) we have from equation (3) that the bolometric luminosity is given by $1.7 \times 10^{45}\ h^{-2}\ \tau_9^{-1}$ ergs/s where τ_9 is the star formation timescale. The ionizing flux would then be about 10% of this value, or approximately $8 \times 10^{54} \tau_9^{-1}$ ionizing photons. Much of this ionizing flux must escape such a galaxy since, to avoid dissipative collapse, the gas collision-time must be long and hence the gas covering factor small. (This is of course also necessary from the arguments of the previous section.) However, we shall adopt the above number as an upper limit to the ionization rate in the galaxy. Now, measured isophotal angular sizes of around $1\overset{''}{.}5$ for members of the faint galaxy population (LCG) imply that, irrespective of cosmology, we are seeing a population with $R < 30\,h^{-1}$ kpc and a mean gas density $> 6 \times 10^{-2}$ cm^{-3}. The recombination rate then exceeds $3 \times 10^{54}\,h^{-3}$ for a gas at 10^4 K. Because the gas is likely to be clumped and centrally condensed we can infer that it must be nearly neutral. The average column density is $> 10^{22}\,h^{-1}$ cm^{-2} and when a clumped portion of the gas is intersected a large damped L_α line will be formed. The interception probability is moderately large because the observations show that the galaxies cover a few percent of the sky.

We can now use Wolfe $et\ al.$'s (1986) damped L_α survey to trace the evolutionary history of the gas in galaxies. The number of confirmed damped L_α absorbers per unit redshift in several wavelength intervals, and the quantity dN_H/dz, can be inferred from the Wolfe $et\ al.$ results. (Inclusion of the unconfirmed systems which have much lower equivalent widths $[W]$ does not significantly change these numbers since for each system the column density is proportional to W^2.) This quantity dN_H/dz may be translated to give the current mass density of material which lay in high-column density neutral gas at epoch z, from

$$n_{H0} = (1+z)^{-3} \frac{dz}{d\ell} \frac{dN_H}{dz} = \frac{(1+2q_0 z)^{\frac{1}{2}}}{(1+z)} \frac{H_0}{c} \frac{dN_H}{dz}. \tag{8}$$

The derived n_{H0} is shown in Figure 4, where we also show (dashed line) the current density of the disks (and roughly of the spheroids) and Twarog's (1980) model of disk-gas density evolution based on the age-metallicity relation in our own disk. (The large uncertainty in n_{H0} reflects the small number of detected systems.) The Twarog model is placed on the graph assuming $H_0 = 50$, $q_0 = 0.02$. A further upper limit at $z = 3.2$ is based on searches for gas around high-z quasars (Cowie and Hu 1987; discussed further in section 6). Figure 4 shows that the average density of high gas column density systems, whether disks or protospheroids, is rising or at most constant as z drops from 3 to 2. Thus the galaxies are not converting gas rapidly to stars at this time but rather seem to be accumulating gas, presumably by infall.

While Figure 4 is, of course, essentially schematic, it is very suggestive of the following picture. Firstly, spheroid formation is mostly complete by $z = 3$, at which point there is very little gas left in the central regions ($R \lesssim 30$ kpc) of the galaxies. The gas may have been fully consumed in star formation or expelled by supernova energy injection. Gas now begins to infall to form the gaseous protodisk which initiates star formation at about $z = 2$ and begins to make the gas into stars. Ongoing gaseous infall continues to fuel the disks until the present time. In particular, Songaila, Cowie and Weaver (1987) have recently used measured distances and crude metal estimates to suggest that the high velocity neutral hydrogen clouds in our own galaxy are metal-deficient material which, infalling at a constant rate, would have formed the dynamical mass of the disk in 10^{10} yr. Thus we see, consistent with

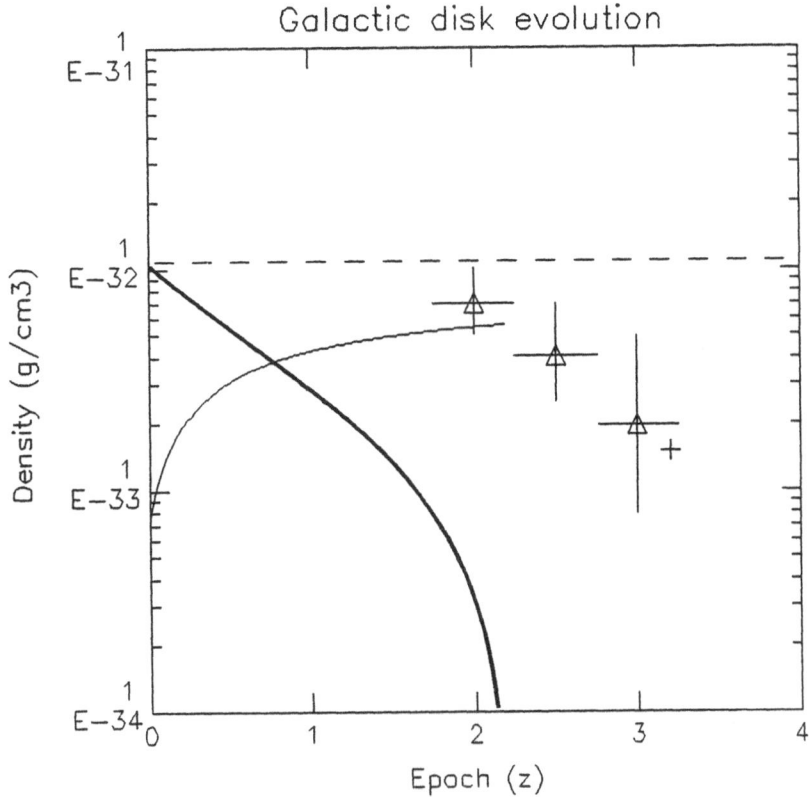

Figure 4 Evolution of average neutral hydrogen density (scaling as h and with q_0 chosen to be small) in high column density systems versus z is shown by triangles with crude error bars based on the number of detected systems. (Points derived from data of Wolfe *et al.* 1986.) Also shown (solid curves) are the evolution of our disk star and gas densities versus z from the age-metallicity relationship (Twarog 1980) with a constant star formation rate, a disk age of 14 billion years, and assuming $H_0 = 50$, $q_0 = 0.2$. The long dashed lines show the assumed average local density in disks (scaling as h^2) from Table 1. The upper limit at $z = 3.2$ is from quasar observations of Hu and Cowie (1987).

our previous discussion, that a protodisk contribution to the blue counts may occur between $z = 0.8$ and $z = 2$ while a protospheroid contribution would have to occur prior to $z = 3$.

4. CONSTRAINTS ON q_0

Let us first postulate that the faint blue counts correspond solely to early disk formation between $z = 0.8$ and $z = 2.5$ (*i.e.* there are no high-z protospheroids in

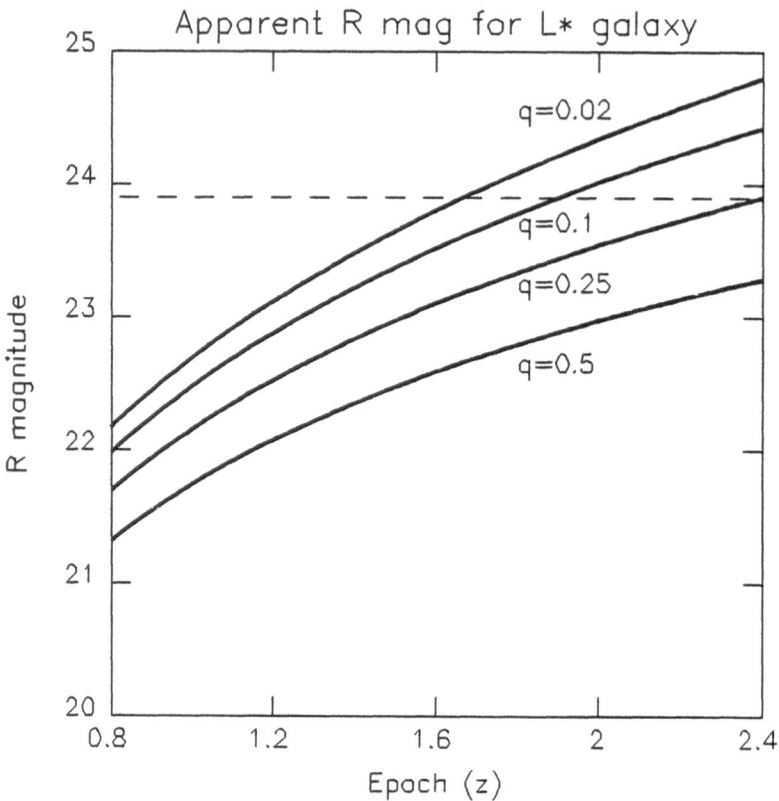

Figure 5 R magnitude for an L_* galaxy forming stars at a constant rate after $z = 2.5$. The dashed line shows the measured $R_* = 23.9$ for the flat blue population.

this population). We can now try to understand the observed surface density and luminosities of the objects. Since the surface brightness is already adjusted by fixing the metal mass to light ratio we need match only one of these parameters. This is most easily considered as the critical magnitude in the Schechter function which we found in section 2 to be $R_* = 23.9$. In the absence of merging, which we would not expect to be important for the disks, this R_* corresponds to a present-day galaxy with a blue $L_* = 3.9 \times 10^{10} h^{-2} L_\odot$. Using a metal mass to blue light ratio of $(MZ/L) = 0.09 h^{-1}$ required to match the observed blue count surface-brightness (Figure 2 and equation [7]), assuming 60% of the disk star formation took place before $z = 0.8$, we obtain R_* from

$$R_* = 2.5 \ \log \left\{ \frac{(1+z) F_{\nu*}}{4 \pi d_\ell^2 f_{R0}} \right\} \tag{9}$$

where we take $f_{R0} = 3.0 \times 10^{-20}$ ergs/cm^2 s Hz. R_* is independent of the Hubble constant and depends primarily on τ and q_0.

Now R_* is maximised and galaxies made as faint as possible by assuming that star formation occurs constantly over the maximum available interval (*i.e.* uniformly

13

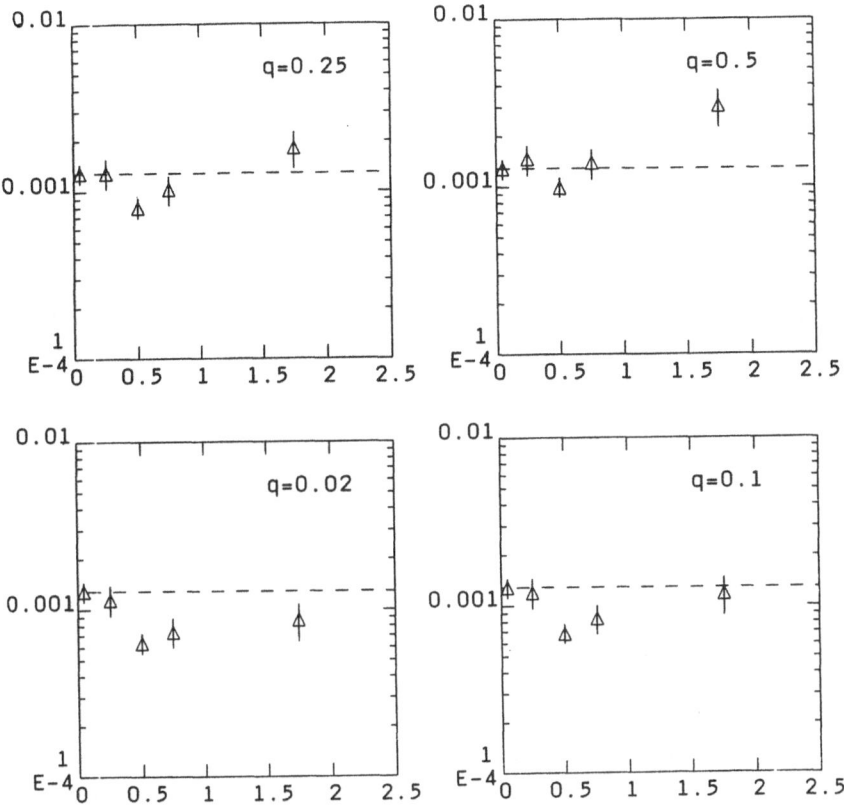

Figure 6 Normalisation of the Schechter function (ϕ_*) as a function of z and q_0. The extreme right-hand point is based on assuming that the blue counts occupy the volume from $z = 0.8$ to 2.5. The remaining points are from Loh and Spillar (1986). The correct geometry should give a constant ϕ_* shown by dashed line normalised to $z = 0$ point.

from $z = 0.8$ to $z = 2.5$) and that τ corresponds to this timescale. The result of such a calculation is shown in Figure 5; it can be seen at once that R_* is extremely sensitive to q_0 and typical values over the z-range are brighter than the measured value for $q_0 \gtrsim 0.1$. The sensitivity to q_0 arises primarily from the smaller luminosity distance and secondarily from the shorter available times at large q_0.

This test is in fact equivalent to the cosmological volume measurement made by Loh and Spillar (1986) though *a priori* it gives a diametrically opposite result for the value of q_0. This may be seen more directly by using the normalization of the Schechter function in Figure 3 together with the q_0-dependent volume between $z = 0.8$ and $z = 2.5$ to obtain the number density of galaxies in this region as a function of q_0. These results are shown in Figure 6 and the ϕ_* determined should equal the local number density and Loh and Spillar's points. This gives the same conclusion that if all the blue flat spectrum sources are protodisks between $z = 0.8$ and $z = 2.5$, then $q_0 \ll 0.5$. From Figure 6 we can see that if $q_0 = 0.5$ less than half the flat blue sources can be such protodisks.

If we now assume instead that half the blue flat spectrum sources are protospheroids between $z = 3$ and 3.5 we can apply the same methods to obtain R_* for this population. Again, a high q_0 would produce a brighter R_* than is observed. In this case, if we wish to have $q_0 = 0.5$, the simplest explanation may be to invoke a small amount of merging to form the bright end of the spheroid population. If an L_* elliptical were the result of a merger of four subunits, we would obtain consistency with $q_0 = 0.5$. (This result fits rather neatly with the break that occurs at roughly this fraction of L_* between lower-luminosity ellipticals and bulges, which have rotationally supported ellipsoidal figures, and giant ellipticals, supported by an anisotropic distribution function [Davies et al. 1983].) In addition, given formation at $z = 3.5$ and $q_0 = 0.5$, protospheroids cannot form stars rapidly on the gravitational collapse time but must burn their gas to stars over cosmological timescales (Baron and White 1987) otherwise R_* is again too bright. If $q_0 \ll 0.5$, faster collapse is possible only if there is significant subsequent merging. To some extent this merging is conceptually equivalent to slow star formation.

5. LYMAN ALPHA EMISSION FROM THE PROTOSPHEROIDS

Let us now assume that there is a significant population of protospheroids at $z_{form} \sim 4$, roughly described by a Schechter function with a continuum f_* flux of 7×10^{-19} ergs/cm^2 s Å at redshifted L_α at around 6000 Å, and a surface brightness of $\sim 10^{-13}$ ergs/cm^2 s Å deg^2 based on the required ionizing flux (cf. section 2). The surface brightness has a multiplicative uncertainty of about a factor of three (Figures 2 and 3). Now the equivalent width of the L_α line is $\sim 45(1 + z_{form})$ Å if half of the ionizing flux is absorbed by the protogalactic cloud (Meier 1976). Hence, in the absence of extinction (either extrinsic or intrinsic) the L_α flux is described by a Schechter function with $\ell_* = 1.5 \times 10^{-16}$ ergs/cm^2 s and a surface brightness of 2×10^{-11} ergs/cm^2 s deg^2. This is shown in Figure 7, where it is compared with our own previously unpublished L_α searches. (Details are given in the caption.) Very similar limits have been obtained by Koo (1986b) and by Pritchet and Hartwick (1987), while similar conclusions would be inferred from recent searches for L_α from quasars (see Spinrad 1987 for a summary).

In order to reconcile the data with the above prediction it is necessary to invoke some extinction or to postulate that our L_α value is too high because a smaller f action of the ionizing flux is absorbed. Only a small amount of dust is required to reduce f_* substantially but this does require that the galaxy have formed significant metals ($Z \sim 10^{-4}$) so that around 1% of the galaxies (the youngest) would still be bright in L_α. One simple model which could be consistent with the data would be to assume that slightly higher-z and younger protospheroids which have not yet formed significant amounts of metals, and hence are dust-deficient, are just slightly too faint to have been seen in L_α, while nearer and therefore brighter protospheroids have higher dust column densities and are suppressing L_α. At this point, intensive spectroscopic study of faint flat-spectrum sources may provide the small increase in sensitivity needed to see the L_α line in the more metal-deficient spheroids.

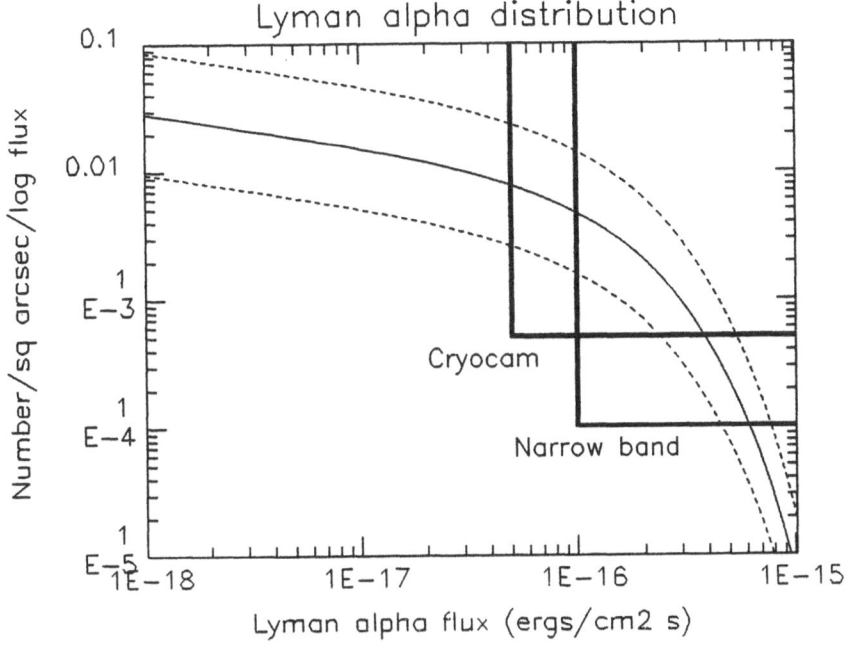

Figure 7 Predicted L_α flux distribution (narrow solid line) and range (dashed lines) as outlined in the text. Thick solid lines show region excluded by observation (to upper right). The 'cryocam' data is based on 2000 $\square''$ of 2 hr exposures with the cryocam spectrograph at the Kitt Peak 4m using a 2.7'' slit width. The covered z range is from 3.5 to 5. The narrow-band data is based on 2 hr exposures of 16 $\square'$ fields taken through 60 Å back-to-back filters with $\lambda \sim 5800$ Å ($z \sim 4$) and $\lambda \sim 7000$ Å ($z \sim 5$) at the prime focus of the Kitt Peak 4m. Two fields were analysed in detail for each z. The area is reduced by the fractional wavelength coverage (120 Å) versus the wavelength spread of the PGs, assumed to be 1200 Å ($\Delta z = 1$) giving approximately 10000 $\square''$ at each of the two redshifts. More recent searches at $\lambda \sim 5200$ Å ($z \approx 3.2$) give similar flux limits over a very much larger area. In all cases the flux limits are subjective estimates of the brightest 4'' × 4'' object which could have escaped detection.

6. HIGH-REDSHIFT EXTENDED OBJECTS

We are now beginning to accumulate observations of a number of $z \gtrsim 2$ galaxies and clouds and, while individually these objects are probably quite peculiar, (and indeed may be highly misleading if naively interpreted) they can, if carefully used, provide valuable diagnostics of galaxy formation history.

In this respect the most interesting cases of all may be Lilly's (1987) studies of faint members of the 1 Jy radio sample. On the basis of IR colors (rest-frame visual) members of this population appear to be very luminous galaxies at $z \sim 2$. However, visual photometry (rest-frame UV) suggests that the objects are forming stars relatively weakly at this epoch. The simplest interpretation of these systems is that they are old spheroidal galaxies which formed at $z \gg 2$. This is consistent with the present picture.

A second object of considerable interest is the quasar companion to PKS 1614 at $z = 3.2$ (Djorgovski *et al.* 1985, 1987, Hu and Cowie 1987). This is a bright L_α cloud with an extremely faint extended underlying continuum. As Hu and Cowie point out, this is almost certainly a gas cloud ionized by the quasar with a small amount of star formation probably induced by the tidal interaction. In our picture this object could be a formed spheroid which is just beginning to reaccumulate disk gas as outlined in Figure 4. More importantly, it already contains a substantial mass ($\sim 1.5 \times 10^{10}$ $M_\odot$ of ionized gas alone) suggesting that mass units are already quite large at this time. This in turn means that only a small amount of subsequent hierarchical merging (perhaps a factor of ten at most) can occur.

It is also of considerable interest to place limits on the galaxies which almost certainly underly high-z quasars. Recently Hu and Cowie (1987) have used very high spatial resolution ($0''.5$) images of a sample of $z = 3.2$ quasars to search for massive star formation in the underlying galaxy, or for L_α emission from gas in the galaxy ionized by the underlying quasar. They find $m_V > 23.5$ in a sample of three quasars and $M_{gas} \lesssim 10^{10}\ h^{-2}\ M_\odot$. If the underlying galaxies are around L_* in luminosity, M_{gas} is less than 10% of the baryonic mass at this point. This constraint has been placed on Figure 4 and is consistent with the Wolfe *et al.* points. The visual magnitude limit when combined with Figure 1 suggests that the star formation rate is less than 100 $M_\odot \mathrm{yr}^{-1}$ for $q_0 = 0.5$ which is not yet a significant constraint. Deeper observations are required.

7. SUMMARY

We can summarize as follows.

(1) Protospheroid formation occurs at $z = 3 \rightarrow 5$ and contributes a flat red spectrum population with very large $B - V$ colors ($z > 3.5$) together with a contribution to flat $B - V$ sources ($z < 3.5$). The latter are easily confused with the protodisks. This population onsets at $I \approx 24$. If, as seems likely, the protospheroids are responsible for ionizing the IGM at this epoch, the sources which are seen contribute only a fraction of the required light and must be the bright-end tail of the population. For $q_0 = 0.5$ we require slow galaxy formation over the local Hubble time together with some merging to form the brightest L_* galaxies.

(2) Disk formation occurs at $z = 0 \rightarrow 2$ and higher z ($\gtrsim 0.8$) protodisks constitute a significant fraction of the blue flat spectrum sources. However, if $q_0 = 0.5$, the protodisks must only be a fraction ($\lesssim 0.5$) of the flat blue sources.

The critical test is now to obtain U band images of the sources since high-z protospheroids should be faint in the U band while protodisks should be bright. On the basis of the present arguments we might expect both populations to be present. If they are not, this will provide a strong constraint on q_0.

If protospheroid gas clouds exist at $z < 4$ they should appear as strong damped L_α signatures in the spectra of high-z ($\gg 3$) quasars. Since the protospheroids cover a few percent of the sky this test requires a large sample of high-z quasars. The protospheroids may be detectable in L_α emission but we know already from L_α surveys that this emission is quite weak.

Finally we note some attractive features of the scenario. The re-formation of the gas disks at $z \sim 2 - 3$ naturally predicts a quasar peak at these redshifts if quasars are formed by the tidal interaction of gas-rich galaxies or by the infall of the gas itself. At higher redshifts the gas has not re-formed in the galaxy while at lower redshifts it has been consumed into stars. However, sporadic quasar activity may be present out to the earliest epoch of galaxy formation at $z = 4 - 5$. A second interesting point is that hot winds at 100 KeV in the protospheroids at $z = 3 - 4$ can naturally explain the observed 20 – 30 KeV temperature of the local diffuse X-ray background (Bookbinder *et al.* 1980). (Note that after subtraction of the quasar contribution the temperature of the residual 'thermal' component is lower than previously assumed [*e.g.* Boldt and Leiter 1987].) The number density of the protospheroids is more than high enough to satisfy existing smoothness constraints (Hamilton and Helfand 1987) but a major difficulty is that the slow formation of the protospheroids results in winds which are too diffuse to release enough of the injected supernova energy. (A few percent of this energy must be radiated.) An alternative possibility is Boldt and Leiter's suggestion of low-level AGNs placed in the protospheroids, but this has no natural explanation for the temperature.

Many of the ideas here were initially raised in conservations with Toni Songaila. Martin Rees raised the question of protogalactic ionization while Jerry Ostriker questioned earlier interpretations of the metal production and made us carefully reconsider the relative contributions of spheroids and disks. Most of the interpretation of the observational material and unpublished data is collaborative work with Esther Hu, Simon Lilly and Jon Gardner. This paper was partially supported by NASA grant NAGW 959.

REFERENCES

Baron, E. and White, S. D. M. 1987, *preprint.*

Bartjlik, S., Duncan, R. C. and Ostriker, J. P. 1987, *preprint.*

Bechtold, J., Weymann, R. J., Lin, Z., and Malkan, M. 1987, *Ap. J.,* **315**, 180.

Boldt, E. and Leiter, D. 1987, *preprint.*

Bookbinder, J., Cowie, L. L., Krolik, J., Ostriker, J. P. and Rees, M. 1980, *Ap. J.,* **237**, 647.

Bruzual, G. 1983, *Ap. J.,* **273**, 105.

Carlberg, R. G., Dawson, P. C., Hsu, T., and VandenBerg, D. A. 1985, *Ap. J.,* **294**, 674.

Coleman, G. D., Wu. C. C and Weedman, D. W. 1980, *Ap. J. Suppl.,* **43**, 393.

Cowie, L. L., Henriksen, M. and Mushotzky, R. 1987, *Ap. J.,* **317**, 593.

Cowie, L. L. and Hu, E. M. 1987, *in preparation.*

Davis, M. 1980, in *Objects at High Redshift,* ed. G. O. Abell and P. J. E. Peebles (Dordrecht: Reidel), p.57.

Davies, R. L., Efstathiou, G., Fall, S. M., Illingworth, G. and Schechter, P. L. 1983, *Ap. J.,* **266**, 41.

Djorgovski, S., Spinrad, H., McCarthy, P. and Strauss, M. A. 1985, *Ap. J. (Letters),* **299**, L1.

Djorgovski, S., Strauss, M. A., Perley, R. A., Spinrad, H. and McCarthy, P. 1987, *A. J.,* **93**, 1318.

Dube, R. R., Wickes, W. C. and Wilkinson, D. T. 1977, *Ap. J. (Letters),* **215**, L51.

18

Ellis, R. S. 1983, in *The Origin and Evolution of Galaxies*, eds. B. J. T. Jones and J. Jones (Dordrecht: Reidel), p. 255.

Feldman, P. D., Brune, W. H. and Henry, R. C. 1981, *Ap. J. (Letters)*, **249**, L51.

Hall, P. and Mackay, C. D. 1984, *M.N.R.A.S.*, **210**, 979.

Hamilton, T. T. and Helfand, D. J. 1987, *Ap. J.*, **318**, 93.

Hu, E. M. and Cowie, L. L. 1987, *Ap. J. (Letters)*, **317**, L7.

Hurwitz, M., Martin, C. and Bowyer, S. 1986, *Adv. Space Res.*, **6**, 69.

Kent, S. M. 1987, *A. J.*, **93**, 816.

King, C. R. and Ellis, R. S. 1985, *Ap. J.*, **288**, 456.

Kirshner, R., Oemler, A. and Schechter, P. 1979, *A. J.*, **84**, 951.

Koo, D. 1986a, *Ap. J.*, **311**, 651.

Koo, D. 1986b, in *Spectral Evolution of Galaxies*, eds. C. Chiosi and A. Renzini (Dordrecht: Reidel), p. 419.

Kron, R. G. 1980, *Ap. J. Suppl.*, **43**, 305.

Larson, R. 1974, *M.N.R.A.S.*, **166**, 585.

Lilly, S. J. 1987, *in preparation.*

Lilly, S. J. and Cowie, L. L. 1987, in *Infrared Astronomy with Arrays*, eds. C. G. Wynn-Williams and E. E. Becklin (Honolulu: U. H. Inst. for Astronomy).

Lilly, S. J., Cowie, L. L. and Gardner, J. 1987, *in preparation.*

Loh, E. D. and Spillar, E. 1986, *Ap. J. (Letters)*, **307**, L1.

Masumoto, T., Akiba, M. and Murakami, H. 1987, *preprint.*

Maucherat-Joubert, M., Deharveng, J. M. and Cruvellier, P. 1980, *Astr. Ap.*, **88**, 323.

Meier, D. 1976, *Ap. J.*, **207**, 343.

Oke, J. B. 1974, *Ap. J. Suppl.*, **27**, 21.

Paresce, F. and Jakobsen, P. 1980, *Nature*, **288**, 119.

Partridge, R. B. and Peebles, P. J. E. 1967, *Ap. J.*, **147**, 868 and *Ap. J.*, **148**, 377.

Pritchet, C. J. and Hartwick, F. D. A. 1987, *Ap. J.*, **320**, 464.

Shapiro, P. R. and Giroux, M. L. 1987, *preprint.*

Songaila, A., Cowie, L. L., and Weaver, H. 1987, *submitted to Ap. J.*

Spinrad, H. and Stone, R. P. S. 1979, *Ap. J.*, **226**, 609.

Spinrad, H. 1987, in *3rd IAP Astrophysics Meeting : 'High Redshift and Primeval Galaxies'.*

Talbot, R. J. and Arnett, W. D. 1973, *Ap. J.*, **186**, 51.

Tinsley, B. 1980, *Ap. J.*, **241**, 41.

Twarog, B. A. 1980, *Ap. J.*, **242**, 242.

Tyson, J. A. 1987, *preprint*

Wolfe, A., Turnshek, D. A., Smith, H. E. and Cohen, R. D. 1986, *Ap. J. Suppl.*, **61**, 249.

THE SEARCH FOR HIGH REDSHIFT QUASARS

David Crampton
Dominion Astrophysical Observatory
5071 W. Saanich Rd.,
Victoria, B.C.,
V8X 4M6, Canada

ABSTRACT. The techniques used for the detection of high redshift quasars are reviewed and compared. There is a gradual decline in the space density of quasars with redshift for z > 2. The space density for quasars with z > 3.4 is estimated to be ~0.2 deg^{-2} for R < 19.5 and the slope of the luminosity function appears to be nearly flat. Surveys of large areas, at least 100 square degrees, are required to obtain the necessary statistics to define the form of the density and luminosity functions at early epochs.

1. INTRODUCTION

The extremely high luminosity of quasars and their existence over a wide range of redshifts make them powerful tools with which to study the post-recombination universe. The early surveys suggested that the space density of quasars of a given luminosity increased dramatically and continuously with look-back time, particularly for those of highest intrinsic luminosity, so that at faint magnitudes and large redshifts, quasars should abound. The first quasars discovered with redshifts greater than 3, OH471, z = 3.40 (Carswell and Strittmatter 1973), and OQ172, z = 3.53 (Wampler et al. 1973) were both relatively bright, m ~ 18, and were both relatively easily discovered, so it was assumed that finding higher redshift objects would not be difficult. The quest to discover and observe objects at an epoch only a few billion years after the Big Bang has consumed considerable time and effort, and has not been as easy or rewarding as was initially expected. Quasars at z ~ 3.5 were discovered in 1973, only 10 years after quasars were first recognized, but to date, 15 years later, only a handful of quasars with higher redshift have been identified, most as a result of very exten- sive surveys in the past few years. There were suggestions from the earlier surveys that the density of quasars with z > 2.2 was declining, but possible observational biases were often raised to dispute this evidence. Recent surveys complete to fainter magnitudes show that high redshift quasars are indeed rarer than initially thought, and that although there is no sharp 'redshift cutoff', there is a gradual

19

N. Kaiser and A. N. Lasenby (eds.), The Post-Recombination Universe, 19–31.

decline in the space density of quasars with $z > 2$.

In this paper the methods of finding high redshift quasars are discussed, the observational biases considered, and the results of recent surveys explored. Since several excellent reviews on similar topics have already been published (e.g. Smith 1983, Veron 1983, Weedman 1987) only recent work is discussed in detail.

2. OBSERVATIONAL DISCOVERY METHODS

2.1. Radio surveys

Quasars were originally discovered as "quasi-stellar" radio sources and consequently large radio surveys were the prime hunting grounds for quasars. It is now recognized that the ratio of radio loud to radio quiet quasars is < 0.1 (Sramek and Weedman 1978) and that quasars discovered by their radio emission may not be representative of the entire population. Nevertheless, radio surveys provide samples of extragalactic sources complete to a fixed flux limit and are unaffected by possible intervening obscuration or other possible biases which may be present in optical surveys. One of the major difficulties in the most of the radio surveys, however, is the time required to get optical identifications and, subsequently, redshifts of the quasars. As Wall and Peacock (1985) point out, the most luminous optically selected quasars are > 2 mag more luminous than the most optically luminous quasars in complete radio samples. Apparently it is the lower bound in luminosity which is common to the radio and optical samples so follow-up spectroscopy of all radio sources is difficult. Because of this, and the natural tendency to observe the brightest and/or the most interesting objects first, the potential inherent in the radio surveys has not yet been fully realized.

High redshift quasars tend to have flatter spectra (Peterson et al. 1982, Dunlop et al. 1986) than lower redshift extragalactic sources so their inclusion in a flux limited survey also depends on the frequency of the survey. The Parkes survey at 2.7GHz (Wall, Shimmins and Merkelin 1971) contains many more flat spectrum sources than the lower frequency 3CR (Bennett 1962) sample, for example. Laing, Riley and Longair (1983) have compiled and discussed the optical identifications for 96% of the 3CR sample ($S > 10$ Jy at 178 MHz). The redshift distribution of the 40 (25%) quasars in the sample which are mostly brighter optically than $V = 19$, is shown in Figure 1a. As might be expected for the apparently brightest sources in the sky, they are confined to relatively low redshifts. Wall and Peacock (1985) have carried out a similar analysis of the brightest ($S > 2.0$ Jy) sources at 2.7GHz. About 45% of the objects with measured redshifts are quasars, again of predominantly low redshift and bright ($m < 19$) apparent magnitudes. From the apparent correlation between the V magnitude and redshift of the quasars (i.e. a "Hubble relation"), Wall and Peacock concluded that the fact that the magnitude distribution peaked at $V \sim 18$ was simply due to the scarcity of quasars with $z > 2$. They also predicted that $z = 4$ radio loud quasars would only become abundant at $V \sim 21$.

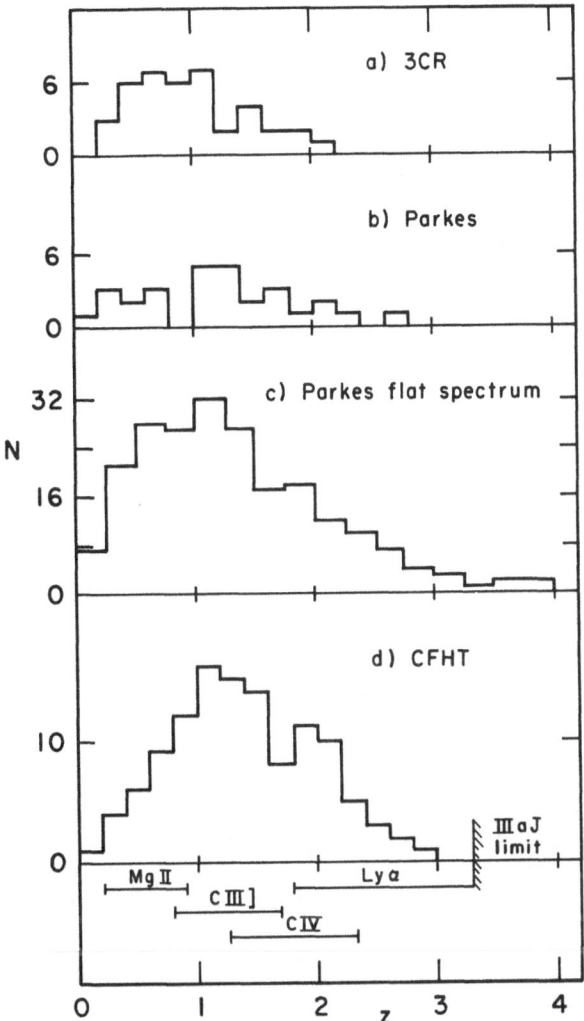

Figure 1. Redshift distributions of a) 3C sources, b) Parkes Selected Area sources, c) Parkes flat spectrum sources, and d) CFHT blue grens quasars. The bars at the bottom show the redshift ranges for which the emission lines indicated are in the IIIaJ passband for slitless spectra surveys. Since Ly α is usually the most easily detectable line, the decline in the numbers of quasars with $z > 2$ in the bottom panel is particularly striking.

Downes et al. (1986) have recently reported initial results of optical identification of the Parkes selected region survey (S > 0.1 Jy at 2.7 MHz). Over half of the 178 sources have been identified to a limit of V ~ 21. The redshift distribution of the 29 quasars discovered (m_b < 20) is shown in Figure 1b. Downes et al. argue that most of the remaining unidentified sources are galaxies and that the comoving density of radio sources begins to drop at z > 2. Peacock (1985) came to a similar conclusion for the compact, flat spectrum sources, and predicted that the density of such sources would be reduced by a factor ≳3 at z = 4 compared to that at z = 2.

Jauncey et al. (1987) recently reported preliminary results of optical identification of flat spectrum sources from the Parkes survey (S > 0.5 Jy at 2.7 MHz). The sample, which should favour inclusion of high redshift quasars, contains over 400 sources and covers 4.5 str or 35% of the entire sky. Over 80% of the sources are quasars, and redshifts have been measured for 225 of them with R < 22. The redshift distribution is shown in Figure 1c. Five of the quasars have z > 3.3. The number of "high redshift" quasars with such radio properties is thus only ~20 over the entire sky. Although this appears to be a successful technique for finding quasars of all redshifts, the number of very high redshift quasars is obviously rather small for statistical analyses.

Thus, although quasars with z > 3 were first discovered through radio surveys, the difficulties in securing optical identifications and redshifts for quasars that are generally not among the most optically luminous has made the quest for high redshift objects by this method more difficult than first anticipated.

2.2. Optical surveys

2.2.1. UV Excess surveys. The first optical surveys for quasars relied upon their ultraviolet excess to distinguish them from the majority of stars. These 'uvx' surveys are very sucessful in identifying candidate quasars with z < 2 but at higher redshifts Ly α enters the passband of the B filter and the U-B colour is no longer a useful criterion. An important survey for bright (B <=16) quasars was carried out with this technique by Schmidt and Green (1983). Their survey of 10,714 deg² yielded 114 quasars, although Wampler and Ponz (1985) suggest that it may be incomplete by as much as 25%.

Boyle et al. (1985) and Shanks et al. (1986) have recently used the 'uvx' method to detect quasar candidates to B ~ 21 on UK Schmidt plates. They find ~100 uvx stars per square degree of which 40% are quasars. The integral space density they derive is thus 40 deg^{-2} at B = 21 and 26 deg^{-2} at B = 20.5 for quasars with z < 2.2. Although this method cannot be used to identify high redshift quasars, it is straightforward and obviously successful at low redshifts and hence serves as an important reference for other techniques. In particular, the completeness of such a survey is not a (strong) function of emission line strength as is the case for many other optical methods.

2.2.2. Multi-colour surveys. Shanks, Fong and Boyle (1983)

demonstrated the potential of a four colour, UBVR, survey for identification of high redshift quasars by quickly discovering a z = 3.6
quasar. Warren et al. (1987) extended the technique to five colours,
UJVRI, and claim excellent separation between stars and quasars (with
stars occupying only ~10% of the 4 colour space). They argue that this
is one of the most effective means of discovering quasars of high
redshift and support this claim by identifying two such objects out of
a sample of 12 of the most likely candidates discovered in a UK Schmidt
field. Although this technique yielded detection of the first quasar
with z > 4, it is not obvious how the completeness of such a survey can
be determined without extensive follow-up spectroscopy of a large
number of candidates. It is also not as independent of emission line
strength as might be expected since strong Ly α emission can significantly affect the colour of a quasar, making it more readily detectable.

Narrow band imaging can be used in conjunction with broad band
images to detect strong emission lines, particularly Ly α, within
selected redshift ranges. Hazard (1987) has begun such a survey with
the UK Schmidt telescope, and Hickson and Kindl (1987) have used a
similar technique with a CCD to probe deeply in a small field. This
technique is likely to become much more important in future surveys.

2.2.3. Slitless Spectra Surveys. This method makes use of the fact
that quasars, by definition, display strong emission lines in their
spectra. Although quasars were first discovered as a class through
their radio emission, it seems certain, in retrospect, that they would
have been soon discovered by deep objective prism surveys at about the
same epoch. Ly α is extraordinarily strong, with equivalent widths
often exceeding 500Å in the observers frame (Hazard et al. 1986), and
so the easiest quasars to detect are those for which Ly α falls in the
passband defined by the earth's atmosphere, the optics of the instrument and the sensitivity of the detector. Thus quasars with z > 1.8
(observed wavelength of Ly α > 3000Å) are the easiest to detect by
ground-based slitless spectra surveys.

There are several variations of this technique, although all
primarily rely on detecting the presence of emission features. The
spectra of all objects in the field are recorded simultaneously but at
the expense of overlapping spectra from nearby objects or other orders
(when a grating is used), and overlying sky background light. The
limiting magnitude and wavelength resolution of the spectra are
strongly dependent on the seeing conditions.

There are three optical instruments used in these surveys:
objective prisms commonly used on Schmidt telescopes, a grating-prism
or 'grism' used near the focus of the many 4m class telescopes, and a
grating-lens-prism or 'grens' used at the prime focus of the CFHT.
There are also three detectors which have been widely used: IIIaJ
plates which are useful for wavelengths less than 5300Å, IIIaF plates
which are sensitive up to ~7000Å, and CCD's which are also sensitive to
at least 7000Å. The latter detectors are often used with filters to
remove the contribution of the bright sky beyond 7100Å, and below 5600Å.
Since spectra are recorded in a slitless mode, the success of this

TABLE 1. RECENT SLITLESS SPECTRA SURVEYS

	Objective prism Resolution is $f(\lambda)$	Grism Constant resolution	Grens
IIIaJ plates			
Field	~25 deg^2	0.3 deg^2	0.8deg^2
m limit	m < 20	m < 21	m < 22
z limits	1.8 < z < 3.3	1.8 < z < 3.3	z < 3.3
W limit	>100Å	>50Å	>30Å
References:	1,2,3	4	5,6
IIIaF plates			
z limits	1.8 < z < 4.7	1.8 < z < 4.7	z < 4.7
References	7	8,9	9,10
CCD			
Field		0.01 deg^2	
m limit		r < 21	
z limits		2.9 < z < 4.8	
W limit		>50Å	
References		11,12	
Scanning CCD			
Field		large	
m limit		faint	
z limits		2.9 < z < 5	
W limit		>50Å	
References		13,14	

REFERENCES to Table 1.
 1. Osmer and Smith (1980)
 2. Clowes and Savage (1983)
 3. Kunth and Sargent (1986)
 4. Gaston (1983)
 5. Weedman (1985)
 6. Crampton, Schade and Cowley (1985)
 7. Hazard, McMahon and Sargent (1986)
 8. Osmer (1982)
 9. Anderson and Margon (1987a)
10. Borra, E. unpublished
11. Koo and Kron (1980)
12. Schmidt, Schneider and Gunn (1986a)
13. Schmidt, Schneider and Gunn (1986b)
14. Schmidt, Schneider and Gunn (1987a)

technique depends on seeing conditions, guiding and the optical perfor-mance of the instrument. Clowes (1981) has discussed these effects on the detection of quasars by slitless techniques, and Schmidt, Schneider and Gunn (1986a) give a detailed description of how they account for these, and other effects, in their CCD survey.

A comparison of the various slitless spectrum techniques is given in Table 1. The limiting magnitudes, equivalent widths and redshift ranges are only approximate, and the parameters of specific surveys may differ considerably. Note the large fields offered by the schmidt telescope objective prism surveys, in contrast to the very tiny fields available with current CCD surveys. The scanning CCD technique, however, allows ~10 deg^2 to be surveyed per night yielding digital spectra to a relatively faint limit ($r \sim 20$). In principle, the same strip of sky could be observed many times to go to fainter limits, or adjacent strips of sky could be scanned to increase the area.

The objective prism surveys offer the possibility of detecting relatively bright, strong-lined quasars over large fields but suffer from the fact that the dispersion produced by prisms is wavelength dependent. In practice this means that the highest redshift quasars are difficult to detect because of the continuum shape and poorer wave-length resolution. In addition, the only large detector available for $z > 3.3$ quasars is the IIIaF plate which has notoriously uneven spectral response. Nevertheless, the large field is a big advantage for the detection of high redshift quasars because of their low space density.

The grism and grens surveys allow extension of the slitless spectra technique to fainter magnitudes. The dispersion is linear with wavelength in these devices and better signal-to-noise is usually achieved allowing weaker emission features to be detected. A further advance in the slitless spectra technique involves the automated machine detection of quasars (Clowes, Cooke and Beard 1984, Hewett et al. 1985, Borra et al. 1987, Schmidt, Schneider and Gunn 1986a). These methods allow quantitative limits to be placed on the strengths of emission features compared to the signal-to-noise of the continuum. The spectra can also be used to determine colours of the objects. Since some quasars have very weak emission lines and also since at some redshifts the strongest lines are not within the observed wavelength region, a combined colour and line strength criterion is important (particularly at low z) if completeness is to be attained. The redshift distribution of a sample believed to be complete for all $z <$ 3.3 and B <= 20.5 (Crampton, Cowley and Hartwick 1987) is shown in Figure 1d. The bars beneath the histogram indicates which emission lines are visible at each redshift on a blue grens plate. The decline in the numbers of quasars with $z > 2$ is particularly striking since Ly α is generally the easiest feature detected.

The recent application of a mosaic of 4 CCD's operated with the telescope in a transit or scanning mode by Schmidt, Schneider and Gunn (1986b) offers exciting prospects for the discovery of high redshift quasars. In theory, such a survey can be pushed both to very large areas and to very faint magnitudes. It's main limitation stems from the fact that only strong-lined quasars are detected. Although Ly α is usually strong, especially in the observers frame due to the (1+z) dependence (Anderson and Margon 1987b), it is sometimes intrinsically weak and, occasionally, almost completely absorbed. Although objective criteria are used to select candidates from the initial CCD grism survey, eye examination is used to reject overlapping orders, CCD flaws, etc., so some subjective judgment is also involved to reduce ~1000 candidates per survey to a manageable number (~50) for slit spectroscopy. The Schmidt, Schneider and Gunn surveys have thus far yielded only ~10 quasars per survey. More sophisticated software making use of pattern recognition techniques to reject late type stars, etc. would likely improve the yield. Nevertheless, this method coupled with a multi-colour approach is likely to provide the best information on the so-called redshift cutoff – the decline in the space density of quasars with z > 2.

2.2.4. <u>Combined techniques</u>. As mentionned above, some of the quantitative slitless spectra surveys offer the possibility of detection by both colour and line strength. The extensive 'APM-MMT' survey (Chaffee <u>et al</u>. 1987) to observe 1000 quasars with m < 18.5 and the Marano, Zamorani and Zitelli (1986) survey to m < 21 are two important examples of the combined use of colour and spectral criteria, although both are restricted to quasars with z < 3.3. In a different approach, Koo, Kron and Cudworth (1986) selected quasar candidates in a 0.3 deg^2 field to B = 22.5 through a combination of UBVI colours, variability and astrometry (lack of proper motion). Although follow-up spectroscopy is not yet complete to this faint limit, no quasars with z > 3.1 have yet been discovered.

3. QUASARS WITH z > 3.3

All published quasars with z > 3.3 are listed in Table 2. The limit of 3.3 was chosen so as to exclude quasars detectable by the extensive IIIaJ surveys (although one exception, 1410+0936, was discovered on a IIIaJ plate). All of the methods (column 2) used to detect these quasars are sensitive to quasars with redshifts up to at least 4.7 in theory, although the objective prism surveys (references 2, 10,16) have decreasing sensitivity for z > 3.9. This bias is likely to be more than offset by the tendency to rush the discovery of any really high redshift object into print. Of the 27 quasars, 8 were discovered by radio surveys, 4 by colour, and the remainder by the slitless spectra technique. The redshift distribution of the quasars in Table 2 is shown plotted in Figure 2a. The hatched area denotes quasars discovered by their radio emission. The distribution appears to be the end of the smooth decline begun near z = 1.8 apparent in the CFHT and the Parkes flat spectrum surveys shown in Figure 1.

Only crude magnitudes have been published for most of the quasars in Table 2, and the V magnitudes must include a significant contribution from Ly α for 3.3 < z < 4. Where Gunn r magnitudes were given, a correction of 0.5 mag was applied to approximate an R magnitude. The apparent magnitude distribution is plotted in Figure 2b. The distribution is surprisingly flat, quite unlike that at lower redshifts. Since the quasars have similar redshifts, Figure 2b also reflects the distribution of absolute magnitudes ($-25 < M_r < -31$ for $H_o = 50$ km s^{-1} Mpc and $q_o = 0.5$).

TABLE 2. LIST OF HIGH REDSHIFT QUASARS (Z > 3.3)

Name	method*	J	V	R	z	W	reference
0014+8118	radio		16.5		3.41	23	1
0042-2627	spec,F			17.0	3.30	50	2
0044-2734	colour	20.7	19.4	19.7	3.42		3
0046-2920	colour	21.0	19.5	19.2	4.01	55	3
0054-2825	colour	19.6	18.2	17.8	3.614	60	4
0055-2659	spec,F			17.1	3.67		2
0105-2634	spec,F			17.3	3.50		2
0131+0120	spec,C			18.9	3.793		5
0135-4239	colour			19	3.97		16
0234+0120	spec,C			19.8	3.301		5
0335-1214	radio		19.8		3.45		6
0345+0130	spec,C			21.2	3.636		5
0642+4454	radio	19.6	18.5		3.402		7
0910+5625	spec,C			20.5	4.04	42	15
1159+1223	spec,F			17.5	3.51		2
1208+1011	spec,F			17.5	3.803		2
1209+0919	spec,F			18.5	3.31		2
1351-0151	radio	20.9		19.3	3.710	35	8
1409+7314	spec,F	21		19.1	3.56	40	9
1410+0936	spec,J	18.8			3.340	55	10
1442+1011	radio	18.6	17.8		3.53	35	11
2000-3300	radio	19.0			3.780		12
2227-3928	spec,F	19	18.8		3.45		13
2239-3608	spec			16	3.5		18
2313-4221	spec,F		19.5		3.36		14
PKS??	radio			19	3.55		17
PKS??	radio			17	3.79		17

* The notation 'spec' refers to the slitless spectra technique, and 'F' to IIIaF plates as detector, 'J' to IIIaJ plates, and 'C' to CCD detectors.

REFERENCES to Table 2.
  1. Kuhr, H. et al. (1983)
  2. Hazard, McMahon and Sargent (1986)
  3. Warren et al. (1987)
  4. Shanks, Fong, and Boyle (1983)
  5. Schmidt, Schneider and Gunn (1987a)
  6. Chu, Zhu, and Butcher (1986)

28

REFERENCES to Table 2 (continued)
 7. Carswell and Strittmatter (1973)
 8. Dunlop et al. (1986)
 9. Anderson and Margon (1987a)
10. Hazard et al. (1986)
11. Wampler et al. (1973)
12. Peterson et al. (1982)
13. Smith et al. (1977)
14. Osmer (1982)
15. Schmidt, Schneider and Gunn (1987b)
16. Irwin, Hazard and McMahon (1987)
17. Jauncey et al. (1987)
18. Hazard and McMahon (1987)

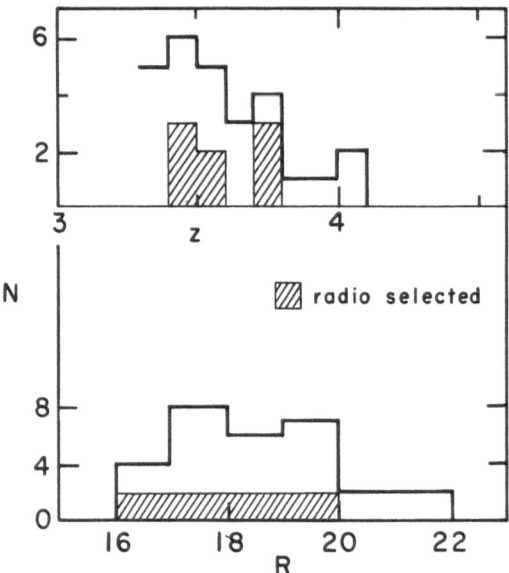

Figure 2. The redshift distribution for the $z > 3.3$ quasars in Table 2
is given in the upper panel, and the magnitude distribution in the
lower panel. Since the quasars are approximately at the same distance
from us, the absolute magnitude distribution has a similar shape to the
apparent magnitude distribution.

    The Ly α rest frame equivalent width of the quasars in Table 2
ranges from ~20Å to 60Å, a range identical to that observed at $z = 2$
(Crampton, Cowley and Hartwick 1987). The average is ~50Å,
corresponding to ~250Å in the observers frame, so that early speculation that Ly α might be intrinsically weaker at high z, or heavily
absorbed by intervening galaxies, appears not to be the case. The Ly α

equivalent widths of the radio-selected quasars is typically ~35Å, somewhat lower than that of the quasars selected by their emission lines, but still easily detectable with slitless spectra surveys.

4. SPACE DENSITY AND LUMINOSITY FUNCTION

The space density of quasars with $z > 3.4$ is difficult to assess because few of the surveys are complete and the numbers are small. Hoag and Smith (1977) discovered one in a 5.1 deg^2 field as did Osmer (1982). Schmidt, Schneider and Gunn (1987) found two relatively strong-lined quasars with $z > 3.4$ in 14 deg^2 although one is very faint (see Table 2). From this meager evidence, we concur with Anderson and Margon (1987b) that the space density of quasars with $z > 3.4$ and R < 19.5 is ~0.2 deg^{-2}. Large areas will have to be surveyed in order to substantially improve our statistics.

At low redshifts the luminosity function is a very steeply rising function with decreasing luminosity (e.g. Schmidt and Green 1983, Crampton, Cowley and Hartwick 1987), but the data also show that there is strong evolution with cosmic time so that high luminosity quasars were relatively much more numerous at high redshift. There is increasing evidence from deep surveys such as that of Koo, Kron and Cudworth (1986) and also from the Schmidt, Schneider and Gunn surveys that large numbers of faint, high redshift quasars do not exist. Of the 9 quasars with $z > 3$ found in the recent Schmidt, Schneider and Gunn (1987a) survey, 7 have r < 19.7, although they obviously can detect quasars up to 2 mag fainter. Based on their experience in the detection of high redshift quasars on Schmidt plates, Hazard and McMahon (1985) argue that the luminosity function for quasars with $z > 3.4$ may be essentially flat. The slope is certainly much less than the ~10 per magnitude observed at very low z. Current indications, including the data in Table 2 (Fig. 2b), are that it may be ~2.

5. CONCLUSION

Obviously, the numbers of quasars with $z > 3.3$ are still insufficient to provide reliable statistics. Evidence from the lower redshift surveys indicates that quasars reached a broad maximum in comoving density, a 'quasar epoch', near $z \sim 1.7$. The 'redshift cutoff' actually appears to be a gradual decline for $z > 2$. Since there is now some evidence that 'primeval galaxies' may also have been forming at $z \sim 1.8$, perhaps the quasar epoch was coeval.

High redshift, $z > 3.3$, quasars are apparently relatively rare, and their luminosity function is relatively flat so that surveys of large areas to moderately faint magnitudes are essential. Since these quasars often have strong emission lines, the slitless spectra technique will continue to be of prime importance. For weaker-lined objects, the multi-colour technique shows great promise. With the advent of the CCD transit type of observation, a multi-colour survey of the identical region to that currently being surveyed by Schmidt,

Schneider and Gunn would be of great interest for the detection of
weaker-lined objects and for comparison. The improved accuracy
possible with a CCD survey should serve to reduce the volume occupied
by normal stars in colour-colour space and improve the reliability of
this technique. The results of such a combined survey should provide
the answers we currently seek.

REFERENCES

Anderson, S.F. and Margon, B. 1987a, Ap.J., 314, 111.
Anderson, S.F. and Margon, B. 1987b, Nature, 327, 125.
Bennett, A. S. 1962, Mem. R. A. S., 68, 163.
Borra, E.F. et al. 1987, Pub. A.S.P., 99, 535.
Boyle, B.J., Fong, R., Shanks, T., and Clowes, R.G. 1985, M.N., 216,
    623.
Carswell, R.F., and Strittmatter, P.A., 1973, Nature, 242, 394.
Chaffee, F.H., Foltz, C.B., Hewett, P.C., MacAlpine, G.M., Turnshek,
    D.A., Weymann, R.J., and Anderson, S.F. 1987, Bull. A.A.S., 19,
    700.
Chu, Y., Zhu, X., and Butcher, H. 1986, Astr. Sp. Sci., 119, 231.
Clowers, R.G. 1981, M.N., 197, 731.
Clowes, R.G., and Savage, A. 1983, M.N., 204, 365.
Clowes, R.G., Cooke, J.A., and Beard, S.M. 1984, M.N., 207, 99.
Crampton, D., Schade, D., and Cowley, A.P. 1985, A.J., 90, 987.
Crampton, D., Cowley, A.P., and Hartwick, D.A. 1987, Ap.J., 314, 129.
Downes, A.J.B., Peacock, J.A., Savage, A., and Carrie, D.R. 1986,
    M.N., 218, 31.
Dunlop, J.S. et al. 1986, Nature, 319, 564.
Gaston, B. 1983, Ap.J., 272, 411.
Hazard, C. 1987, preprint Wide Field Telescopes and High Redshift
    QSO Surveys.
Hazard, C., and McMahon, R.G. 1985, Nature, 314, 238.
Hazard, C. and McMahon, R.G. 1987, private communication.
Hazard, C., McMahon, R.G., and Sargent, W.L.W. 1986, Nature, 322, 38.
Hazard, C., Morton, D.C., McMahon, R.G., Sargent, W.L.W., and
    Terlevich, R. 1986, M.N., 223, 87.
Hickson, P., and Kindl, E. 1987, private communication.
Hewett, P.C. et al. 1985, M.N., 213, 971.
Hoag, A.A., and Smith, M.G. 1977, Ap.J., 217, 362.
Irwin, M.J., Hazard, C., and Mcmahon, R.G., 1987 private communication.
Jauncey, D.L., White, G.L., Savage, A., Peterson, B.A., Gulkis, S.,
    and Batty, M.J. 1987, in QSO Absorption Lines: Probing the
    Universe, STScI Colloq., Blades, J.C., Norman, C. and Turnshek,
    D.A. eds.
Koo, D.C., and Kron, R.G. 1980, Pub. A.S.P., 92, 537.
Koo, D.C., Kron, R.G., and Cudworth, K.M. 1986, Pub. A.S.P., 98, 285.
Kuhr, H., Liebert, J.W., Strittmatter, P.A., and Schmidt, G.D. 1983,
    Ap.J., 275, L33.
Kunth, D., and Sargent, W.L.W. 1986, A.J., 91, 761.
Laing, R.A., Riley, J.M., and Longair, M.S. 1983, M.N., 204, 151.

Marano, B., Zamorani, G., and Zitelli, V. 1986, in Structure and Evolution of Active Galactic Nuclei, Giuricin et.al. eds., (Reidel), p. 339.

Osmer, P.S. 1982, Ap.J., 253, 28.

Osmer, P.S., and Smith, M.G. 1980, Ap.J.Suppl., 42, 523.

Peacock, J.A. 1985, M.N., 217, 601.

Peterson, B.A., Savage, A., Jauncey, D.L., and Wright, A.E. 1982, Ap.J., 260, L27.

Schmidt, M., and Green, R.F. 1983, Ap.J., 269, 352.

Schmidt, M., Schneider, D.P., and Gunn, J.E. 1986a, Ap.J., 306, 411.

------. 1986b, Ap.J., 310, 518.

------. 1987a, Ap.J., 316, L1.

------. 1987b, preprint

Shanks,T., Fong, R., and Boyle, B.J. 1983, Nature, 303, 156.

Shanks,T., Fong, R., Boyle, B.J., Peterson, B.A. 1986, in Quasars (ed. G.Swarup and V.K. Kapahi; Reidel), p. 37.

Smith, M.G. 1983, in Quasars and Gravitational Lenses, Proc. of the 24th Liège Int. Ap. Colloq., (Université de Liège), p.4.

Smith, M.G., Boksenberg, A., Carswell, R.F., and Whelan, J.A.J. 1977, M.N., 181, 67p.

Sramek, R.A., and Weedman, D.W. 1978, Ap.J., 221, 468.

Veron, P. 1983, in Quasars and Gravitational Lenses, Proc. of the 24th Liège Int. Ap. Colloq., (Université de Liège), p. 210.

Wall, J.V., and Peacock, J.A. 1985, M.N., 216, 173.

Wall, J.V., Shimmins, A.J., and Merkelijn, K.J. 1971, Austr. J. Phys. Ap. Suppl., 19, 1.

Wampler, E.J., and Ponz, D. 1985, Ap.J., 298, 448.

Wampler, E.J., Robinson, L.B., Baldwin, J.A., and Burbidge, E.M. 1973, Nature, 243, 336.

Warren, S,J. et al. 1987, Nature, 325, 131.

Weedman, D.W. 1985, Ap.J.Suppl., 57, 523.

Weedman, D.W. 1987, Quasar Astronomy, Cambridge Astrophysics Series, Cambridge U. Press.

COSMOLOGICAL EVOLUTION OF EXTRAGALACTIC SOURCES

L. Danese and A. Franceschini
Dipartimento di Astronomia
Vicolo dell'Osservatorio, 5
I-35122 Padova, Italy

Abstract We review all the relevant information on the cosmological evolution of extragalactic sources. We show that the evolution of Active Galactic Nuclei selected at radio, optical and X-ray wavebands is well represented by simple Luminosity Evolution models. A population of Actively Star Forming galaxies is also found to evolve on time scales similar to those of AGNs. Finally, we touch on the problem of the physical interpretation of the cosmological evolution for these source populations.

1. INTRODUCTION

The concept of a cosmological evolution for some populations of extragalactic sources is quite natural in the context of an expanding universe. Cosmological evolution simply means that global properties such as the number density and/or the typical luminosity of the population undergo significant changes with cosmic time.

Number counts of galaxies as a function of their apparent magnitude have been proposed some 50 years ago by Hubble and Tolman (1935) as a test for cosmologies. Under the assumption of a constancy of source properties in the space/time, such counts would depend on the geometry of the universe only. Radioastronomers soon discovered objects, like powerful extragalactic radiosources, which are visible up to high redshifts and hence suitable for an efficient application of the test. The well known result was that these sources are not uniformly distributed in the comoving volume, they should have been vastly more numerous or more luminous at earlier epochs. More recently, a similar behaviour has been been found to be shared also by optically selected QSOs, X-ray selected Active Galactic Nuclei (AGNs) and, lastly, by some of the source populations selected by the IRAS satellite in the far-IR. Thus, if the idea to use powerful sources to test the geometry of the universe failed, a new research field opened, aimed at the understanding of the phenomenological patterns and the physical reasons of the cosmological evolution of extragalactic sources. The main features in this scenario are the birth epoch of the objects and their evolutionary time scale. It is worth to notice that the relevant conclusions do not depend on the adopted Friedman model ($H_0 = 50$ and $q_0 = 0.5$ will be assumed, anyway).

The ultimate motivation for studying the cosmological evolution is to under-

N. Kaiser and A. N. Lasenby (eds.), The Post-Recombination Universe, 33–50.

stand the physical link between the characteristic evolution timescales of the various populations considered as statistical ensembles and those of the phenomena ruling the birth, the life and the death of the single objects. Therefore, it is extremely important to keep a close relation among "kinematical" studies and basic ideas on the physical mechanisms that rule the emission of the individual sources. We will show, in this context, that a simple model of luminosity evolution is able to reproduce a large body of data at radio, far-IR, optical and X-ray wavelengths on various classes of extragalactic objects.

We will also demonstrate that the cosmic evolution is not an exclusive property of AGNs, but that a population of Actively Star Forming (ASF) galaxies does also substantially evolve with cosmic time.

More specifically, in Section 2 we will touch on some technical aspects of modelling the cosmological evolution and comparing models with the data. Section 3 will be devoted to a brief presentation of the Local Luminosity Functions (LLFs) of extragalactic populations. In Section 4 we shall review results on the evolution of powerful radiogalaxies and QSOs and of faint source populations in the radio band. In Section 5 we will see that the recent surveys of IRAS at far-IR wavebands do also imply evolution and discuss the efficiency of far-IR observations in revealing primeval galaxies. In Section 6 the evolution properties of optically selected AGNs will be reviewed. Also the relevance of ASF galaxies for the counts of galaxies in the optical band will be briefly addressed. Section 7 is devoted to the X-ray selected AGNs and Section 8 to the discussions and conclusions.

2. EVOLUTION MODELS AND CONFRONTATION WITH THE DATA

Evolution studies require large observational efforts to carry out complete surveys down to various limiting fluxes and with sky coverages as large as possible. Moreover, for surveys in wavebands other than the optical, optical identifications are mandatory.

Although a wealth of spectroscopic redshifts have been obtained in the last few years using multi-object spectroscopy tecniques, the coverage of the luminosity-redshift plane (the 'evolution plane') is still meagre.

2.1. The observables

The most immediate result of a survey is the log N - log S plot. The differential $N(S)$ can be expressed as a simple integral of the Luminosity Function (LF) $\rho(L, z)$:

$$N(S) = \frac{1}{4\pi} \frac{dS}{S} \int_{z_1}^{z_2} dz \frac{\rho(L, z)}{ln10} \frac{dV}{dz} \quad sources/sr \quad (1)$$

z_1 and z_2 being the effective upper and lower limits of the redshift distribution and dV/dz the differential comoving volume.

The spectral intensity of the diffuse background radiation in a given e.m. band and its cell-to-cell fluctuations also provide relevant constraints on models of cosmological evolution. They are simply the first and second moment of the N(S) distribution: $I = \int S \cdot N(S)dS$ and $(\Delta I)^2 = \int S^2 \cdot N(S)dS$. Clearly, number counts, but even more the background intensity and fluctuations, are rather integrated pieces of information.

Whenever redshifts for all sources of a complete sample are available, other statistics can be built up, such as the V/V_{max} test (see Avni and Bahcall, 1980) and the source luminosity or redshift distributions (e.g. Danese et al., 1985).

An extended coverage of the evolution plane would allow the direct computation of the LFs of a class of objects at different cosmological epochs. However, this ideal coverage is hampered by the increasing difficulty of achieving the redshift measurements at low flux densities and by the finiteness of the sampled volumes.

2.2 The continuity equation in the L-z plane

Although direct testing of the LFs as a function of redshift is still difficult at present, it is not hopeless to produce a scenario linking the statistical behaviour of a population – as reflected in the LF $\rho(L, z)$ – with the physics ruling the individual sources via a sort of continuity equation

$$d\rho/dt + \partial/\partial L(\dot{L}\rho) = S(L, z) \qquad (2)$$

(Cavaliere et al., 1983), where $\dot{L} = dL/dt$ is the rate of change of L and $S(L, z)$ the source term.

Actually, the statistical behaviour of a population is the result of the convolution of the individual fates, which can be affected by a lot of peculiar situations. Moreover, the above equation may also not always apply. For instance, this should be the case when at a given L there exhist sources both in a brightening ($\dot{L} > 0$), as well as in a dimming ($\dot{L} < 0$) phase (Peacock, 1986; see however Cavaliere & Vagnetti, 1987).

Although we should avoid using this equation in a crude way (the data are far from giving us informations on the derivatives contained in [2]), nevertheless the continuity equation models many interesting features in a convenient way.

2.3 Evolution models

An inspection of eq. (2) suggests two extreme ways of modelling the evolution : a Density Evolution (DE) mode applies when $\dot{L} = 0$, while a Luminosity Evolution (LE) mode works wherever $S(L, z) = 0$. For long time the scanty coverage of the $L-z$ plane hampered the discrimination between the two extreme scenarios (already proposed by Oort, 1961) of Pure Density Evolution (PDE) and Pure Luminosity Evolution model (PLE). In both cases the LF at any epoch keeps the same shape of the LLF $\rho(L, z = 0)$: the LLF shifts uniformly at increasing z, towards higher densities and higher luminosities for PDE and PLE models, respectively.

More recent data sets, from both radio and optical bands, have shown that PDE models significantly fail to account for the data at the faint flux levels. They implied, in particular, that the AGN evolution rates, for the case of DE, but probably also of LE, should dependent on the luminosity. A Luminosity Dependent version of the Density Evolution (LDDE) model was proposed by Wall et al. (1980) to interprete the radio counts. In this case the LF writes:

$$\rho(L, z) = \rho(L, z = 0) \cdot exp[\tau \cdot f(L)] \qquad (3)$$

where $f(L)$ is a function of L which is constant at $L > L_2$, zero at $L < L_1$ and linearly increasing in between (L_1 and L_2 beeing free parameters) , and $\tau = 1 - t(z)/t_0$ is the fractional look back time. A similar model with $logL_2 \rightarrow \infty$ has been adopted by Schmidt and Green (1983) when discussing optical counts of quasars.

It has been shown by several authors (Mathez, 1978; Cheney and Rowan-Robinson, 1981; Mathez and Nottale, 1982; Marshall et al., 1984; Cavaliere et al., 1983, 1985; Danese et al., 1985) that simple PLE models are remarkably successful. Cavaliere et al. (1983, 1985) also pointed out that moderately luminous sources can be fuelled for times $\sim H_0^{-1}$ by stellar mass loss in a large galaxy and proposed a differential luminosity evolution of the form

$$dL/dt = \begin{cases} -\kappa H_0(L - L_S) & \text{for } L > L_S \\ 0 & \text{for } L \leq L_S. \end{cases} \qquad (4)$$

For $L < L_S$ the evolutionary timescale is longer than H_0^{-1}, for $L \gg L_S$ it is equal to $(\kappa H_0)^{-1}$. A reasonable spread in the threshold luminosity L_S avoids discontinuities at $L \simeq L_S$ in the evolved LFs.

A mixed model have been suggested by Condon (1984), in which the LF undergoes both density and luminosity evolution: $\rho(L, z) = g(z) \cdot \rho[L/f(z), z = 0]$, with $g(z)$ and $f(z)$ arbitrary functions of z.

Free–form polynomial expansions of the LF have also been proposed (Robertson,1977; Peacock and Gull, 1981). These representations are more flexible than the above discussed canonical parametric models, but their results are of a much less immediate interpretation.

A further relevant parameter is the high redshift cutoff of the source population. Actually, a redshift cutoff is a crude way to represent the changing behaviour of the LF when approaching the epoch of object formation. DE representations of the LF suffer in general serious divergences in the number density of objects at high redshifts, whereas this is not the case for LE, so that a z–cutoff is quite more dramatically required for the former models than for the latter.

2.4 Statistical tools

As already mentioned, except than in a few cases of rich sampling, direct testing of the LFs as a function of z is not feasible. The observations provide some more integrated statistics (such as the $N(S)$ and the L or z distributions) to be compared with the model predictions by means of trial–and–error techniques.

The parameter optimization is in general performed through the χ^2 test for the case of binned data or the maximum likelihood in the unbinned case. The χ^2 also provides the goodness–of–fit testing of the models, while the Kolmogorov-Smirnov is usually adopted for the case of unbinned data.

The optimal use of the information from a complete survey down to a given flux limit would require a measurement of the redshifts and luminosities for all objects, so that a direct sampling of the evolution $(L - z)$ plane is obtained. Peacock (1983) and Fasano & Franceschini (1987) have discussed a two dimensional KS test suitable to analyse the LF modelling on such plane.

A comprehensive discussion of the commonly used statistical methods and of the associated problems can be found in Peacock (1985).

3. THE LOCAL LUMINOSITY FUNCTIONS

As apparent from the above discussion, the investigation of the cosmological evolution of a class of extragalactic sources strongly demands a correct and precise determination of the LLF $\rho(L, z = 0)$. Contrariwise, a scanty determination of the LLF badly affects any conclusions on the cosmological evolution. This entails the necessity of complete surveys with redshift determinations over large areas of sky, to avoid local fluctuations and to obtain large sampled volumes also with moderately deep or local surveys. This is particularly relevant for strongly evolving populations, in which case even a moderately deep survey can be affected by the presence of a relevant fraction of relatively distant and evolved objects.

When truly local, complete samples of objects are available, the LLFs can be computed using the generalized Schmidt's estimator (Felten, 1976; see also Toffolatti *et al.*, 1987).

A different procedure is required for the computation of the LLF in a given e.m. band using a sample selected in other wavebands. A typical case is the computation of the LLF of low luminosity radiogalaxies: radio selected samples at fairly bright flux limits are largely dominated by distant powerful radiosources. In such case one needs resort to a local, optically selected sample whose objects have been observed in the radio down to faint flux limits. The usual technique is to compute the bivariate radio/optical luminosity distribution and then to derive the radio LLF by convolving the bivariate with the optical LLF of the sample (see Toffolatti *et al.*, 1987, and references therein). Often the secondary survey has not enough sensitivity to detect all objects of the primary optical sample, so that an important fraction of the data is in the form of upper limits. Survival analysis techniques (Feigelson *et al.*, 1985; Schmitt, 1985) are particularly powerful in exploiting data containing detections and upper limits.

In the radio band the LLF of flat and steep spectrum sources at 2.7 GHz have been computed by Peacock (1985) and, more recently, by Toffolatti *et al.* (1987).

The LLFs at optical, radio and far-IR (60 μm) wavelenghts of various galaxy populations have been determined by Franceschini *et al.* (1987a) on the basis of a complete sample of 1671 galaxies selected from the UGC Catalogue (Nilson, 1973), with full radio coverage taken from the Arecibo survey at 2.4 GHz (Dressler & Condon, 1978). Far-IR data have been obtained with a positional cross–correlation with the IRAS Point Source Catalogue. The global sample was divided into four well distinct classes of sources: E+S0, Spiral+Irregulars, Seyfert galaxies and a new class characterized by enhanced Star Formation Activity (herein ASF galaxies), comprising non-Seyfert Markarians and galaxies with peculiar, distorted, strongly interacting morphologies.

The optical LLF of AGNs (Seyfert nuclei, in practice) in the range $-23.5 < M_B < -18.5$ has been derived by Cheng *et al.* (1985), using a homogeneous sample selected from the Markarian lists. The nuclear magnitudes were estimated by means of two independent (a "colour-given" and a "galaxy-given") methods. This LLF well matches that of optically selected QSOs at $M_B \simeq -24$ by Schmidt & Green (1983) and Crampton *et al.* (1987). Under the assumption of a continuity between properties of Seyfert galaxies and QSOs, it provides a fundamental tool for the investigation of the evolution of optically selected AGNs.

Piccinotti *et al.* (1982) derived the LLF of X-ray selected AGNs from the HEAO-1 A2 all sky survey in the 2-10 keV band. Their LLF extends down to $log L_x (erg\ s^{-1}) = 42.5$. More recently, Persic *et al.* (1987) showed that downward $L_x = 10^{42}$ the LLF clearly flattens. The X-ray LLF is an important boundary condition not only in studying the evolution, but also in connection with the problem of the origin of the X-ray background.

4. COSMOLOGICAL EVOLUTION OF RADIOSOURCES

4.1 Classical powerful radiosources and low luminosity populations

Wall (1983) and van der Laan (1987, this volume) vividly illustrate the status of the observations and our understanding of the cosmic evolution of radiosources. Here we shall concentrate on the recent ultradeep counts obtained through aperture synthesis telescopes (Fomalont *et al.*, 1984; Condon & Mitchell, 1984; Windhorst *et al.*, 1984; Partridge *et al.*, 1986). These observations extended the radio counts over almost 7 decades in flux.

The exciting result of these surveys is that around and below 1 mJy at 1.4 and 5 GHz the normalized differential counts clearly flatten. A large fraction of these faint sources are optically identified with blue galaxies, quite different from the red galaxies associated with brighter radiosources.

To interpret these results, a crucial point was understanding the role of low luminosity radiosources, which is expected to be relevant at these faint flux levels. Thus, instead of confining ourselves to the classical distinction between steep and flat spectrum sources, we have singled out the following classes of sources, characterized by well distinct astrophysical properties (as revealed by morphological and spectroscopic characters and broad-band spectral indices), hence also probably endowed with different evolution rates: *i)* evolving flat spectrum sources, typically the nuclei of active elliptical and S0 galaxies and QSOs; *ii)* evolving steep spectrum E+S0 galaxies and QSOs, the classical powerful double–lobed radiosources; *iii)* non evolving normal spirals and irregulars; *iv)* evolving Active Star Forming galaxies (see Sect.3); *v)* galaxies with bright Seyfert nuclei.

The radio LLFs of the above classes have been derived by Franceschini *et al.* (1987a; cf. also Danese *et al.*, 1987). For the evolving classes we have adopted the generalized LE model described in Sect.3 (eq.[4]).

The data base to be fitted include number counts at various frequencies from 178 MHz to 5 GHz, luminosity and redshift distributions and also some constraints on the fraction of red to blue galaxies in the milliJansky domain (see Danese *et al.*, 1987), for a total of more than 300 data bins.

Fig. 1a and 1b illustrate the best-fit models by Danese *et al.* (1987) and highlight the fact that luminous E+S0+QSOs dominate the bright radio counts, while at deeper flux levels relevant contributions from fainter source populations are indicated.

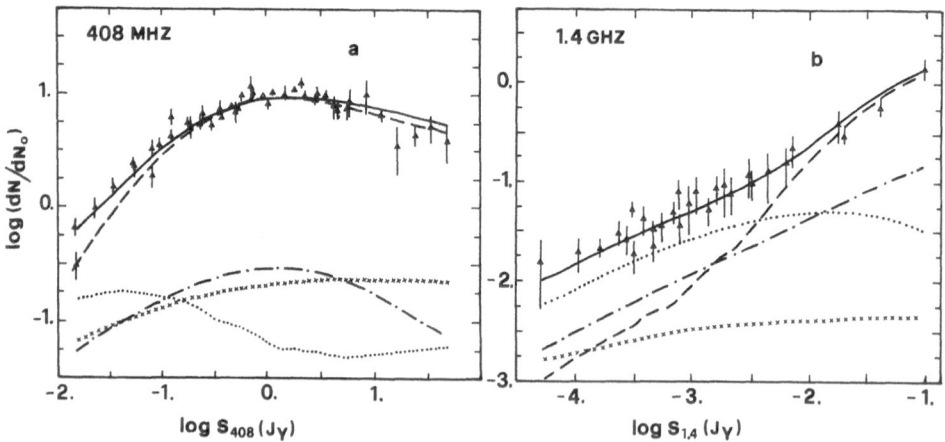

Figure 1. (a) Differential counts at 408 MHz. Dashed and dot-dashed are the contributions of stee and flat spectrum E+S0+QSOs; dotted lines and crosses are evolving ASF and unevolving spiral galaxies, respectively. (b) Same as in (a), but at 1.4 GHz.

A first interesting result is that the cosmological evolution of E+S0+QSOs with steep spectra has a markedly differential character: in keeping with results of LDDE models (Wall *et al.*, 1980; Peacock and Gull, 1981; Peacock, 1985), also LE models require that low luminosity ($L_{2.4} < 10^{23}(W/Hz/sr)$) radiogalaxies should not evolve, as implied in particular by the 408 MHz counts (Fig.1a). This entails a strong convergence of the counts of powerful E+S0s and radio loud QSOs below 10 mJy also at 1.4 GHz.

Flat spectrum sources still contribute significantly to the 1.4 and 5 GHz counts at very faint flux levels, in agreement with the finding of a number of flat spectrum sources at very low flux levels by Fomalont *et al.* (1984). In this case, the available data do not seem to require a lower limit to the luminosity of the evolving sources.

The most striking outcome of this analysis was that a cosmological evolution of the ASF galaxies on time scales of the order of those of powerful radiosources is required to explain the upturn of radio counts at $S_{1.4} < 10 mJy$ (see Fig.1b). The ultra deep counts are equally contributed by flat spectrum sources and ASF galaxies, while normal spirals are a factor five or more below the observed counts.

The model by Danese *et al.* (1987) well accounts also for the fraction of red galaxies and QSOs to blue galaxies as observed by Kron *et al.*, (1985), Windhorst *et al.* (1985), Windhorst *et al.* (1987).

Wall *et al.* (1986) pointed out that it is very difficult to produce pronounced features, such as the flattening in the deep counts, with sources dominating at high flux densities. They proposed that the flattening in the counts is due to non evolving normal spirals at $log P_{408} (W Hz^{-1} sr^{-1}) \simeq 20 \div 21$, whose radio LLF is then required to be almost a factor 10 higher than the current estimates. However, the fairly small uncertainties of the radio LLF by Franceschini *et al.* (1987b) in this power range seem to exclude such a possibility. Moreover, the photometric and spectroscopic work performed by Koo, Windhorst and collaborators shows that about 75% of the blue radio galaxies have $z > 0.3$.

On the other hand, Condon (1984) suggested that the blue radio galaxies could be normal spirals evolving both in luminosity and in density. This model is able to reproduce the deep counts but it does not predict the correct portion of blue galaxies at fluxes $> 1 mJy$, as already noted by van der Laan *et al.*, (1986). This conclusion only relies on the radio LLF of the spirals, which is now well defined.

More recent spectroscopic observations by Windhorst *et al.*, (1987) do actually suggest that the majority of the sub-mJy sources are identified with actively star forming galaxies.

All the above arguments imply that galaxies with enhanced star formation (ASF galaxies), evolving on timescales $\sim t_0/4$, are natural candidates to be the primary constituents of the faint radio galaxy population. We will see that the latter conclusion is also supported by IRAS observations in the far-IR bands.

Once again, radio observations turn out to be a fundamental cosmological tool: many years ago they allowed to evidenciate the cosmological evolution of the powerful radio sources, now they strongly suggest that much fainter ASF galaxies do also evolve. Thus strong cosmological evolution is no more an exclusive property of AGNs.

4.2. The decline of powerful radiosources at large redshifts

A possible fall of the comoving density of powerful flat spectrum radiosources and steep spectrum QSOs at high redshifts has been pointed out by Peacock (1985) and Peacock & Dunlop (1987). It is unclear how this result compares with the findings that only faint optically selected QSOs seem to show a decline at $z > 2$, whereas powerful QSOs do not (see Sect.6). Of course, selection effects and biases can affect the optical surveys, while enviromental effects may play a relevant role in determining the nature of high-z radiosources (Barthel, 1986).

The presence of a redshift cutoff is very interesting in connection with many fundamental problems, such as galaxy formation and the central engine (Black Hole?) formation. Hopefully, a large effort will be devoted in the future to investigate the high-redshift radiosources.

5. IRAS COUNTS

The great success of the IRAS mission has opened a new e.m. band (12 to 100 μm) to the cosmological investigations. This waveband is particularly sensitive to the presence of dust heated by young massive stars: typical ratios of far-IR ($60\mu m$) to radio (5 GHz) fluxes are of the order of 250 for disc and ASF galaxies, whereas they are $\sim 1/20 - 1/30$ for radiogalaxies and radio loud QSOs (Neugebauer et al., 1986). Consequently, while at radio wavelengths the counts are mainly contributed by radio loud QSOs and E/S0 radiogalaxies – normal or actively star forming galaxies significantly appear only at very faint fluxes – in the far-IR band the latter objects are expected to largely dominate also at bright flux levels (Franceschini et al., 1987b). Indeed, the star formation phenomenon (quiet as it is in normal spirals or active as in ASF galaxies) shows up strikingly at IRAS wavelengths. This fact, coupled to the well established tight correlation between radio and far-IR emission observed in disc galaxies (De Jong, 1985; Helou et al., 1985), implies that far-IR surveys can provide a powerful, independent check of the radio model for the ASF galaxies.

5.1. Analysis of bright $60\mu m$ samples
Particularly relevant for cosmological studies are the IRAS Point Source Catalogue and the samples at $60\mu m$ derived from it by a number of authors (e.g. Rowan-Robinson et al., 1986; Soifer et al., 1987; Lawrence et al., 1986; Smith et al., 1987), and the Hacking & Houck's (1987) deep survey complete to $S_{60} = 50mJy$).

A first, though marginally significant, indication of evolution for the high luminosity IRAS galaxies has been suggested by Soifer et al. (1987) on the basis of their direct evaluation of the $60\mu m$ LLF, whose values at the highest luminosity bins keep somewhat below the LF by Lawrence et al.. This difference was explained as due to an evolution effect, starting form the consideration that objects of the Lawrence et al.'s sample at $L_{60} > 10^{12}L_\odot$ have redshifts in the range $0.1 < z < 0.26$. Franceschini et al. (1987b) confirm and strenghten this finding. Using the UGC selected sample, they have derived the $60\mu m$ LLF of all source populations considered in the radio analysis (Sect.4). The global LLF turns out in quite good agreement with those of IRAS selected galaxies. Their highest luminosity bin ($L_{60} > 10^{12}L_\odot$), which is dominated by ASF and Seyfert galaxies, agrees with the Soifer's et al. LLF, thus confirming the evolution effect.

The cosmological evolution of ASF galaxies, as implied by the radio data, can easily reconcile the truly "local" LLFs with that of Lawrence et al.. Franceschini et al. (1987a) showed that the correlation between radio (at 2.4 GHz) and $60\mu m$ luminosities of ASF galaxies is non-linear: $L_{60} \propto L_{2.4}^{0.82}$. Thus, if their LE time scale is $\tau_{ev} = t_0/\kappa_R$ in the radio band, it will be $\tau_{ev} = t_0/(0.82\kappa_R)$ at $60\mu m$. A LE of ASF galaxies with that time scale accounts for the highest luminosity bins of the Lawrence et al. LF, as well as for their luminosity and redshift distributions.

5.2. Deep IRAS counts: confirmation of cosmological evolution for the ASF galaxies

Hacking & Houck (1987), using a set of scans over an IRAS primary calibrator within an area of 5 sq. deg., have derived a sample of 99 sources at $S_{60} \geq 50 mJy$. Almost all sources brighter than 100 mJy have been identified with POSS galaxies ($m < 18 \div 19$), while below this limit the identifications drop to 30-40%.

The predicted counts of non evolving spiral+irregulars and E+S0 galaxies have been obtained using the far-IR LLFs. Far-IR K-corrections have been determined from the 60 to 25 μm spectral indices.

Normal spiral+irr galaxies turn out to dominate the $60\mu m$ counts at least down to 1 mJy and the ASF galaxies appear to be the other relevant population, as detailed in Fig.2. Seyfert galaxies, even evolving, are only marginally important, while radio loud QSOs and powerful radiogalaxies are completely negligible in this analysis, due to their very low ratios of radio to far-IR luminosities.

This illustrates the main point we address here: ASF galaxy evolution fully explains the deep far-IR counts. We stress that no free parameter was allowed in predicting the counts.

On the other hand, the far-IR counts add futher difficulties to some alternative models proposed to fit the radio counts. The non evolving spiral galaxies proposed by Wall et al. (1986) are in trouble, due to the flat shape (independently confirmed by the IRAS selected samples) of the LLF in the range $30.5 < logL_{60} < 31.5$ relevant for the deep counts. On the other hand, Condon's (1984) model, assuming evolution of all spiral galaxies, predicts too large surface densities at faint fluxes (Hacking, Condon & Houck, 1987).

Although many uncertainties are still present, we can conclude that the cosmic evolution of ASF galaxies is able to quantitively account for a wealth of far-IR data.

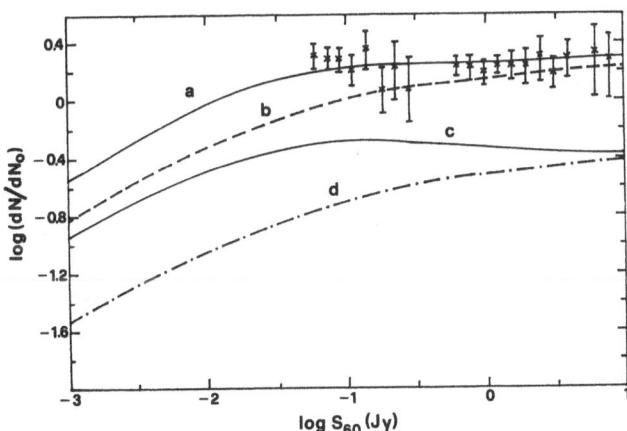

Figure 2. Differential counts at 60 μm from the IRAS survey. Curves a and b are the predicted total counts for the cases of evolution and no-evolution of the ASF galaxies, curves c and d the separate contributions of the latter.

5.3. Primeval galaxies in far-IR bands
A number of authors already noticed the difficulty of finding primeval galaxies with surveys at optical wavebands (Ellis, 1986; Koo, 1986; Tyson & Seizer, 1987; Cowie, this volume). This elusivity might be explained in a cold dark matter dominated

universe, in which the luminous cores of galaxies can form long before the associated halos (Silk, 1987).

A different possibility is that the disks of the spiral galaxies were still forming at $z \simeq 3$ (Rees, 1986). Actually, if dark halos around galaxies extend out to 100 kpc, the galaxy disks could have formed from material falling from distances of $\geq 100 kpc$, whose infall times are of the order of $1 - 2 \, 10^9$ years, or $z \simeq 2.5 - 3$ ($q_0 = 1/2$). This may be supported by the observation of Hunstead et al. (1987) of an absorption line system at $z_{abs} = 3.1723$ in the direction of the quasar 2000-330, with an inferred metal abundance ~ 100 lower than the solar value. The low ionization level is compatible with an origin of the line system in the disc of an intervening galaxy.

Therefore, it is not implausible that at $z \simeq 3$ the galaxy disks started a very rapid phase of metal enrichment through enhanced star formation. In this case in few 10^7 years the metal abundance became high enough that a large part of the enhanced star formation activity has been shrouded by the dust (see e.g. Koo, 1986). Hence the optical wavelengths might not be optimal to search for enhanced phases of SF or even for primordial galaxies.

Contrariwise, the far-IR bands are very promising. Referring for instance to the $60\mu m$ wavelength (but longer wavelengths could be even more informative), we have roughly modelled an enhanced phase of SF activity assuming that at $z \sim 2.3 - 3$ all spiral galaxies were brighter in the far-IR band by a factor 100 during $2 \, 10^8$ years, the time necessary to roughly produce a solar metal abundance in the disc. The result is that the primeval disks of galaxies would clearly show up at about 10 mJy. Interestingly enough, this flux level can be achieved by ISO, whose launch is foreseen in 1993.

Extrapolating to the faintest fluxes the above count predictions, it turns out that the contributions to the extragalactic background at $60\mu m$ of evolving ASF galaxies is about $2 \, 10^4 Jy/sr$, of spirals/irr again $2 \, 10^4$; $0.5 \, 10^4 Jy/sr$ are contributed by Seyferts, while primeval galactic disks can add a further $10^4 Jy/sr$. The overall contribution at $100\mu m$ is then probably not exceeeding $0.2 MJy/sr$, a small fraction of the intensity estimated by Rowan-Robinson (1986).

6. OPTICAL COUNTS AND THE LUMINOSITY EVOLUTION OF AGNs

A dramatic improvement of our knowlwdge on optically selected QSOs has occurred in the recent past, thanks in particular to the use of multi-object spectroscopy and grism techniques on 4 meter class telescopes.

Two main problems have been addressed by various authors on the basis of the new data: i) modelling the cosmological evolution of objects at $z \leq 2$ and ii) looking at a possible decline of the cosmological evolution for $z > 2$. Here, rather than to discuss the details, we simply show that Luminosity Evolution models are able to reproduce the main features of the data at $z < 2$. On the other hand, what is happening beyond that epoch is still unclear.

6.1 The data base

A non exhaustive list of the data sets is presented below. A large fraction of them derive from quasar samples selected according to the UV excess criteria. They comprise: a) The Bright Quasar Survey (BQS) by Schmidt and Green (1983), including 114 sources over ~ 10000 sq. deg. down to an effective limiting magnitude $B \simeq 16.16$. Among them, 92 quasars have absolute magnitudes $M_B < -23$. The authors warn that for intrinsically fainter AGNs the sample is not likely to be complete since, as the nuclear luminosity diminishes, the nebulosity becomes relatively stronger and the integrated colours redder. This severe incompleteness is a typical limitation of all quasar samples. *b)* The AB and BF samples, including 22 quasars with $B \leq 18.25$ over 37 sq. deg. (Marshall *et al.*, 1983) and 35 quasars to $B \leq 19.8$ over 1.72 sq. deg. (Marshall *et al.*, 1984); *c)* The samples by Usher *et al.* (1983) and Mitchell *et al.* (1984), including 43 confirmed quasars with $16 \geq B \leq 18.3$; *d)* The sample by Boyle *et al.* (1987) of 170 confirmed quasars over ~ 500 colour selected objects down to B=21; *e)* The sample ol local Seyfert 1 and 1.5 nuclei by Cheng *et al.* (1985) at $B \leq 17$ (see Sect.3).

Quasar samples selected by ultraviolet excess criteria are reasonably complete only to $z \simeq 2.2$ (for higher redshifts the Ly α enters the B band). Multicolor techniques are in principle able to detect quasars beyond this limit. The most important samples based on such techniques are: *f)* The QSO sample selected by Koo *et al.* (1986) and Koo and Kron (1987) in the Selected Areas 68 and 57, comprising 151 quasar candidates down to B=23; *g)* the multicolor+grism selected sample by Zamorani *et al.* (1986) at $B \leq 22$.

High redshift quasars, in any event, are more easily selected from objective-prism surveys: *h)* the Curtis-Schmidt CTIO surveys, as summarized by Schmidt & Green (1983); *i)* the CFHT blue grens quasar survey by Crampton *et al.* (1985, 1987) with 117 QSOs at $B \leq 20.5$; *j)* the grism survey in the J band by Weedman (1985), which has allowed to investigate the QSO number counts and LF within the small redshift interval $2 < z < 2.5$.

The continuation of the QSO space distribution at $z > 2.5$ has been investigated on IIIaF plates and CCD detectors. Osmer (1982) has found no Ly α QSOs in the redshift range $3.7 < z < 4.7$ over 5 sq. deg. to an effective $B_{lim} \simeq 21$. Schmidt *et al.* (1986), using the PFUEI machine at Palomar over a 0.91 sq. deg. field, found 6 quasars in the range $2.1 < z < 2.6$, but no Ly α QSO at $z > 2.6$ to an estimated limiting magnitude $B \simeq 22$. This evidence of a paucity of faint ($B > 21$) quasars at $z > 2.6 \div 3$ from objective-prism surveys is in full agreement with findings by Koo *et al.* (1986) and Zamorani *et al.* (1986) based on colour selected surveys. On the contrary, at brighter limiting magnitudes and over more extended fields, Hazard & Mc Mahon (1985) and Hazard *et al.* (1986) were able to find quasars up to $z = 3.8$. They claim that no redshift cutoff (at least by $z \sim 4$) is apparent for quasars with $B \sim 18.5$.

These data seemed to favour a luminosity dependent redshift cutoff of QSOs, but they have been challenged by more recent data. Warren *et al.* (1987) reported the discovery of two QSOs with $z = 4.01$ and 3.42 significantly fainter ($m_R > 19$) than previously known for high redshift QSOs. Nine further QSOs with $3 < z < 3.8$ and $18.5 < m_R < 21.7$ have been detected by Schmidt *et al.* (1987) in a grism survey searching for Ly α emission with the Palomar 4 Shooter. Thus, a drastic redshift cutoff of the faint QSOs seems now excluded by the data, which rather favour a slow decline of the evolution, probably sharper for the fainter objects.

To summarize, three main general conclusions can be drawn from the data: *i)* there is a dramatic flattening of QSO counts at $B > 20$; *ii)* the Luminosity Functions at different redshifts are similar (Boyle et al, 1987; Koo & Kron, 1987); *iii)* the number density of QSOs at redshifts $z > 2.5$ is no more increasing at the same rate at increasing redshift, it may well be decreasing.

6.1 The cosmological evolution of optically selected AGNs

Several different attempts have been tried to model the cosmological evolution of optical AGNs and sometimes contradictory results have been claimed. We stress, however, that the QSO statistics in this waveband is likely to suffer severely from photometric errors, errors in the line emission subtraction and survey incompleteness (Wampler & Ponz, 1985; 1986). It is, hence, quite dangerous to confine to a few particular data sets and to draw detailed inferences from a strict application of some powerful statistical tests. We judge much safer to maintain some more general results complying with all constraints imposed by the available data sets.

Using all the above information, we have found that a LE model with an exponential dependence of L on τ (see e.g. eq.[4]) provides the best representation of the data ($\chi_\nu^2 \simeq 1.4$), while a p.l. dependence $L \propto L(z = 0)(1 + z)^\alpha$ gives a poorer fit ($\chi_\nu^2 \simeq 2.15$) (Franceschini et al., in preparation). A generalized LDDE model with free-form $f(L)$ (see eq.[3]) is only marginally consistent with the data ($\chi_\nu^2 \simeq 1.6$)

6.2 Few remarks on galaxy counts

Although the likely presence of dust could partly mask the evolution of Actively Star Forming galaxies in the optical (particularly in the blue) band, some evolutionary effects should also come out here, though possibly milder than in radio or far-IR bands.

Is there any hint of galaxy evolution in the optical band? Koo (1986), Ellis (1987) and Tyson (1987) have found that in the galaxy counts there is an extra component of faint blue galaxies at $m_B > 21$. In particular, Koo's data show that the fraction of field galaxies intrinsecally bluer than $B - V = 0.7$ has increased significantly from a local value of 40% to a 70% at z=0.4. Ellis (1987) found a surprising proportion of galaxies showing [OII] emission among objects with redshifts $0.2 < z < 0.4$. This suggests that galaxies at $z \simeq 0.4$ were undergoing much more prominent bursts of star formation than local objects do. Notice also that Windhorst et al., (1987) optically identified the complete sample of radio sources down to 70 μJy in the Hercules field and found that the magnitude distribution peaks at $m_B \simeq 21$. They also confirm that the fraction of emission line galaxies in the sample is significantly higher than the 30% found in local field galaxy samples.

All this is in qualitative agreement with an evolution in the optical band of the ASF galaxies, which are locally about 20-30% of the field galaxies.

7. EVOLUTION OF THE X-RAY SELECTED ACTIVE GALACTIC NUCLEI

The data sets relevant for discussing the statistical properties of X-ray selected AGNs are: *i)* the number counts and LLF in the 2-10 keV band derived from the HEAO-1 A2 experiment (Piccinotti et al., 1982); *ii)* counts, luminosity and redshift distributions from the Einstein Medium Sensitivity Survey (MSS), derived from a collection of several IPC frames in the 0.3-3.5 keV band (e.g. Gioia et al., 1984); and *iii)* the counts at the Einstein Deep Survey limit (Griffith, this volume). Although these samples can only provide rather loose limits to the space distribution of X-ray AGNs, a cosmological evolution is definitely indicated in this case also. The fact that the implied evolution rates (Table 1) are remarkably lower than those of radio and optical AGNs (e.g. Maccacaro, Gioia & Stocke, 1984) aroused the suspect that X-ray surveys would not sample the same AGN population as optical surveys do. Indeed, a clear deficit of X-ray AGNs resulted from predicting the number counts at X-ray frequencies by means of the available statistical information on the optical AGNs, together with the observed distribution of their X-ray to optical luminosity ratios (Franceschini, Gioia & Maccacaro, 1986; Setti, 1987 and references therein).

Although possible ways out have been proposed – either in terms of broadening effects in the observed distribution of the X-ray to optical luminosity ratios, or in terms of a substantial upward scaling of the MSS flux scale (Danese *et al.*, 1986; Setti, 1987) – the issue is not yet fully settled and awaits for the forthcoming extension of the MSS survey (Gioia *et al.*, 1987).

Important constraints on the X-ray AGN evolution are provided by the XRB's spectral intensity and fluctuations. In particular, De Zotti *et al.* (1982) pointed out that the smoothness of the XRB spectrum (well fitted in the 3 to 50 keV range by a thermal bremsstrahlung with $T \simeq 45 keV$; Marshall *et al.*, 1980), strongly argues against a > 30% contribution from objects with "wrong" spectra, such as intermediate luminosity AGNs, whose average spectral index is $\alpha_x \simeq 0.7$ (Mushotzky, 1982), or QSOs, which exhibit, at least in the few observed cases, an even steeper slope (Elvis *et al.*, 1985).

There is now a fairly general consensus that bright QSOs cannot contribute more than 20% of the XRB at 2 keV (Cavaliere *et al.* 1981; Schmidt & Green, 1986). On the other hand, the contribution of low luminosity AGNs is more difficult to assess. Persic *et al.* (1987), exploiting the HEAO-1 A2 data base and the published data of the Einstein Observatory, concluded that low luminosity AGNs $(L_x < 10^{42} erg/s)$ do not seem able to saturate the XRB. This conclusion seems also confirmed by the analysis of the arcminute scale fluctuations at the Deep Survey limit (Hamilton & Helfand, 1987).

Using the above information, Danese *et al.* (1986) have studied the evolution of X-ray AGNs, showing that these data are fitted by a luminosity dependent LE model, with sources at $log L_X(2 - 10 keV) < 43.5$ (corresponding to $M_B \geq -20$) evolving much more slowly than those with higher luminosities. (see Table 1).

Incidentally, the possible failure of the X-ray AGNs to account for all the XRB intensity suggested that a new astrophysical component should be devised, endowed with definite spectral properties, different from those of the known sources (Cavaliere *et al.*, 1981). Persic *et al.* (1987) have shown, in particular, that dwarf Seyfert Nuclei, such as those considered by Keel (1983), very likely cannot be relevant contributors to the XRB, whereas the contribution of ASF galaxies deserves further investigation.

8. DISCUSSIONS AND CONCLUSIONS

We have seen that cosmological evolution is required to explain observations of AGNs in all accessible wavebands, the radio, optical and X-ray. In addition, we have seen that a population of Actively Star Forming galaxies also evolves on similar time scales. It is interesting that such large amount of data can well be described by a simple model requiring Luminosity Evolution for the intrinsically bright sources and slower or no evolution for the faint ones. The evolution timescales are within the 22% of the Hubble time observed for optically selected AGNs and the 32% for the X-ray selected AGNs ($q_0 = 0.5$, see Table 1).

The circumstance, suggested in particular by radio and X-ray data, that low luminosity sources do not undergo relevant cosmological evolution, may be related to the existence of a number of processes able to fuel the central powerhouse and to garantee a basal luminosity level ($L_s \simeq 10^{43} erg/s$ or $M_B \simeq -20$) for times intervals of the order of the age of the Universe (Cavaliere *et al.*, 1985).

Many authors correctly warn against an overinterpretation of LE models. On the other hand, their success in depicting the main aspects of the data suggests that the average number of AGNs and the shape of their LF are approximately preserved at least during the cosmological epochs more accessible to observations ($z \lesssim 2$). The most obvious possibility is that since $z \simeq 2$ we are looking at the dimming phase of the already born AGNs (Cavaliere et al., 1985; Cavaliere and Vagnetti, 1987). In particular, Cavaliere et al. (1985) pointed out that in a large variety of models, based on the hypothesis that the power supply of AGNs is primarly gravitational, the time derivative $\dot{L}$ is a simply parametrised function of L itself: $dL/dt \propto -L^p$. In a pure LE model the time scale $\tau = L(dL/dt)^{-1}$ is a constant ($p = 0$). The assumption $p = 0$, together with the allowance for no evolution of low luminosity AGNs, results in the minimal model of eq. 4, widely used in our previous discussions and showing such a widespread success. This scenario requires $dL/dt < 0$ and $S = 0$ since $z \sim 2.5$.

On the other hand, Phinney (1983) pointed out that this class of models must tackle the problem of the accreted mass, which is obviously very large particularly for objects of high luminosity. As a consequence, the local objects must irradiate at luminosity levels much lower than their Eddington limit, the most luminous objects being also the most sub-Eddington emitters with $L/L_E \ll 10^{-2}$. However, the data (Wandel & Mushotzky, 1986) indicate average local values $L/L_E \geq 10^{-2}$.

Table 1. Parameters of the best-fit luminosity evolution models ($H_0 = 50, q_0 = 0.5$)

class	of	sources	(evol. time scale/t_0)$^{-1}$ κ	threshold luminosity of the evolution
RADIO		Steep-spectrum E/S0/QSOs	4.5	$\log L_{408} = 23.5 \ (W/Hz/sr)$
		Flat-spectrum E/S0/QSOs	3.4	–
		ASF galaxies	4.3	–
$60\mu m$		ASF galaxies	3.5	–
OPTICAL		QSOs	4.6	$M_B \simeq -20$
X-RAY		AGNs	3.2	$\log L_{2-10} \simeq 43.5 \ (erg/s)$

A rather different scenario has been proposed by Blandford (1986, and this book), in which the cosmological evolution of AGNs is ruled by a more complex pattern of births-and-deaths of the sources. The model assumes that: i) all galaxies harbor a black hole in their nucleus; ii) there is a local gas supply independent on the mass of the hole; iii) the gas is accreted at a critical rate until the gas supply falls well below the Eddington limit and the object virtually disappears as an AGN. It is interesting that this model is not dissimilar from the former scenario by Cavaliere et al. (1985) as for the ability to reproduce the data, while getting the advantage of not requiring very massive B.H. ($M \gg 10^9 M_\odot$) in a few percent of the galaxies, but rather a more spread distribution of B.H. masses among the galaxies.

In a compromise scenario the activity is discontinous (Cavaliere & Vagnetti, 1987) and possibly triggered by interactions among active galaxies and companions.

The net result would be that the number of potential AGNs should increase with cosmic time and the Eddington constraint alleviated. This possibility is indeed supported by many obsevations (Heckman *et al.*, 1984; Yee and Green,1984, Stockton & McKenty, 1987; Hutchings, 1987) suggesting that gravitational interactions actually play an important role in triggering the activity of galaxies (see also Roos, 1985).

To summarize, LFs at different cosmological epochs, as well as observations of L/L_E ratios and B.H. mass distributions in local AGNs and QSOs are strongly needed to meaningfully constrain the models and to bridge the gap between kinematical studies and the physics of the evolution. The HST will provide a unique opportunity for this purpose.

Radio and optical surveys provide clear evidences of a decline of the evolution at high redshifts, albeit its precise behaviour is still unclear. If dust obscuration becomes severe at $z > 3 - 4$, as envisaged by Ostriker & Heisler (1984), then the true evolution of optically selected QSOs would be masked. This would not explain, in any case, the decline of the evolution in the radio band, wherein the dust has no effect. In addition, relevant differences in the high-z decline are seen at radio and optical frequencies. If optical data suggest that the slowing down of the evolution concerns the low luminosity QSOs, while high luminosity objects seem to keep a fast evolution (Crampton, 1987, and Mc Mahon, 1987, this volume), in the radio domain even the most powerful radio sources seem to decline at relatively low redshifts. Of course, different physical conditions in the environment can play an important role (Barthel, 1986), whereas gravitational lensing can affect in particular the optical counts of high luminosity QSOs (Peacock & Dunlop, 1987). We see that our knowledge of the evolution of AGNs at epochs earlier than $z \sim 3$ is still quite uncertain.

We have also shown that a strong cosmological evolution is not an exclusive property of AGNs. ASF galaxies are also evolving, as a population, on very short timescales, comparable with those of the AGNs. Unless we admit that this is simply due by chance, we should try to infer reasons for this similarity. There are two physical situations that could be in common between AGNs and ASF galaxies: *i)* the gravitational interactions as a triggering mechanism, and *ii)* the presence and the amount of the gas available to form stars and/or to fuel the nucleus. It could be that the similarity of the timescales is related to the fact that one or both the above mentioned circumstancies dictate the evolution of both populations.

There is actually also an increasing evidence (Lawrence, 1987) that a distinction between starburst galaxies and galaxies with active nuclei is difficult, particularly for high luminosity starburst galaxies. Sanders *et al.* (1987) proposed that the ultrabright far-IR galaxies are the initial stage in the QSO evolution. An extreme point of view on the subject is the starburst model for AGNs proposed by Terlevich *et al.* (1987). To sum up, radio and far-IR observations have opened a new interesting field that could be crucial to understand the the formation and evolution of AGNs and galaxies in general.

Aknowledgments: Our understanding of these subjects has benefited of many years of collaboration and a number of enlightening discussions with A. Cavaliere and G. De Zotti. L. Danese thanks also the organizers of the Meeting and the staff of the IOA for their warm ospitality.

REFERENCES

Avni, Y. & Bachall, J.L., 1980, *Ap. J.*, **235**, 694.

Bartel, P.D., 1986, in *Quasars*, eds. G. Swarup and B.K. Kapahi, Reidel, Dordrecht, 181.

Blandford, R.D., 1986, in *Quasars*, eds. G. Swarup and B.K. Kapahi, Reidel, Dordrecht, 359.

Boyle, B.J., Fong, R., Shanks, T. & Peterson, B.A., 1987, *M.N.R.A.S.* **227**, 717.

Cavaliere, A., Danese, L., De Zotti, G. & Franceschini, A., 1981, *Astron. Astroph.*, **97**, 269.

Cavaliere, A., Giallongo, E., Messina, A. & Vagnetti, F. 1983, *Ap. J.*, **269**, 57.

Cavaliere, A., Giallongo, E., and Vagnetti, F. 1985, *Ap. J.*, **296**, 402.

Cavaliere, A. & Vagnetti, F. 1987, to be published in the Proceedings of the Conference *Supermassive Black Holes*, George Mason University.

Cheney, J.E. & Rowan-Robinson M. 1981, *M.N.R.A.S.*, **197**, 313.

Cheng, F.Z., Danese, L., De Zotti, G. & Franceschini, A. 1985, *M.N.R.A.S.* **212**, 857.

Condon, J.J. 1984, *Ap. J.*, **284**, 44.

Condon, J.J. & Mitchell, K.J. 1984. *Astr. J.*, **89**, 610.

Crampton, D., Cowley, A.P. & Hartwick, F.D.A. 1987, *Ap. J.* **314**, 129.

Crampton, D., Schade, D. & Cowley, A.P., 1985, *Ap. J.* **90**, 987.

Danese, L., De Zotti, G., & Franceschini, A. 1985, *Astron. Astroph.*, **143**, 277.

Danese, L., De Zotti, G., Fasano, G. & Franceschini, A. 1986, *Astron. Astroph.*, **161**, 1.

Danese, L., De Zotti, G., Franceschini, A. & Toffolatti, L. 1987, *Ap. J. Lett.*, **318**, L15.

de Jong, T., Klein U., Wielebinski R. & Wunderlich, E., 1985. *Astron. Astroph. Lett.*, **147**, L6.

De Zotti, G. *et al.*, 1982, *Ap.J.*, **253**, 47.

Dressel, L.L. & Condon, J.J., 1978. *Ap. J. Suppl.*, **36**, 53.

Ellis, R., 1987, in *Observational Cosmology*, eds. Hewitt A., Burbidge, G. and Fang L.Z., Reidel, Dordrecht, 367.

Elvis, M., Wilkes, B. & Tananbaum, H., 1985, *Ap. J.*, **292**, 357.

Fasano, G. & Franceschini, A. 1986, *M.N.R.A.S.* **225**, 155.

Feigelson, E.D. & Nelson, P.I., 1985. *Ap. J.*, **293**, 192.

Felten, J.E., 1977. *Astr. J.*, **82**, 861.

Fomalont, E.B., Kellermann, K.I., Wall, J.V. & Weistrop, D. 1984, *Science*, **225**, 23.

Franceschini, A., Gioia, I.M., Maccacaro, T., 1986, *Ap. J.*, **301**, 124.

Franceschini, A., Danese, L., De Zotti, G. & Toffolatti, L. 1987a, *M.N.R.A.S.*, in press.

Franceschini, A., Danese, L., De Zotti, G. & Xu, C., 1987b, submitted to *M.N.R.A.S.*

Gioia, I.M., Maccacaro, T., Schild, R.E., Stoke, J.T., Liebert, J.W., Danziger, I.J., Kunth, D., Lub, J., 1984, *Ap. J.*, **283**, 495.

Gioia, I.M., Maccacaro, T., Wolter, A., in *Observational Cosmology*, eds. Hewitt A., Burbidge, G. and Fang L.Z., Reidel, Dordrecht, 593.

Hacking, P. & Houck, J.R., 1987 *Ap. J. Suppl.*, **63**, 311.

Hacking, P., Condon, J.J. & Houck, J.R., 1987 *Ap. J. Lett.*, **316**, L15.

Hamilton, T.T. & Helfand, D.J., 1987, *Ap. J.*, **318**, 93.

Hazard, C., and Mc Mahon, R. 1985, *Nature*, **314**, 238.
Hazard, C., Mc Mahon, R., and Sargent, W.L.W. 1986, *Nature*, **322**, 38.
Heckman, T.M., Mailey, G.K. & Green, R.F., 1984, *Ap. J.*, **281**, 525.
Helou, G., Soifer, B.T. & Rowan-Robinson, M., 1985 *Ap. J. Lett.*, **298**, L7.
Hubble, E. & Tolman, R., 1935, *Ap. J.*, **82**, 302.
Hunstead, R.W., Pettini, M., Blades, J.C. & Murdoch, H.S., 1987, in *Observational Cosmology*, eds. Hewitt A., Burbidge, G. and Fang L.Z., Reidel, Dordrecht, 799.
Hutchings, J.B., 1987, in *Observational Evidence of Activity in Galaxies*, eds. Khachikian Y.E. et al.
Keel, W.C., 1983, *Ap. J. Suppl.*, **52**, 229.
Koo, D.C. 1986, in *Spectral Evolution of Galaxies*, eds. Chiosi C. & Renzini A., Reidel, Dordrecht, 419.
Koo, D.C., and Kron, R.G., Cudworth, K.M. 1986, *P.A.S.P.*, **105**, 107.
Koo, D.C., and Kron, R.G., 1987, STSI preprint.
Kron, R.G., Koo, D.C. & Windhorst, R.A. 1985, *Astron. Astroph.*, **146**, 38.
Lawrence, A., 1987, to be published in *'Comets to Cosmology*, prodeedings of 3rd IRAS Conference, Springer-Verlag.
Lawrence, A., Walker, D., Rowan-Robinson, M., Leech, K.J. & Penston, M.V., 1986. *M.N.R.A.S.*, **219**, 687.
Maccacaro, T., Gioia, I.M. & Stocke, *et al.*, 1984, *Ap. J.*, **283**, 486.
Marshall, F.E. *et al.*, 1980, *Ap. J.*, **235**, 4.
Marshall, H.L., Tananbaum, H., Zamorani, Huchra, J.P., Braccesi, A., and Zitelli, V. 1983, *Ap. J.*, **269**, 42.
Marshall, H.L., Avni, Y., Braccesi, A., Huchra, J.P., Tananbaum, H., Zamorani, G., and Zitelli, V. 1984, *Ap. J.*, **283**, 50.
Mathez, G. 1978, *Astron. Astroph.*, **68**, 71.
Mathez, G. & Nottale, L., 1982, *Astron. Astroph.* **113**, 336.
Mitchell, K.J., Warnock III,A., and Usher, P.D., 1984, *Ap. J. Lett.*, **287**, L3.
Mushotzky, R.F., 1982, *Ap. J.*, **256**, 92.
Neugebauer, G., Miley, G.K., Soifer, B.T., Clegg, P.E. 1986. *Ap. J.*, **308**, 815.
Nilson, P., 1973, *Uppsala General Catalogue of Galaxies*, Uppsala Astronomical Observatory, Uppsala.
Osmer, P.S. 1982, *Ap. J.*, **253**, 28.
Ostriker, J.P., & Heisler, J., 1984, *Ap. J.*, **278**, 1.
Partridge, R.B., Hilldrup, K.C. & Ratner, M.I., 1986. *Ap. J.*, **308**, 46.
Peacock, J.A. & Gull, S.F. 1981, *M.N.R.A.S.* **196**, 611.
Peacock, J.A. 1983, *M.N.R.A.S.* **202**, 605.
Peacock, J.A. 1985, *M.N.R.A.S.* **217**, 601.
Peacock, J.A. 1986, *M.N.R.A.S.* **217**, 601.
Peacock, J.A. & Dunlop, J.S. 1986, in *Quasars*, eds. G. Swarup and B.K. Kapahi, Reidel, Dordrecht, 455.
Persic, M., De Zotti, G., Danese, L., Palumbo, G., Franceschini, A., Boldt, E.A. & Marshall, F.E. 1987, submitted to *Ap. J.*
Phinney, E.S., 1983, Ph.D., Cambridge University, England.
Piccinotti, G., Mushotzky, R.F., Boldt, E.A., Holt, S.S., Marshall, F.E., Serlemitsos, P.J. & Shafer, R.A., 1982. *Ap. J.*, **253**, 485.
Rees, M., 1986, in *Quasars*, eds. G. Swarup and B.K. Kapahi, Reidel, Dordrecht, 1.
Robertson, J.G. 1977, *Austr. Jour. of Phys.*, **30**, 241.

Roos, N., 1985, *Ap. J.* **294**, 486.

Rowan-Robinson, M., Walker, D., Chester, T., Soifer, T. & Fairclough, J., 1986. *M.N.R.A.S.*, **219**, 273.

Rowan-Robinson, M., 1986, *M.N.R.A.S.*, **219**, 737.

Sanders, D.B. *et al.*, 1987, in *Star Formation in Galaxies*, eds. Persson C.J., U.S. Government Printing Office.

Schmidt, M., and Green, R.F. 1983, *Ap. J.*, **269**, 352.

Schmidt, M., and Green, R.F. 1986, *Ap. J.*, **305**, 68.

Schmidt, M., Schnider, D.P., and Gunn, J.E. 1986, *Ap. J.*, **306**, 411.

Schmitt, J.H.M.M, 1985. *Ap. J.*, **293**, 178.

Silk, J., 1987, in *Observational Cosmology*, eds. Hewitt A., Burbidge, G. and Fang L.Z., Reidel, Dordrecht, 391.

Smith, B.J., Kleinmann, S.G., Huchra, J.P. & Low, F.J., 1987, *Ap. J.*, **318**, 161.

Soifer, B.T., Sanders, D.B., Madore, B.F., Neugebauer, G., Danielson, G.E., Elias, J.H., Persson, C.J., & Rice, W.L., 1987. *Ap. J.*, in press.

Stockton, A. & McKenty, J.W., 1987, *Ap. J.* **316**, 584.

Terlevich, R., Melnick, J., Moles, M., 1987, in *Observational Evidence of Activity in Galaxies*, eds. Khachikian Y.E. et al., 499.

Toffolatti, L., Franceschini, A., De Zotti, G. & Danese, L., 1987. *Astron. Astroph.*, **184**, 7.

Tyson, J.A. & Seizer, P., 1987, in *169th A.A.S. Meeting*, 1047.

Usher, P.D., Green, R.F., Huang, K.L., and Warnock, III.A. 1983, in *Quasars and Gravitational Lenses*, Proceeding of the 24th Liege Astrophysical Symposium, 245.

van der Laan, H., Katgert, P. & Oort, M.J.A. 1986, in *Structure and Evolution of Active Galactic Nuclei*, eds. G. Giuricin, F. Mardirossian, M. Mezzetti and M. Ramella, Dordrecht, Reidel, 437.

Wall, J.V., Pearson, T.J. & Longair, M.S., 1986, *M.N.R.A.S.*, **193**, 683.

Wall, J.V. 1983, in *Origin and Evolution of Galaxies*, eds. B.J.T. Jones and J.E. Jones, Reidel, Dordrecht, 295. 683.

Wall, J.V., Benn, C.R., Grueff, G. & Vigotti M. 1986, *Highlights Astr.*, **7**, 345.

Wampler, E.J. & Ponz D., 1985, *Ap. J.*, **298**, 448.

Wampler, E.J. & Ponz D., 1986, in *Quasars*, eds. G. Swarup and B.K. Kapahi, Reidel, Dordrecht, 439.

Wandel, A. & Mushotzky, R.F., 1986, *Ap. J. Lett.*, **306**, L61.

Warren, S.J., Hewitt, P.C., Irwin, M.J., McMahon, R.G., Bridgeland, M.T., Bunclark, P.S., & Kibblewhite, E.J. 1987, *Nature* **325**, 131.

Weedmann, D.W. 1985, *Ap. J. Suppl.*, **57**, 423.

Windhorst, R.A., Dressler, A. & Koo, D.C. 1987. *'Observational Cosmology', IAU Symp. N° 124*, eds. Burbridge, G. & Fang, L.Z., Reidel, Dordrecht, Holland, 573.

Windhorst, R.A., van Heerde, G.M. & Katgert, P. 1984, *Astron. Astroph. Suppl.*, **58**, 1.

Windhorst, R.A., Miley, G.K., Owen, F.N., Kron, R.G. & Koo, D.C., 1985, *Ap. J.* **289**, 494.

Yee, H.K.C. & Green, R.F., 1984, *Ap. J.*, **280**, 79.

Zamorani, G., Zitelli, V., and Marano, B. 1986, in *Quasars*, G. Swarup and V.K. Kapahi Eds., 33.

THE EVOLUTION OF EXTRAGALACTIC X-RAY SOURCES

A.C. Fabian
Institute of Astronomy
Madingley Road
Cambridge CB3 OHA
U.K.

ABSTRACT. The various classes of extragalactic X-ray source and their possible contributions to the X-ray Background are reviewed. Particular attention is paid to recent limits to the source density from fluctuation analysis and to the problem of spectral mismatch. A new class of X-ray source is probably needed to explain the Background. Bremsstrahlung from a hot Intergalactic Medium is a strong possibility if enough energy can be found.

1 INTRODUCTION

An X-ray survey of the extragalactic sky reveals clusters of galaxies, normal galaxies, active galaxies and, if either the beam is large or the detector sensitive enough, the X-ray background. This background radiation may be dominated by the integrated contribution of many active galaxies, of intergalactic gas, or of some new class of X-ray source. All of the contributing sources must evolve and the X-ray background provides us with an observational envelope for evolutionary models. Surveys and source counts, too, show the evolution of X-ray sources, although it is still early days in this work and most observed objects have redshifts less than one. The potential for future X-ray studies is enormous. It is clear that we shall be able to define the evolution of clusters and active galaxies and, through fluctuation analyses of the X-ray background, the extent and growth of clustering on scales up to 1000 Mpc. Perhaps also, we can discover new objects or events at large redshifts and witness dissipation during galaxy formation.

In this review, I first discuss observations of the well-known classes of source such as clusters, Seyfert galaxies and QSO and outline their X-ray properties. The surveys and source counts are then described before an account is given of the X-ray background. A recent fluctuation analysis provides a strong constraint here on the evolution of distant active galaxies. The contributions of each class of X-ray source to the X-ray background are still fairly uncertain and some estimates are given.

The possible evolution of clusters and active galaxies suggested by theory is outlined. The spectrum of the X-ray background is a strong constraint and may argue for a new class of X-ray source. The current situation is that no nearby or known source has clearly the same spectral shape as the X-ray background or the required number density and so either there is some strong spectral evolution in known sources or there is a new source.

There is some confusion created by the broad wavelength coverage of X-ray astronomy, which corresponds to about four decades. A similar range is obtained in going from millimetre-wave radio techniques to the shortest wavelengths of IUE. Surveys made in

51

N. Kaiser and A. N. Lasenby (eds.), The Post-Recombination Universe, 51–68.
© *1988 by Kluwer Academic Publishers.*

Table 1

THE X-RAY WAVEBANDS

Energies	0.1-1 Soft	1-10	10-100 Hard	100-1000keV Soft γ-ray
Objects	Clusters AGN	Clusters AGN	AGN	AGN
Instruments	Einstein IPC/HRI EXOSAT (ROSAT)	HEAO-1 A2 EXOSAT TENMA/GINGA	HEAO-1 A4	
Background	Galactic	Extragalactic well-measured	Extragalactic	MeV bump
Flux limits AXAF & XMM	$\sim 5.10^{-14}$ $\mathrm{erg\,cm^{-3}\,s^{-1}}$ $\sim 10^{-15}$	$\sim 5.10^{-12}$ $\mathrm{erg\,cm^{-3}\,s^{-1}}$ $\sim 10^{-15}$		

soft X-rays are often combined with measurements of the X-ray background in the 'ordinary' and hard X-ray bands without stressing the extrapolations that are necessary. No one would be particularly convinced if submillimetre observations were used to predict the UV background without a strong theoretical basis for doing so! Table 1 is a guide to these X-ray wavebands and some recent instruments.

2 CLASSES OF X-RAY SOURCE

2.1 Clusters and Groups of Galaxies

All clusters and groups of galaxies appear to be X-ray sources of luminosity $10^{42}\,\mathrm{erg\,s^{-1}} < L_X < 10^{46}\,\mathrm{erg\,s^{-1}}$ due to thermal bremsstrahlung and line radiation from hot gas. The temperature of the gas is close to the virial temperature of the cluster, $2 < kT < 10\,\mathrm{keV}$. Individual early-type galaxies also contain much hot gas at $kT \sim 1\,\mathrm{keV}$. The mass of X-ray emitting gas in a cluster is comparable to the mass of galaxies, and together they constitute $\sim 10 - 20$ per cent of the total binding mass.

Many of the lower luminosity clusters are irregular in appearance - the non-XD clusters of Jones & Forman (1984). There is also a significant fraction (~ 10 per cent) of clusters which appear double (Forman et al. 1981). This is further evidence for the clustering of clusters. In clusters of all luminosities with a central dominant galaxy, the X-ray emission is usually strongly peaked on that galaxy. The radiative cooling time of the gas there is less than a Hubble time and it is likely that a cooling flow operates. There will be more discussion of cooling flows later, but for the moment it is worth noting that this situation holds in most nearby clusters (e.g. Virgo, Hydra, Centaurus, Perseus ...). Coma is the obvious exception, but it does have 2 large central galaxies.

An important sample of extragalactic X-ray sources was obtained with the HEAO-1 A2 instrument by Piccinotti et al. (1982). Clusters form about one half of that sample. Detailed X-ray images of clusters can be found in the works of Jones & Forman (1984) and of Abramopoulos & Ku (1983). X-ray luminosity functions have most recently been published by Piccinotti et al. (1982), Soltan & Henry (1983), Kowalski et al. (1984) and by Johnson et al. (1983). Correlations of the various X-ray and other cluster properties can be found in the paper of Mushotzky (1984). An excellent review of all X-ray aspects of clusters of galaxies is given by Sarazin (1986).

2.2 Normal Galaxies

As already mentioned, early-type galaxies are X-ray luminous through their hot gas. Late-type galaxies contain point X-ray sources which are mostly binary X-ray sources, supernovae and active stars, in an order of decreasing luminosity. Since most so-called 'normal' galaxies are of relatively low luminosity ($10^{38} < L_X < 10^{41}\,\mathrm{erg\,s^{-1}}$), they are only detected within the nearest few tens of Mpc and little is known of their evolution. Recent summaries of their X-ray properties are given by Forman, Jones & Tucker (1984), Fabbiano & Trinchieri (1985), Canizares, Fabbiano & Trinchieri (1987).

Some starburst galaxies have interesting ratios of X-ray to optical luminosities L_X/L_{opt} (Fabbiano et al. 1982). In normal galaxies this ratio is $\sim 10^{-3} - 10^{-4}$ but it can reach up to $\sim 10^{-1}$ as the metal abundance decreases (Clark et al. 1978, Stewart et al. 1982). It is likely that X-ray binaries are responsible by comparison with our Galaxy, M31 and the Magellanic Clouds.

2.3 Seyfert 1 Galaxies

These form the major class of active galaxy at the current epoch. (Some) Seyfert 2 galaxies may be obscured Seyfert 1s, but we shall ignore them here. The Seyfert 1 galaxies constitute roughly the other half of the Piccinotti sample. Their luminosities are in the range $10^{42} < L_X < 10^{45}\,\mathrm{erg\,s^{-1}}$ and their spectra are well-characterized by a power-law of energy index $\alpha \approx 0.7$ (Mushotzky 1982; Petre et al. 1984; Turner & Pounds, private communication) over the energy range $0.5 < \epsilon \lesssim 100\,\mathrm{keV}$. At the lowest energies, some Seyfert 1 galaxies show evidence of a 'soft excess' of emission (e.g. Mkn 841, Arnaud et al. 1985; Mkn 335, Pounds et al. 1987). This may be modified blackbody radiation from the inner regions of an accretion disc. At the highest energies, the spectra may extend to a few MeV before dropping (e.g. NGC 4151; Perotti et al. 1982). Rapid X-ray variability is a further characteristic of Seyfert galaxies (see e.g. Barr & Mushotzky 1986).

The luminosity function of Seyfert 1 galaxies has a slope of 2.75 above a luminosity of $\sim 10^{43}\,\mathrm{erg\,s^{-1}}$ and flattens below (Elvis, Soltan & Keel 1984). The point where the luminosity function flattens provides the dominant contribution of these sources to the X-ray background. Some Seyfert galaxies show intrinsic photoelectric absorption (e.g. NGC 4151 Holt et al. 1980). This is probably prevalent in sources of luminosity below $\sim 10^{43}\,\mathrm{erg\,s^{-1}}$, otherwise they would feature much more strongly in soft X-ray surveys (Fabian, Kembhavi & Ward 1982).

2.4 BL Lac Objects

These objects have steep X-ray spectra and are often highly variable. The luminosity function is flat (Schwartz & Ku 1983) and explained by beaming (Urry & Shafer 1984).

2.5 QSO

Only 3C273 was well-studied before 1978. Now there are data on several hundred QSO, mostly from the Einstein Observatory (Tananbaum *et al.* 1979; Zamorani *et al.* 1981). QSO are often considered to be a strong candidate for the origin of the X-ray background and show clear evolution, so a little more time will be devoted to them here. I shall concentrate on 6 recent samples, which cover most aspects of QSO detected and observed with the Einstein Observatory. It should be noted that these observation cover only the soft energy range of $\sim 0.3 - 3 \, \text{keV}$.

a) *Radio-loud Quasars* (Worrall *et al.* 1987). Observations of 114 radio-loud quasars were considered, of which 102 were detected, a success rate of 89 per cent. Radio-loud quasars clearly have a higher L_X than radio-quiet QSO and at a given L_{opt}, L_X increases as the radio luminosity L_r increases. This correlation is most notable when core emission (as opposed to diffuse or lobe emission) is used for L_r.

b) *Bright Quasar Sample* (Tananbaum *et al.* 1987). 57 quasars were detected out of a sample of 66 with $B < 16.6$, a success rate of 86 per cent.

c) *Braccesi Faint Sample* (Marshall *et al.* 1984). 15 quasars were detected out of a UV excess sample ($B < 19.8$) of 35 (43 per cent).

d) *Optically Selected Sample* (Anderson & Margon 1987). In this case the quasar was not the target of the X-ray observation. Instead, ~ 400 quasar candidates were obtained from slitless spectra of 17 square degrees which had been imaged deep in X-rays. Two samples were formed; a 'complete' one with $1.8 < z < 3, B < 19.5$ and 78 in a 'narrow' sample in redshift and L_{opt} space. Only 25 per cent were actually detected in these samples. Anderson & Margon found that the 'effective' $\alpha_{ox} = 1.50 \pm 0.03$, where α_{ox} is the energy spectral index of a (hypothetical) power-law joining monochromatic flux measurements of a quasar at 2500 Å and 2 keV (in the rest-frame of the quasar).

e) *Heterogeneous Sample* (Avni & Tananbaum 1987b). 53 QSOs of various types were observed with 24 detections (45 per cent success). The QSO were chosen as the X-ray targets (as was also the case in samples a, b, and c). When combined with the bright quasar and Braccesi faint samples, Avni & Tanabaum conclude that any X-ray quiet population is less than 8 per cent of all QSO.

f) *Optically Selected Sample* (Kriss & Canizares 1985). They consider 178 optically-selected quasars (and some Seyfert 1 galaxies) of which 77 are detected (43 per cent). This sample is interesting as they make a point of including some 'high' redshift ($z > 1$) quasars. They find that their upper limits are consistent with a sample drawn from the detected population and so conclude that there is no obvious X-ray-quiet quasar population.

A general result from many studies is that $L_X \propto L_{opt}^{0.7}$. This correlation is apparently stronger than any redshift dependence. This together with the low mean-redshift of the detected QSO, shows that their evolution is weaker in the X-ray band than in the optical.

Although it appears from most studies and samples that all quasars are X-ray 'loud', the low detection rate of optically faint, $z < 1$ quasars - which after all are the dominant quasar population - means that this statement is open to question. Whether the 'effective' α_{ox} for the detected population applies to most quasars is also questionable. Most of the samples do not emphasize the facts that the majority of X-ray detected quasars have $\bar{z} \ll 1$ and that most 'serendipitous' X-ray quasars have $\bar{z} < 0.5$. The Anderson & Margon and Kriss & Canizares samples are the only ones to seriously concentrate on 'typical' quasars. To be fair, however, it should be noted that if $\bar{\alpha}_{ox}$ is greater than 1.5, then few 'typical' quasars were detectable with the Einstein Observatory. How valid the answers for 'typical' quasars are depends upon whether you believe that upper limits can be safely combined with detections in a sample of low detection rate. They obviously can if they really are the same population. The samples studied so far strongly suggest that all quasars are X-ray loud with, say, $\alpha_{ox} < 1.7$, but it is still possible that there is a much broader range. Only the 'tip of the iceberg' has been examined so far.

Few X-ray spectra of QSO have been obtained in detail. 3C273 has been studied in detail (Worrall *et al.* 1979; Turner *et al.* 1987) over a wide energy range. Wilkes & Elvis (1987) have estimated spectral indexes from Einstein Observatory Imaging Proportional Counter (IPC) spectra of 24 quasars ($\sim 0.3 - 3$ keV). They obtain $\alpha_X \sim 0.2 - 1.8$ with $\bar{\alpha}_X \sim 0.5$ for radio-loud quasars and $\bar{\alpha}_X \sim 1$ for radio-quiet QSO. The errors on individual spectral indexes are fairly large. Canizares & White (private communication) find that radio-loud quasars have α in the range 0.2 to 0.8. There are a few results from EXOSAT (e.g. McGlynn *et al.* 1985) and GINGA (Tanaka, private communication) indicating that in the 1 - 10 Kev band, $\alpha_X \sim 0.7$, as for Seyfert galaxies. As discussed later, X-ray surveys are providing further indications of $\alpha_X \gtrsim 0.7$, which is significantly steeper than the spectral index of the X-ray background in the 3 - 20 KeV band of 0.4.

The X-ray spectrum clearly cannot be a simple power-law at lower energies as $\alpha_X \neq \alpha_{ox}$. Where the break occurs is not yet known. There could even be soft excesses (cf. PG1211+143 as reported by Bechtold *et al.* 1987). The lack of any dramatic redshift dependence argues against any abrupt spectral change above a few keV*.

In summary, the currently available evidence suggests that most quasars are X-ray sources with spectra significantly steeper than the X-ray background. It should be remembered, however, that few high redshift, radio-quiet quasars have been detected (we shall return to this when discussing the deep surveys) and the energy band where most detections have occurred is at a lower energy than that where the extragalactic X-ray background is observed.

3 X-RAY SURVEYS

Mention has already been made of the Piccinotti sample (1982) which remains the most complete X-ray sample in the 2-20 keV range. This was assembled from the HEAO-1 A2 data and covers the whole sky above $|b| = 20$ degrees. The next all sky survey will not occur before that made by ROSAT (see J. Truemper's contribution). In the meantime, some fractions of the sky have been surveyed at varying depths by the Einstein Observatory and by EXOSAT.

The *Einstein Medium Sensitivity Survey* (MSS; Gioia *et al.* 1984) produced 112 sources out of 40 square degrees of images searched to a (0.3 to 3 keV) flux level of $7.10^{-14} < S < 2.10^{-12}$ erg cm^{-2} s^{-1}. The sources have been identified and 68 per cent found to be extragalactic. Most of those are relatively modest luminosity AGN at low $\bar{z}$, providing further evidence for weaker evolution of AGN in X-rays when compared to the optical and radio bands (Maccacaro *et al.* 1984).

The MSS has been followed up by the *Extended Medium Sensitivity Survey* (EMSS; Gioia *et al.* 1987a) yielding about 550 sources from 780 square degrees. The task of source identification is still underway but can be summarized as;

a) at least 67 clusters with $\bar{z} \sim 0.2$ and 12 per cent with $z > 0.3$. The cluster source counts give $N(> S) \propto S^{1.04 \pm 0.23}$ (Gioia *et al.* 1987b; Morris' contribution to these Proceedings), i.e. significantly flatter than the Euclidean slope.

b) of 403 identifications so far, 171 are AGN and 16 are BL Lacs (many of the rest are active stars in our Galaxy). The source counts for the AGN are $N(> S) \propto S^{-1.71}$, which is steeper than Euclidean and indicates evolution, (Gioia *et al.* 1987c). The BL Lacs

---

* There is at present no evidence that the spectra of QSO flatten above a few keV. This would allow them to match the background.

have a flat log N - log S, probably due to the flat luminosity function (Cavaliere et al. 1986). For the MSS, $\bar{z}$ for AGN is ~ 0.4, whereas the EMSS has > 18 with $z > 1$. $\bar{M}_V = -23.7$. Again it should be noted that very few AGN with $L_X < 10^{42}$ erg s^{-1} were detected, perhaps because of intrinsic absorption. Problems in the inter-comparison and -calibration of the various source counts and instrument sensitivities are reviewed by Setti (1987).

Some progress has been made on determining the X-ray spectra of the extra-galactic MSS sources (Maccacaro et al. 1984) by using IPC hardness ratios and the source detection rate as a function of latitude (and thus Galactic photoelectric absorption column, N_H). The result is $\alpha_X = 0.93 \pm 0.04$, with a dispersion of ~ 0.4. The AGN have $\alpha_X = 0.99 \pm 0.06$. These values apply to the $0.3 - 3$ keV band.

Giommi & Tagliaferri (1987) have carried out an EXOSAT high galactic latitude survey yielding 130 sources in 570 square degrees of which ~ 60 per cent are identified so far. 16 of the identification are extragalactic and the remaining 56 are also probably extragalactic (they are the optically fainter objects). From the MSS, they can predict the expected number of sources and compare with the observed number (knowing N_H) and so obtain an estimate of the spectrum. Giommi & Tagliaferri (1987) find $\bar{\alpha} > 1$. The soft X-ray spectra of serendipitous AGN thus appear to be steep. In this respect, it is worth noting that Branduardi-Raymont et al. (1985) found 6 QSO in the EXOSAT image of Coma, several of which were not previously seen in the Einstein image. This may indicate that at the softest X-ray energies, where EXOSAT was more sensitive than the Einstein Observatory, the spectra are yet steeper.

The deepest X-ray surveys so far were made with the Einstein Observatory and reached $\sim 10^{-14}$ erg cm^{-2} s^{-1} in the 1-3 keV band (Giacconi et al. 1979; Griffiths et al. 1983). 19 ± 8 sources per square degree were found above 2.6×10^{-14} erg cm^{-2} s^{-1}. Most of these appear to be faint QSO (Griffiths et al. 1983; Griffiths contribution to these Proceedings).

The total source counts of the MSS and deep surveys are roughly compatible with a Euclidean extrapolation of the $1 - 3$ keV source counts down to a level of $S - 10^{-14}$ erg cm^{-2} s^{-1}.

4 THE X-RAY BACKGROUND

The spectrum of the X-ray background is well-measured in the 3-30 keV range (Marshall et al. 1980). It fits a 40 keV thermal bremsstrahlung spectrum well, indeed better than the microwave background fits a blackbody spectrum (De Zotti et al. 1982)! As already noted, the energy spectral index below 20 keV is about 0.4.

Below 1 keV, the spectrum is dominated by Galactic emission. There is also a small Galactic component, probably of different origin, detectable up to about 10 keV (Warwick et al. 1980; Iwan et al. 1982). Between 1 and 3 keV where the Einstein Observatory results apply, the intensity is poorly determined. Garmire & Nousek (1980) have reported an excess there which has not been confirmed (or denied). The background spectrum rolls over at about 30 keV and then flattens again at ~ 100 keV to give the so-called 'MeV bump'.

Fluctuations are detected in the X-ray background intensity, mostly due to those sources which are just not detected (i.e. P(D) or 'confusion noise'). Source counts can be used to predict the level of these fluctuations and, when this contribution is accounted for, any 'excess' fluctuations give $\delta I/I < 2$ per cent on a scale of a few degrees (Shafer 1983; Shafer & Fabian 1983).

The cosmic dipole has been observed at 95 per cent confidence (Shafer & Fabian 1983). It is confirmed by the energy dependence of the Compton-Getting effect measured by HEAO-1 (Boldt 1987). There could possibly be other large scale deviations at a level

< 0.5 per cent on large scales.

Deep IPC fields have been used to assess the small-scale isotropy of the background in the 1-3 keV band (Stewart and Fabian, reported in Fabian 1981; Hamilton & Helfand 1987). These studies find that the P(D) noise is less than predicted on the basis that the deep survey source counts continue to extrapolate to lower fluxes and conclude that the counts must flatten or turn over just below that level.

The careful and detailed work of Hamilton & Helfand (1987) suggests a flux level cut-off (or flattening) to the source counts of $\sim 6.5 \times 10^{-15}$ erg cm^{-2} s^{-1}. This has the interesting conclusion that sources like those detected in the deep survey cannot produce more than $\sim$ 50 per cent of the 1-3 keV background. The intensity of the background in that band is estimated by extrapolating the 3-30 keV spectrum to lower energies and may, of course, be underestimated. This is a very firm conclusion for it is difficult to think of any way of reducing fluctuations. The only loophole that occurs to me is the small pixel size (one arcmin) used by Hamilton & Helfand (1987), which is similar to (if not even smaller than) the point-spread-function of the detector. This could result in source smearing and a reduction in the spread of the predicted P(D) distribution. Stewart and I, however, used a 2 arcmin pixel and drew a similar conclusion from a much more preliminary analysis, so smearing is not a likely way out. It should be testable by retrieving the deep survey counts from a fluctuation analysis of MSS fields. Given that X-ray sources may enjoy clustering, it is likely that the above estimate of 50 per cent is an upper limit (Barcons & Fabian 1987) and that deep-survey-like sources contribute only 20-30 per cent of the extrapolated X-ray background. Note that this method does not require that the background be actually measured in order to limit the source count behaviour. It does require that the background intensity is known when fractional contributions are estimated. If Garmire & Nousek's factor of 2 increase in the extragalactic intensity by 1 keV is confirmed, then sources such as those found in the deep survey need contribute only 10 per cent or so of the 3-30 keV background at an energy of, say, 5 keV. I shall have more to say on this issue later.

5 THE X-RAY EVOLUTION OF GALAXIES, CLUSTERS AND AGN

The X-ray background places strong constraints on the density, luminosity and spectral evolution of X-ray sources.

5.1 Galaxies

The high specific X-ray luminosity of some starburst objects (e.g. NGC 5408 at 5.10^{47} erg s^{-1} $M_{\odot}^{-1}$) means that young galaxies could produce the X-ray background (Stewart et al. 1982; see also Bookbinder et al. 1979). X-ray binaries do have appropriately hard spectra, but they do not extend to 100s of keV, which would be necessary if they are at redshifts of 3 or more. Little more can be done at the present time without a clear theoretical model.

5.2 Clusters of Galaxies

Perrenod (1978; 1980) has modelled the (one-dimensional) evolution of gas in developing cluster potential wells. Clusters brighten as the wells develop and strong to moderate evolution is predicted. The subclustering evident in clusters indicates that the real picture is more complicated and Cavaliere et al. (1986) have had some success with crude 3D hydrodynamic simulations. Kaiser (1986) has considered cluster evolution in a hierarchical fashion and predicts strong density evolution.

Henry et al. (1979; 1982) find some weak evidence for cluster evolution in Einstein Observatory data. This could just be Malmquist bias (see also White et al. 1981, Soltan & Henry 1983, Henry & Lavery 1984 and Henry & Henriksen (1986) for discussions of observed X-ray emission from distant clusters). More distant clusters could be cooler (Perrenod & Henry 1981). The contribution of clusters to the X-ray background at 3 keV is less than 10 per cent.

The prevalence of cooling flows now in clusters of galaxies and their presence in poor clusters (e.g. MKW3s, in which gas is cooling at $\sim 100 \, M_\odot \, yr^{-1}$, Canizares et al. 1983) suggests, on a hierarchical clustering picture, that they could have been more common and stronger in the past (Stewart et al. 1984; Fabian et al. 1986). One is certainly present at $z \sim 0.5$ in 3C295 (Henry & Henriksen 1986). The X-ray images and particularly the X-ray spectra show that gas is cooling at rates of $\sim 10 - 1000 \, M_\odot \, yr^{-1}$ in many clusters. What the cooled gas condenses into is not clear, although low-mass stars are a plausible sink (Fabian, Nulsen & Canizares 1982; Sarazin & O'Connell 1983). This means that at least some dark matter around massive galaxies is baryonic. Strong cooling flows in the past may then be associated with the formation of galaxy haloes. Indeed, if a significant fraction of the mass of a galaxy was ever at its virial temperature, then it must have been a bright X-ray source.

The merging of subclusters to form the present day rich clusters and groups would have led to the disruption and mixup of central cooling flows (McGylnn & Fabian 1984). It is therefore possible that the distant X-ray sky is littered with peaked soft X-ray sources.

Only a small fraction of the flow is needed to power even the most luminous quasar and again it is likely that the evolution of radio-loud quasars and radio galaxies is implicated in the growth of clustering (Fabian et al. 1986). The relatively recent decline in the source activity is then due to the recent formation of clusters. This means that many distant cooling flows may have highly luminous active cores, i.e. quasars. It is likely moreover that the X-ray emission from the quasar will outshine that from the cluster or group and so, even with a positive density evolution of clusters, fewer clusters may be recognized by their X-ray emission at larger redshifts in instruments of modest spatial resolution (e.g. the IPC). If, as the surveys of Yee & Green (1984, 1987) indicate, radio-loud quasars lie in poor clusters, then the existing surveys will find quasars at the expense of clusters. This effect may already be involved in the MSS counts. Future, higher spatial-resolution instruments (ROSAT and AXAF) will sort out this problem. For the time being we must rely on indirect optical methods. We have already shown that 3C48 is probably surrounded by a luminous cooling flow (Fabian et al. 1987).

Some of the evolution of radio-quiet QSO, which are presumably in spirals, could also be due to the growth of clusters, as ram pressure stripping will increasingly affect cluster spirals.

5.3 AGN and QSO

Danese et al. (1986 and these Proceedings), Cavaliere et al. (1985) and Tucker & Schwartz (1986) have constructed evolution models for the X-ray emission from QSO. The strongest constraint on their evolution is, of course, the X-ray background and this is complicated by uncertainties in the spectrum of QSO. As stressed already, where spectra and counts are available they apply principally to photon energies of 1 to 2 keV, whereas the background is only well-measured above 3 keV. Most workers assume $\alpha_X \sim 0.4$, which is only relevant to radio-loud quasars. This last class can at most only provide 10 per cent of the background intensity (Kembhavi & Fabian 1982).

The clearest direct result is the summed emission of the deep survey sources

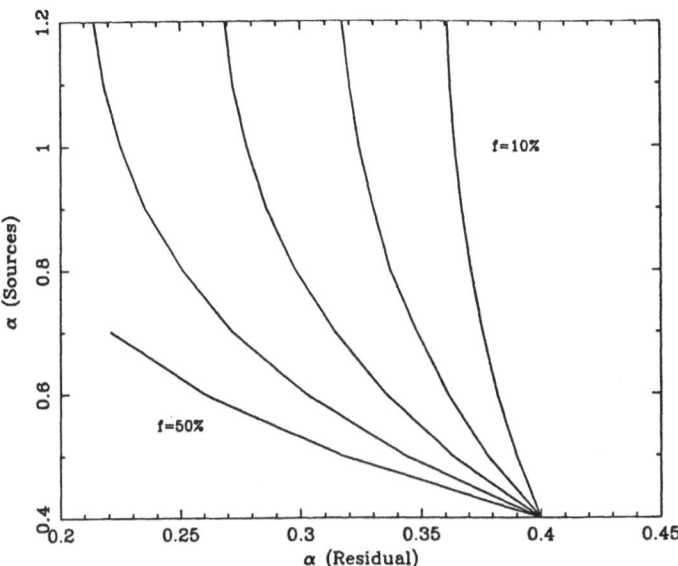

Figure 1a. Residual 3-10 keV spectral index obtained after subtracting a source fraction equal to 10, 20, 30, 40 and 50 per cent of the 3 keV intensity from a background of spectral index 0.4. The source spectra shown range from 0.4 to 1.2.

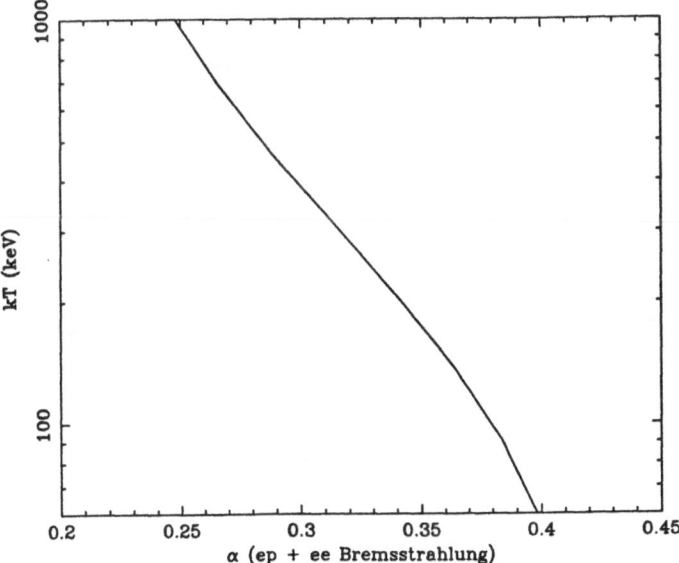

Figure 1b. Effective 3-10 keV spectral index of ep and ee thermal bremsstrahlung from gas at $kT = 60$ to 1000 keV, when redshifted so that $kT_o = 40(1 + z)$ keV.

Table 2

(SOME) ESTIMATES FOR f_{QSO}% AT 2 keV
from Giacconi & Zamorani (1987)

	Setti 85	Schmidt & Green 86	G & Z 87
QSO	> 20	8-13	> 40
Sey 1	34	29	
BL Lac	< 7	1	
Galaxies	~ 13	4-8	~ 13
Clusters	< 10	10	
Σ	~ 70	53-62	> 53

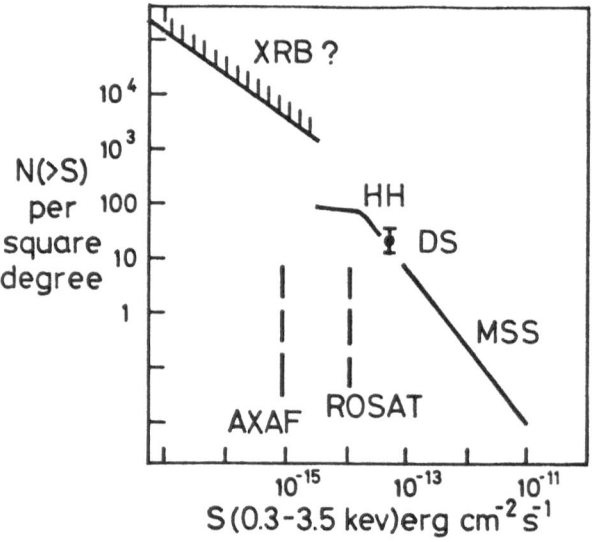

Figure 2. 0.3-3 keV source counts from the *Einstein Observatory* Deep Survey, the Medium Sensitivity Survey and inferred from the IPC fluctuation analysis. The level at which an intensity equal to the extrapolated spectrum of the 3-30 keV X-ray background is reached by sources with a Euclidean (3/2) slope is shown. This figure has been adapted from a similar one in the paper of Giacconi & Zamorani (1987).

(Giacconi *et al.* 1979) of 26 ± 11 per cent at 2 keV (estimated assuming $\alpha_X = 0.4$ for the individual sources). This was initially taken as strong evidence that QSO make the whole X-ray background. Indirect methods rely on optical QSO counts * and estimates of α_{ox}. More recently the estimates for the summed contributions of point sources have settled

---

* Early attempts using power-law optical QSO counts and single values of α_{ox}

around 50 to 100 per cent (Table 2). A careful breakdown of the QSO contribution by Anderson & Margon (1987) shows that from QSO with $B < 19.5$ is 21 ± 8 per cent. A QSO contribution of up to 70 per cent is consistent with the data, but it does rely heavily on the faint, $z > 1$ population for which detection rates are low (~ 25 per cent).

The Hamilton/Helfand (1987) constraint on a 50 per cent point source contribution (of the kind contributing to the known source counts) means that the estimates in Table 2 are wrong, especially if we treat the 50 per cent as an upper limit on the assumption that QSO cluster. Why? Firstly, it should be noted that Seyfert 1 galaxies are usually assumed to make a substantial contribution. This is estimated on the basis that Seyfert 1s do not evolve. Perhaps they have negative X-ray evolution or QSO evolve into Seyferts Either way, that fraction attributed to Seyfert 1 galaxies is highly questionable, especially since observers might classify a luminous Seyfert 1 at $z \sim 0.2$ as a QSO. Secondly, as belaboured in Section 2.5, the X-ray properties of faint, radio-quiet QSOs at redshifts greater than one are not known. Thirdly, many parameters (> 21) are required in the calculations (Maccacaro, Gioia & Stocke 1984). Finally, as also belaboured at regular intervals in this paper, the spectra are not well-known, either of the QSO or of the X-ray background in this energy range. There is indeed a problem in making 34 per cent of the X-ray background with a spectral index of $\alpha_X = 0.7$, as found for Seyfert 1 galaxies. The remaining X-ray background would then have an unusually flat spectrum, requiring an even more exceptional spectrum from QSO.

To clarify this point, I have calculated the 3-10 keV spectral index of the residual spectrum after removing a fraction, f, for sources of spectral index between 0.4 and 1.2 from the X-ray background, assumed to be of slope 0.4 (Fig. 1a) Most of the estimates in Table 2 would require QSO to have $\alpha_{ox} \sim 0.3$ which is completely at odds with the Wilkes & Elvis (1987) result that most radio-quiet QSO studied (which are the brightest ones) have $\alpha_{ox} \sim 1$.

The Hamilton/Helfand and Wilkes/Elvis results show that sources similar to those observed have neither the numbers nor the spectral shape to account for most of the X-ray background. This leads to the conclusion, reached earlier by Cavaliere et al. and by Leiter & Boldt (1982), that the X-ray background is due to a new class of source. A corollary to this is that, unless this source fortuitously drops in flux below 3 keV, the X-ray background must have an excess below 3 keV to accommodate Seyfert 1 galaxies, QSO and so on.

5.4 A New Population

The new source population must exceed 5000 sources per square degree in order that their Poisson fluctuations are not obviously detected (Fig.2 and Hamilton & Helfand 1987). This seems to rule out optically-selected QSOs and means that the individual sources are not particularly powerful (Boldt & Leiter 1987). What the sources actually are is not at all clear. Boldt & Leiter (1987) favour primordial active galaxies, noting that their number density is comparable to that of Seyfert 1 galaxies. Of course the density could be much higher. They also favour thermal Comptonization to give the flat spectrum. However, as α_X depends upon the Compton y-parameter rather sensitively in this range, it is not obvious why the parameters are so-tuned in such sources. Zdziarski (1987) has found one solution.

---

soon showed that the optical counts must flatten around $B \sim 20 - 21$ in order that the X-ray background not be exceeded. This result has since been confirmed optically. In this context, the Hamilton/Helfand result means that the QSO counts must flatten still further beyond $B \sim 23$, which is just fainter than the magnitude corresponding to many deep survey identifications (Griffiths, these Proceedings).

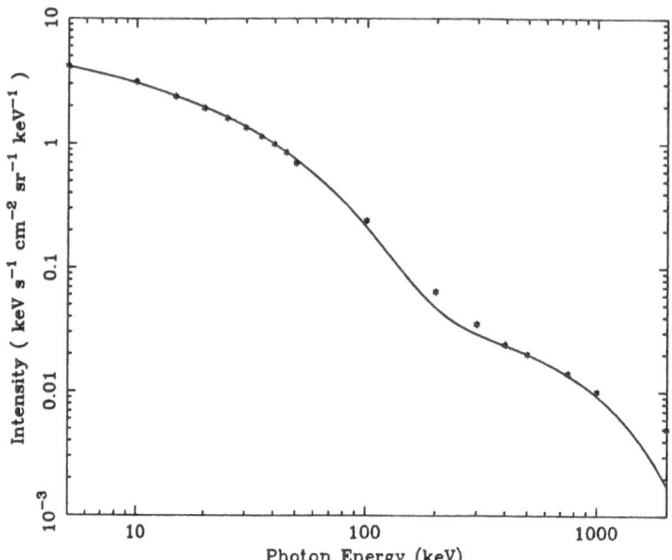

Figure 3. Comparison of the spectrum of the X-ray background (star points from Marshall *et al.* (1980) and Gruber *et al.* (1984)) with the emission of a 2-phase hot gas (from Barcons & Fabian 1987). The error bars on the spectral intensity are very small below ~ 50 keV but exceed the deviations of the fit above ~ 100 keV.

I have investigated nonthermal sources with Comptonization in which there is much more power injected in relativistic electrons than in soft photons available for up-scattering. This automatically produces a hard spectrum. The electrons are allowed to cool and are injected with $\gamma_{max} = 600$. The model is similar to one for gamma-ray bursts by Zdziarski & Lamb (1986) which are also observed to have a flat bremsstrahlung-like spectrum. The spectral turnover around 30-40 keV is particularly difficult to produce as the only natural turnover occurs around $m_e c^2$, so that the sources have to be at $z \sim 50$. This will continue to be a significant problem in synthesizing the X-ray background spectrum from point sources. Unless they have very similar parameters, including redshift, it may be difficult to obtain a sharp enough break at 30-40 keV. Of course, the 'MeV bump' may be due to the integrated emission from AGN, as has been discussed for the past decade.

6 A HOT INTERGALACTIC MEDIUM

This has a venerable pedigree, dating back to Hoyle's interpretation based on the steady-state theory in 1964. His idea was soon refuted by Gould & Burbidge (1964) and was not revived until 1972 by Cowsik & Kobetich. Since then it has been considered by Field & Perrenod (1979), Sherman (1979), Marshall *et al.* 1980, Fabian (1981), Guilbert & Fabian (1986) and Barcons (1987). A variant is discussed by Daly (1987).

As already noted, thermal bremsstrahlung gives an excellent fit to the 3-100 keV spectrum of the X-ray background (Marshall *et al.* 1980), even when evolution and some ($<$ 20 per cent) point source contribution are included (Fig. 3 and Guilbert & Fabian 1986).

It is usually assumed that the gas is heated at some redshift z_H and then cools adiabatically so that $n \propto (1 + z)^3$ and $T \propto (1 + z)^2$. As the spectrum required fits $kT \approx 40\,\text{keV}, kT_H \approx 40(1 + z_H)\,\text{keV}$ Then, if $z_H > 3$, the gas is mildly relativistic and electron-electron bremsstrahlung is important. The bremsstrahlung emissivity $\propto n^2$, so most of the background originates from close to z_H, which means that the resultant spectrum is fairly close to that of a single-temperature gas.

At moderate z, Compton cooling on the microwave background is important and exceeds adiabatic cooling for $z > 3.5$. The resultant spectrum is steepened unacceptably by Compton cooling if $z_H > 6$ (Guilbert & Fabian 1986). Compton scattering also distorts the microwave background, particularly at the shortest wavelengths. The distortion is smaller (in the 5-10 mm band) than that reported by Matsumoto et al. (1987).

The degree to which some point source contribution can be included is seen by combining the curves in Figs.1a and 1b. This last figure shows the 3-10 keV spectral index of electron-proton and electron-electron bremsstrahlung from gas of temperature kT. Clearly this has to correspond to the index of the residual spectrum (Fig. 1a) if emission from diffuse gas dominates the X-ray Background. $\alpha(residual) < 0.3$ if $z_H < 6$, so $f < 35$ per cent when $\alpha(sources) = 0.7$. A more stringent result would be obtained if we were to choose, say, 3-5 keV spectral indexes, but I believe that such an exercise should involve fitting HEAO-1 A2 pulse height data (proportional counters only record the energy of incoming photons to about 20 per cent accuracy around 5 keV, so model-fitting is the appropriate course).

The Compton limit on z_H could be relaxed if some extended heating were available, but it seems pointless to pursue this at present without some good theoretical guide to the origin of the heat. The heat source is certainly prodigious (too much for some, see Field 1980) and involves $\sim$ 33 per cent of the energy in the microwave background at z_H. However, so does the 'bump' on the microwave spectrum of Matsumoto et al. (1987) and that is presumably not primordial. No one seems to think that is too much energy. On this 'philosophical' basis, I will not reject a hot IGM as the source of the X-ray background. Incidentally, the energy is only $\sim$ 0.1 per cent of the rest-mass energy of the IGM, about right for nucleosynthesis or Hoyle's decaying neutrons (which were created in too large a number, not energy).

The biggest problem is the required number density of baryons, corresponding to $\Omega_b > 0.22$ (Guilbert & Fabian 1986; Barcons 1987). Current deuterium (and Li) measurements require $\Omega_b < 0.19$ (Young et al. 1984). Baryon inhomogeneities from the quark-hadron era could relax this problem (Applegate, Hogan & Scherrer 1987) but I understand that lithium may prevent this (see Olive 1987). The required Ω_b can be reduced by clumping the gas, although the isotropy of the X-ray and microwave backgrounds are a severe constraint. Guilbert & Fabian (1987) invoke pressure confinement of small blobs by a still hotter gas which also gives most of the 'MeV' bump (see also Barcons & Fabian 1987) and Daly (1987) uses gravitational confinement by very deep potential wells formed by exotic particles which have since decayed.

A final problem is that of electron-ion coupling, for which there is insufficient time by 2-body interactions. Barcons (1987) has computed the spectrum assuming that the protons remain cold. It may even be unnecessary for the electron distribution to be a close approximation to a Maxwellian in order that the resultant emission fits the spectrum.

An origin for the X-ray background in a hot-intergalactic medium * still remains attractive to me, despite all the problems. Perhaps abundances measured from small parts

---

* It is possible that most of the IGM (apart from the tiny fraction now in HI) has always been ionized. In that sense, there never has been a 'Post-Recombination Universe'!

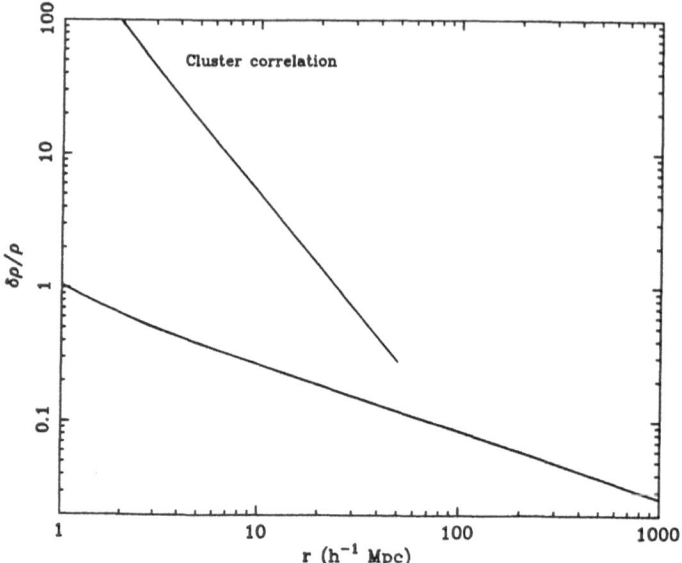

Figure 4. Possible fluctuations in the large-scale X-ray emissivity distribution (and presumably the matter distribution) detectable by the next-generation of X-ray telescopes (AXAF & XMM) in a few hours. The cluster-cluster correlation function is indicated. Note that $\delta\rho/\rho$ is scaled to current values. See Barcons & Fabian (1987) for details.

of the 5 (or less) per cent of baryons in visible galaxies are unrepresentative and maybe there was a major energy event at $z \sim 5$. The first explanation (i.e. bremsstrahlung) for the first cosmic background discovered may have some validity yet. (Even the decay of spontaneously created neutrons seems no more bizarre to me than the flailing of superconducting strings or the decay of exotic particles!). More of the Universe may be due to baryons than is generally imagined. The heat source is most probably some form of nucleosynthesis.

7 FINAL COMMENTS

X-ray sources do, of course, evolve. The evolution of AGN appears to be less in the X-ray band than at other wavelengths. The X-ray background provides a strong constraining envelope to estimates of any such evolution. Such a constraint becomes very strong if, as seems likely, the 3-30 keV X-ray background is due to some new class of source.

The evolution of clusters will be particularly well-studied by sensitive soft X-ray telescopes. Hierarchical clustering may be implicated, through cooling flows and the pressure of the intracluster medium, in the recent evolution of quasars and QSO. The X-ray properties of a typical, $z \sim 2$, QSO are poorly-determined at present. It is probably X-ray loud and will be louder still if it has active core radio emission. Whether the evolutionary histories of Seyfert 1 galaxies and QSO are separate or joint is unclear.

The recent IPC fluctuation analysis of Hamilton & Helfand (1987) is very important in showing that point sources such as those already observed cannot make all the

X-ray background. It should be checked using a larger bin size in order to eliminate possible source smearing problems. If QSO have even a small amount of clustering (see Shaver, these Proceedings) then known classes of source can only make 20 – 30 per cent of the 1-3 keV background (Barcons & Fabian 1987). Studies of the spectrum and intensity of the background in this energy range will help to define the evolution of QSO.

Finally, it is worth remarking again that fluctuation analyses of the X-ray background are a powerful method for measuring the homogeneity and isotropy of the recent Universe. (Rees 1980; Fabian 1981; Kaiser 1982; Shafer & Fabian 1983; Barcons & Fabian 1987). One deep AXAF or XMM image should measure the fractional X-ray emissivity on scales of hundreds of Mpc down to a few Mpc to between ten and one hundred per cent (Fig. 4). If this corresponds to $\delta\rho/\rho$ (as most analyses assume) then we shall be able to trace the growth of clustering. Unfortunately, the day when we can do this does not seem to come any closer.

Acknowledgements

I thank Xavier Barcons and Ajit Kembhavi for discussions and help, Isabella Gioia and Tommaso Maccacaro for providing preprints and advance information on the EMSS, and The Royal Society for support.

References

Abramopoulos, F. & Ku, W-M., 1983. *Astrophys. J.*, **271**, 446.
Anderson, S.F. & Margon, B., 1987. *Astrophys. J.*, **314**, 111.
Applegate, J.H., Hogan, C.J. & Scherrer, R.J., 1987. *Phys.Rev.D*, **35**, 1151.
Arnaud, K.A. *et al.*, 1985. *Mon. Not. R. astr. Soc.*, **217**, 105.
Avni, Y. & Tananbaum, H., 1986. *Astrophys. J.*, **305**, 83.
Barcons, X., 1987. *Astrophys. J.*, **313**, 547.
Barcons, X. & Fabian, A.C., 1987. *Mon. Not. R. astr. Soc.*, in press.
Barr, P. & Mushotzky, R.F., 1986. *Nature*, **320**, 421.
Bechtold, J., Czerny, B., Elvis, M., Fabbiano, G. & Green, R.F., 1987. *Astrophys. J.*, **314**, 699.
Boldt, E.A., 1987. *Phys. Rep.*, **146**, 215.
Boldt, E.A. & Leiter, D.L., 1987. *Astrophys. J.*, **322**, L1.
Bookbinder, J., Cowie, L.L., Krolik, J.H., Ostriker, J.P. & Rees, M.J., 1980. *Astrophys. J.*, **237**, 647.
Branduardi-Raymont, G., Mason, K.O., Murdin, P.G. & Martin, C., 1985. *Mon. Not. R. astr. Soc.*, **216**, 1043.
Canizares, C.R., Stewart, G.C. & Fabian A.C., 1983. *Astrophys. J.*, **272**, 449.
Canizares, C.R., Fabbiano, G. & Trinchieri, G., 1987. *Astrophys. J.*, **312**, 503.
Cavaliere, A., Santangelo, P., Tarquini, G. & Vittorio, N., 1986. *Astrophys. J.*, **305**, 651.
Cavaliere, A., Giallongo, E. & Vagnetti, F., 1985. *Astrophys. J.*, **296**, 402.
Cavaliere, A., Giallongo, E. & Vagnetti, F., 1986. *Astr. Astrophys.*, **156**, 33.
Clark, G., Doxsey, R., Li, F., Jernigan, G. & Van Paradijs, J., 1978. *Astrophys. J.*, **221**, L37.
Cowsik, R. & Kobetich, E.J., 1972. *Astrophys. J.*, **177**, 585.
Daly, R.A., 1987. *Astrophys. J.*, **322**, 20.
Danese, L., De Zotti, G., Fasano, G. & Franceschini, A., 1986. *Astr. Astrophys.*, **161**, 1.

De Zotti *et al.*, 1982. *Mon. Not. R. astr. Soc.*, **253**, 47.

Elvis, M., Soltan, A. & Keel W.C., 1984. *Astrophys. J.*, **283**, 479.

Fabbiano, G., Feigelson, E. & Zamorani, G., 1982. *Astrophys. J.*, **256**, 397.

Fabbiano, G. & Trinchieri, G., 1985. *Astrophys. J.*, **296**, 430.

Fabian, A.C. 1981. *Ann. N.Y. Acad.Sci.*, **375**, 235.

Fabian, A.C., Nulsen, P.E.J. & Canizares, C.R., 1982. *Mon. Not. R. astr. Soc.*, **201**, 933.

Fabian, A.C., Arnaud, K.A., Nulsen, P.E.J. & Mushotzky, R.F., 1986. *Astrophys. J.*, **305**, 9.

Fabian, A.C., Crawford, C.S., Johnstone, R.M. & Thomas, P.A., 1987. *Mon. Not. R. astr. Soc.*, **228**, 963.

Fabian, A.C., Kembhavi, A.K. & Ward, M.J., 1981. *Space Sci. Rev.*, **30**, 113.

Field, G.B., 1980. In: *Some Strangeness in the Proportion*, ed Woolf, H., Addison-Wesley.

Field, G.B. & Perrenod, S.C., 1977. *Astrophys. J.*, **215**, 717.

Forman, W., Schwarz, J., Jones, C., Liller, W. & Fabian, A.C., 1979. *Astrophys. J.*, **234**, L27.

Forman, W., Bechtold, J., Blair,W., Giacconi, R., Van Speybroeck, L. & Jones, C., 1981. *Astrophys. J.*, **243**, L133.

Forman, W., Jones, C. & Tucker, W., 1985. *Astrophys. J.*, **293**, 102.

Garmire, G. & Nousek, J., 1981. *B.A.A.S*, **12**, 853.

Giacconi, R. *et al.* 1979. *Astrophys. J.*, **234**, L1.

Giacconi, R. & Zamorani, G., 1987. *Astrophys. J.*, **313**, 20.

Gioia, I.M. *et al.* 1984. *Astrophys. J.*, **283**, 495.

Gioia, I.M., Maccacaro, T., Schild, R.E., Wolter, A., Stocke, J.M., Morris, S.L. & Danziger, I.J., 1987. in *Proc. I.A.U. 121*, (eds. Kachikian *et al.*), 329.

Gioia, I.M., Maccacaro, T., Morris, S.L., Schild, R.E., Stocke, J.T. & Wolter, A., 1987. Preprint.

Gioia, I.M., Maccacaro, T. & Wolter, A., 1987. *in Proc. I.A.U. 124*.

Giommi, P. & Tagliaferri, G., 1987. *in Observational Cosmology*, (eds Hewitt *et al.*), Reidel.

Gould, R.J. & Burbidge, G.R., 1963. *Astrophys. J.*, **138**, 969.

Griffiths, R.E. *et al.*, 1983. *Astrophys. J.*, **269**, 375.

Gruber, D.E., Rothschild, R.E., Matteson, J.L. & Kinzer, R.L., 1984. *MPE Rep.*, **184**, 129.

Guilbert, P.W. & Fabian, A.C., 1986. *Mon. Not. R. astr. Soc.*, **220**, 439.

Hamilton, T.T. & Helfand, D.J., 1987. *Astrophys. J.*, **318**, 93.

Henry, J.P. *et al.*, 1979, *Astrophys. J.*, **234**, L15.

Henry, J.P., Soltan A., Briel, U. & Gunn, J.E., 1982. *Astrophys. J.*, **262**, L1.

Henry, J.P. & Lavery, R.J., 1984. *Astrophys. J.*, **280**, 1.

Henry, J.P. & Henriksen, M.J., 1986. *Astrophys. J.*, **301**, 689.

Holt, S.S. *et al.*, 1980. *Astrophys. J.*, **241**, L13.

Hoyle, F., 1963. *Astrophys. J.*, **137**, 993.

Iwan, D. *et al.* 1982. *Astrophys. J.*, **260**, 111.

Johnson, M.W., Cruddace, R.G., Ulmer, M.P., Kowalski, M.P. & Wood, K.S., 1983. **266**, 425.

Jones, C. & Forman, W., 1984. *Astrophys. J.*, **276**, 38.

Kaiser, N., 1982. *Mon. Not. R. astr. Soc.*, **198**, 1033.

Kaiser, N., 1986. *Mon. Not. R. astr. Soc.*, **222**, 323.

Kembhavi, A.K. & Fabian, A.C., 1982. *Mon. Not. R. astr. Soc.*, **198**, 921.

Kowalski, M.P., Ulmer, M.P., Cruddace, R.G. & Wood, K.S., 1984. *Astrophys. J. Suppl.*, **56**, 403.

Kriss. G.A. & Canizares, C.R., 1985. *Astrophys. J.*, **297**, 177.

Leiter, D. & Boldt, E.A., 1982. *Astrophys. J.*, **260**, 1.

Maccacaro, T., Gioia, I.M. & Stocke, J.T., 1984. *Astrophys. J.*, **283**, 486.

Maccacaro, T., Gioia, I.M., Wolter, A., Zamorani, G. & Stocke, J.T., 1987. *Astrophys. J.*, Submitted.

Marshall, F.E., Boldt, E.A., Holt, S.S., Miller, R., Mushotzky, R.F., Rose, L.A., Rothschild, R. & Serlemitsos, P.J., 1980. *Astrophys. J.*, **235**, 4.

Marshall, H.L. *et al.*, 1984. *Astrophys. J.*, **283**, 50.

Matsumoto, T., Hayakawa, S., Matsuo, H., Marakami, H., Sata, S., Lange, A.E. & Richards, P.L., 1987. *Astrophys. J.*in press.

McGlynn, T.A., Tennant, A.F., Shafer, R.A. & Stewart, G.C., 1985. *Space. Sci, Rev.*, 40, 633.

McGlynn, T.A. & Fabian, A.C., 1984. *Mon. Not. R. astr. Soc.*, **208**, 709.

Mushotzky, R.F., 1982. *Astrophys. J.*, **256**, 92.

Mushotzky, R.F. 1984. *Phys. Scripta*, **T7**,157.

Olive, K.A., *Nature*, **330**, 700.

Perotti, F. *et al.*, 1981. *Astrophys. J.*, **247**, L63.

Perrenod, S.C., 1978. *Astrophys. J.*, **226**, 566.

Perrenod, S.C., 1980. *Astrophys. J.*, **236**, 373.

Perrenod, S.C. & Henry, J.P., 1981. *Astrophys. J.*, **247**, L1.

Petre, R., Mushotzky, R.F., Krolik, J.H., & Holt, S.S., 1984. *Astrophys. J.*, **280**, 499.

Piccinotti, G., Mushotzky, R.F., Boldt, E.A., Holt, S.S., Marshall, F.E., Serlemitsos, P.J. & Shafer, R.A., 1982. *Astrophys. J.*, **253**, 485.

Pounds, K.A., Stanger, V., Turner, T.J., King, A.R. & Czerny, B., 1987. *Mon. Not. R. astr. Soc.*, **224**, 443.

Rees, M.J., 1980. In: *Proc. IAU 92*, eds Abell, G.O. & Peebles, P.J.E., Reidel.

Sarazin, C.L., 1986. *Rev. Mod. Phys.*, **58**, 1.

Sarazin, C.L. & O'Connell, R.W., 1983. *Astrophys. J.*, **258**, 552.

Setti, G., 1987.*in Observational Cosmology*,(eds Hewitt *et al.*), Reidel.

Schwartz, D.A., Schwarz, J. & Tucker, W.H., 1980. *Astrophys. J.*, **238**, L59.

Schwartz, D.A. & Ku, W.H-M., 1983. *Astrophys. J.*, **266**, 479.

Shafer, R.A., 1983. *PhD thesis*, Univ. of Maryland.

Shafer, R.A. & Fabian, A.C., 1983. In: *Proc. IAU Symposium 104*, eds Abell G.O. & Chincarini, G., Reidel.

Sherman, R.D., *Astrophys. J.*, **232**, 1.

Soltan, A. & Henry, J.P., 1983. *Astrophys. J.*, **271**, 442.

Stewart, G.C., Fabian, A.C., Terlevich, R.J. & Hazard, C., 1982. *Mon. Not. R. astr. Soc.*, **200**, 61P.

Stewart, G.C., Fabian, A.C., Jones, C. & Fabian, A.C., 1984. *Astrophys. J.*, **285**, 1.

Tananbaum, H. *et al.*, 1979. *Astrophys. J.*, **234**, L9.

Tananbaum, H., Avni, Y., Green, R.F., Schmidt, M. & Zamorani, G., 1987. *Astrophys. J.*, **305**, 57.

Tucker, W.H. & Schwartz, D.A., 1986. *Astrophys. J.*, **308**, 53.

Turner, M.J. *et al.*, 1987. Preprint.

Urry, C.M. & Shafer, R.A., 1984. *Astrophys. J.*, **280**, 569.

Warwick, R.S., Pye, J.P. & Fabian, A.C., 1980. *Mon. Not. R. astr. Soc.*, **190**, 243.

White, S.D.M., Silk, J. & Henry, J.P., 1981, **251**, L65.

Worrall, D.M., Mushotzky, R.F., Boldt, E.A., Holt, S.S. & Serlemitsos, P.J., 1979. *Astrophys. J.*, **232**, 683.

Worrall, D.M., Giommi, P., Tananbaum, H. & Zamorani, G., 1987. *Astrophys. J.*, **313**, 596.

Wilkes, B.A. & Elvis, M., 1987. *Astrophys. J.*, **323**, 243.

Yang, J., Turner, M.S., Steigman, G., Schramm, D.N. & Olive, K.A., 1984. *Astrophys. J.*, **281**, 493.

Yee, H.K.C. & Green, R.F., 1984. *Astrophys. J.*, **280**, 79.
Yee, H.K.C. & Green, R.F., 1987. *Astrophys. J.*, **319**, 28.
Zamorani, G. *et al.*, 1981. *Astrophys. J.*, **245**, 357.
Zdziarski, A.A. & Lamb, D.Q., 1986. *Astrophys. J.*, **309**, L79.
Zdziarski, A.A., 1987. Preprint.

HII GALAXIES

Roberto Terlevich
Royal Greenwich Observatory,
Herstmonceux, BN27 1RP,
U.K.

ABSTRACT. Models of the stellar population in "primaeval galaxies" indicate that these objects are similar in many respects to nearby HII galaxies. As HII galaxies they have UV, optical and IR spectra which are dominated by the young O-B stellar population and the associated narrow emission lines. Hence besides providing information about the physics of massive star formation and evolution at very low abundances, studies of HII galaxies can also be used to test primaeval galaxy model. Observations of HII galaxies can also indicate how a primaeval galaxy will look to observers when one is found. Some of the more important statistical properties of a sample of more than 400 HII galaxies are reviewed.

1 - INTRODUCTION

Starbursts or Violent Star Forming Regions(VSFRs) are regions where thousands of massive stars (masses larger than $20 \, M_{\odot}$) have recently been formed in a very small volume (few tens of parsecs in diameter) and over a time scale of only a few million years. The term "Violent Star Formation Region" seems appropiate to characterize the extremely short time scales, small dimensions and large luminosities involved.

VSFRs are ubiquitous in a variety of extragalactic systems, the best known examples being the giant extragalactic HII regions associated with normal late Hubble type spirals or irregular galaxies like 30-Doradus in the LMC and NGC604 in M33, the extragalactic detached HII regions or HII Galaxies like IIZW40 and IZW18 (these are very luminous burst of star formation inside a low luminosity galaxy) and the Starburst Galaxies, i.e. VSFRs in the central region of early Hubble type luminous spiral galaxies.

Galaxies with VSFR are commonly separated into three groups according to their optical morphology:

1 - Starburst nuclei: These are galaxies where the young stellar component and its associated HII region is located in the nuclear region of an otherwise normal early Hubble type spiral galaxy. Starburst nuclei have been studied by Weedman and coworkers (Balzano 1983 and references therein).

2 - Clumpy irregular galaxies: Irregular galaxies with widespread star formation (Heidmann 1982).

3 - HII galaxies or detached extragalactic HII regions: This class of galaxy has small dimensions, spheroidal shape, a young and hot stellar

N. Kaiser and A. N. Lasenby (eds.), The Post-Recombination Universe, 69–83.

population and spectroscopic properties indistinguishable from those of giant extragalactic HII regions (Sargent and Searle 1970).

HII galaxies are found very frequently in objective prism surveys. Most of the emission line galaxies in the Tololo and Michigan Surveys belong to this type. The Markarian and Zwicky lists also include several of these galaxies although their proportion is much lower than in the Tololo and Michigan Surveys due to the different selection criteria. In many cases these systems are associated with low surface brightness amorphous galaxies with no indication of spiral structure. Detailed studies (Melnick, Terlevich & Eggleton 1985; Campbell, Terlevich & Melnick 1986; Melnick 1987) show that they are in fact dwarf galaxies with one or more luminous giant HII regions which dominate their spectral characteristics. The presence of an extended component or galaxy is evident in about half of these objects for which deep CCD pictures have been obtained. The rest are stellar-like objects with no evidence for any extension or fuzz indicating that they may be truly intergalactic giant HII regions (see below). We will refer to these systems as HII galaxies. HII galaxies are not to be confused with "blue compact galaxies", which exhibit qualitatively different properties. Blue compact galaxies are systems selected by colour and compacticity and they are usually old bursts with weak emission lines superimposed in a relatively old stellar component.

HII galaxies are among the most luminous narrow emission line objects in the sky. The observed H_β luminosities are frequently in the range 10^{40} to 10^{41} erg s^{-1} with some galaxies reaching almost 10^{43} erg s^{-1} (for $H_0 = 100$ km s^{-1} Mpc^{-1} and uncorrected for reddening). The absolute blue magnitude of the stellar continuum (i.e. not including the emission lines) ranges from -14^m to -24^m. The analysis of the emission line spectrum indicates that these objects are normally underabundant in heavy elements and photoionized by normal hydrogen-burning hot young stars (Bergeron 1977; French 1980). The optical and ultraviolet continuum is dominated by the O and B star population. The remarkable stellar composition of these compact and isolated objects, combined with the very low heavy-element abundance deduced from their emission-line spectra, lead to the conclusion that some of them can be truly "young" galaxies forming their first generation of stars. In any case they represent the youngest galaxies that can be studied in any detail and therefore the study of their systematic properties can provide important model constraints and clues for theories of formation and evolution of massive stars and for searches of primaeval galaxies.

From their emission line spectrum it is relatively straightforward to derive precise abundances of He, N, O, Ne and S. This and the fact that in HII galaxies we observe massive star formation and evolution in conditions that do not exist any longer in the Galaxy, makes the study of HII galaxies also very important for theories of chemical evolution. From the comparison between the helium and metal content the primordial abundance of helium is deduced. This is one of the most important parameters to select among the many possible cosmological models describing the early phases of the Universe. Moreover, within the canonical big-bang model it could provide some insight into the physics of elementary particles.

I will review here some of the statistical properties of HII galaxies based on the "Spectrophotometric Catalogue of HII Galaxies", hereafter SCHG (Terlevich et al. 1988). Most of the work reported here has been done in collaboration with Jorge Melnick and Mariano Moles.

2 - STATISTICAL PROPERTIES

The SCHG contains about 920 spectra of 432 HII galaxies, Giant HII regions, and Star-burst nuclei found in objective prism surveys using IIIa-J emulsion. The SCHG also includes about 30 new Seyfert galaxies and QSOs. All objects from the University of Michigan and Tololo surveys classified as emission line galaxies were observed in low resolution. About 100 HII galaxies with detected [OIII] λ4363 Å were further selected for higher resolution and high S/N observations to study their abundances and gas kinematics.

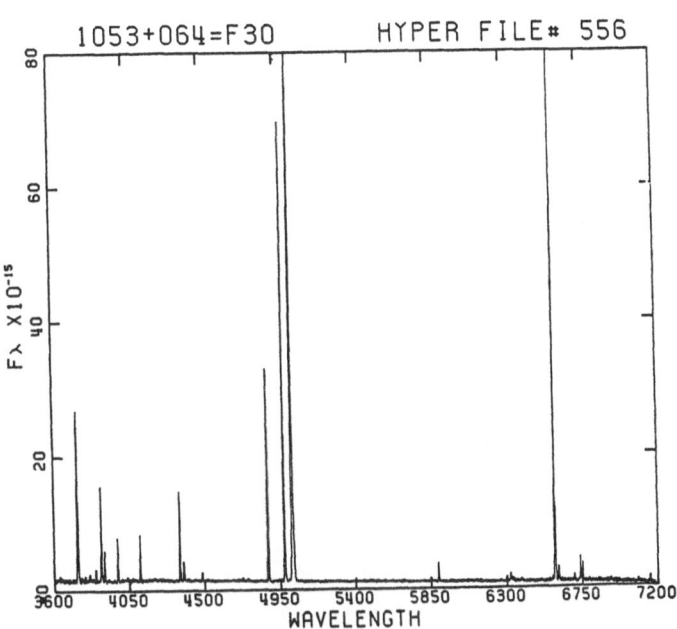

Figure 1 - Spectrum of a typical HII galaxy. Both H_α and [OIII] λ5007 Å are out of scale.

To illustrate the spectral similarities between Giant HII regions and HII galaxies I plotted in figure 1 the spectrum of a bright HII galaxy. This spectrum is basically identical to those of typical giant HII regions such as 30 Doradus in the LMC.

2.1. Redshift and Luminosity distribution

In figure 2 the redshift distribution of the complete sample is presented. Although the sample is selected primarily by the presence of the [OIII] $\lambda\lambda$4959,5007 Å doublet and therefore strongly biased towards redshift < 0.08, some higher redshift objects discovered by the presence of strong [OII] λ3727 Å are present in the sample.

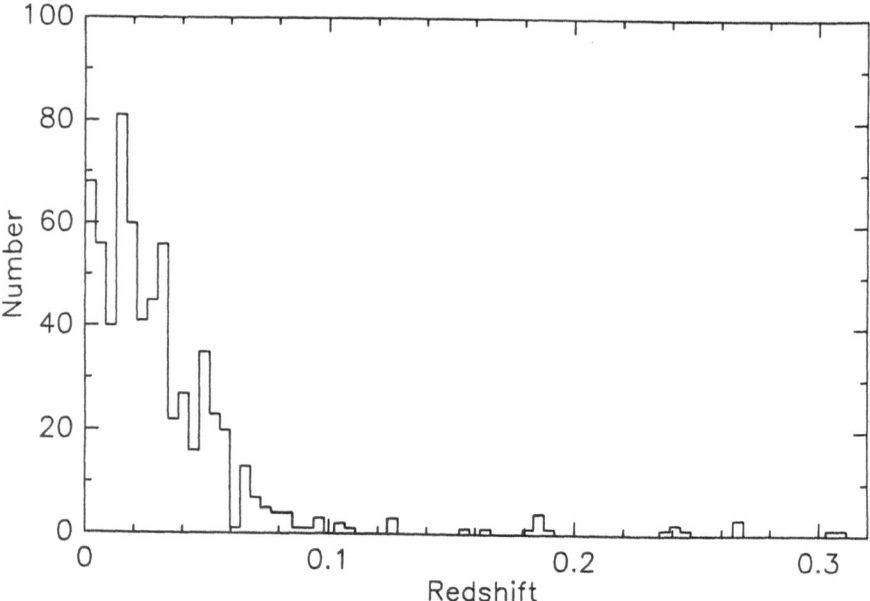

Figure 2 - Redshift distribution for 432 HII galaxies from the SCHG. Some of the HII galaxies have redshift in excess of 0.3.

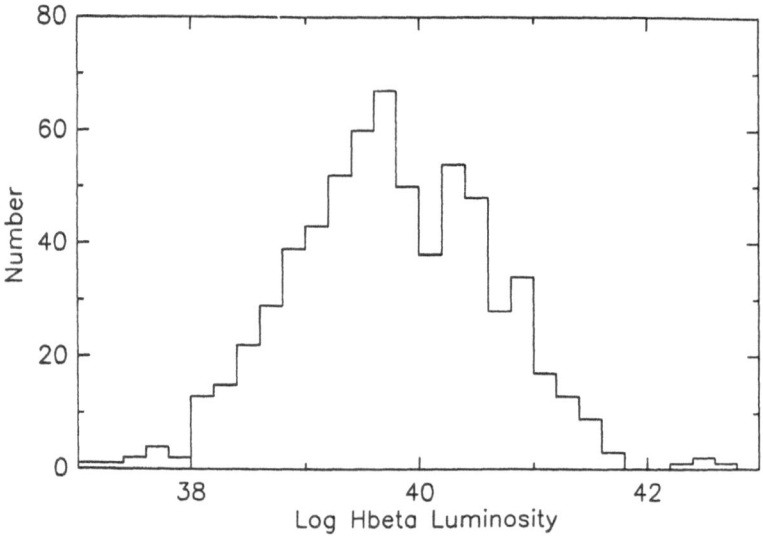

Figure 3 - Distribution of observed H_β luminosity (H_0 =100). Log L_{H_β} = 43 corresponds to a bolometric luminosity of $2.0*10^{12} L_\odot$.

Figure 3 shows the H_β luminosity distribution. The large luminosities of HII galaxies reflect their large star formation rate. 10^{43} erg s^{-1} can be considered as a reasonable upper limit for the H_β luminosity of HII galaxies. This in turn implies a maximum bolometric luminosity of HII galaxies (a conservative estimate can be done for an average extinction coefficient of 0.3 at H_β and assuming that 25% of the luminosity is emitted in the ionizing continuum) of $8 * 10^{45}$ erg s^{-1} or $2 * 10^{12}$ L$_\odot$ ($H_0 = 100$).

It is very important to emphasize that this maximum bolometric luminosity is the same as that found in the infrared luminous IRAS galaxies. The brightest and best known examples are Markarian 231 and Arp 220 with bolometric luminosities of $2 * 10^{12}$ L$_\odot$ and $9 * 10^{11}$ L$_\odot$ respectively (Soifer et al. 1986; Soifer et al. 1987). The fact that HII galaxies do not appear to be strong IRAS sources is probably related to their low metal content (Gondhalekar et al. 1986).

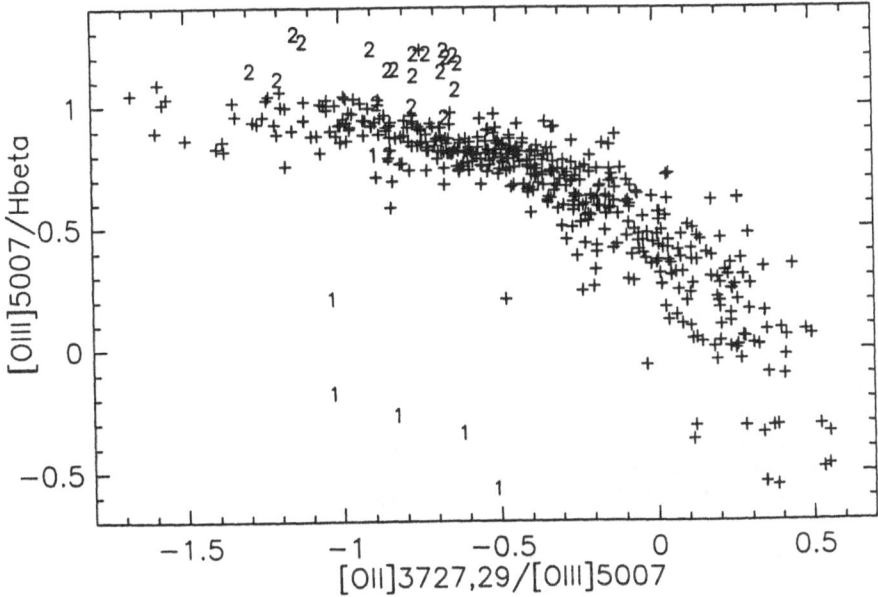

Figure 4 - BPT diagnostic diagram for all objects in the SCHG; +=HII galaxies; 1= type 1 Seyferts; 2=type 2 Seyferts. Note the tight relation for HII galaxies.

2.2. Emission line ratios and metallicity distribution

In order to investigate the homogeneity of the sample I plotted in figure 4 one of BPT diagnostic diagrams (Baldwin, Phillips & Terlevich 1981) using two line ratios [OIII] $\lambda5007$/H_β and [OII] $\lambda3727,29$/[OIII] $\lambda5007$ for all objects in SCHG. There is a tightly defined relationship regarding the HII galaxies (crosses) in both these diagrams. So tight is this relation that anything more than about 0.3 in the logarithm away from the locus would immediately be suspected of falling into a different category. All but one of the HII galaxies fall clearly inside the line ratio region characteristic of normal HII regions. This object near the upper limit of the figure is in fact a missclassified Seyfert type 2. Thus a reasonable approach to classifying a new object as a star-forming or primaeval galaxy would be to plot the object's position in diagrams such as that of figure 4.

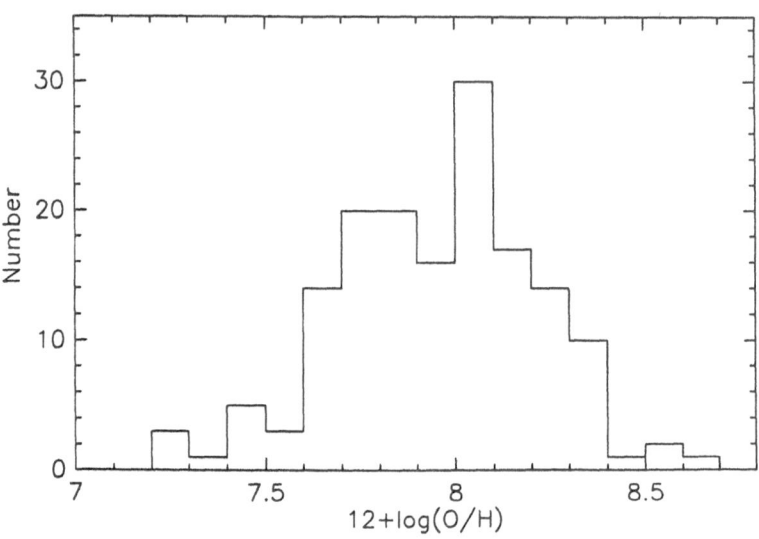

Figure 5 - The metallicity distribution for 102 HII galaxies with detected [OIII] λ4363 Å. The solar abundance is 8.75.

One of the main aims of producing the SCHG was the search for low metallicity systems. The importance of such systems has been emphasized for several problems including chemical evolution of galaxies (Lequeux et al. 1979) and primordial nucleosynthesis (Yang et al. 1984). It was quite unexpected that the result of the analysis of more than 400 galaxies provided no object with abundances below or equal to that of I Zw 18. This is illustrated in figure 5 where the distribution of oxygen abundance of 102 HII galaxies with detected [OIII] λ4363 Å is shown.

The selection criteria used guarantee that this subset of HII galaxies represents the low metallicity tail of the distribution; (in these units solar abundance is 8.75 and I Zw 18 would be at 7.1). It remains possible that objective prism searches for high excitation and luminous intense bursts of star formation were not optimized for finding extremely low abundance galaxies. Searching in local group gas rich dwarf irregulars Skillman and co-workers (Skillman et al. 1988) found evidence supporting a good correlation between total galaxy mass and abundance. Furthermore they found that in the extremely low luminosity dwarf galaxy GR 8 the oxygen abundance is 0.03 of the solar value, which is, within errors, equal to that of I Zw 18. Figure 6 reproduces figure 2 of Skillman et al. (1980); a plot of metallicity versus the total galaxian mass. The metallicity Z was computed assuming that oxygen represents 45% of the total mass of metals. The solid line shows the correlation derived by Talent(1980) without GR 8 and I Zw 18. The position of GR 8 in this diagram lies very close to that predicted by the relationship of Talent (1980). The low mass and low metal content of GR 8 confirms the discoveries of Lequeux et al. (1979) and Talent (1980) that there is a good correlation metallicity and total galaxy mass for dwarf galaxies.

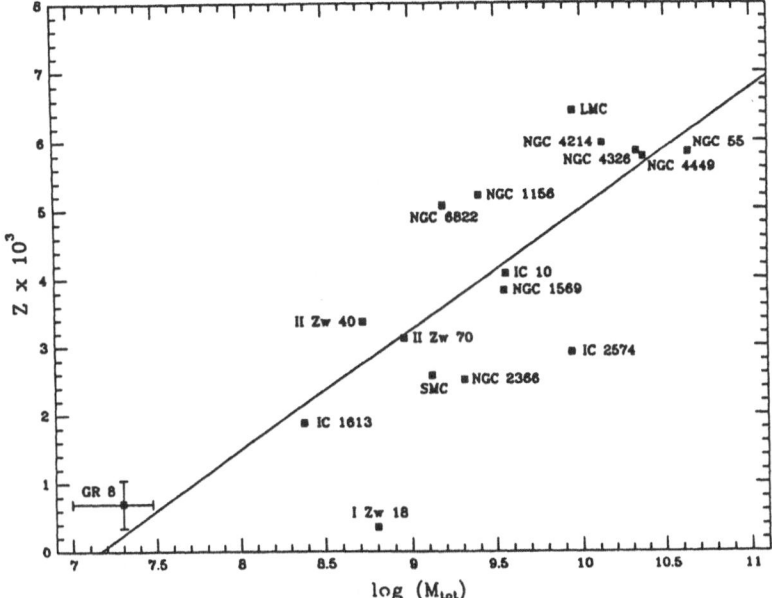

Figure 6 Plot of metallicity versus the logarithmic total galaxian mass from Skillman *et al.* (1988). The solid line represents the relationship found by Talent (1980).

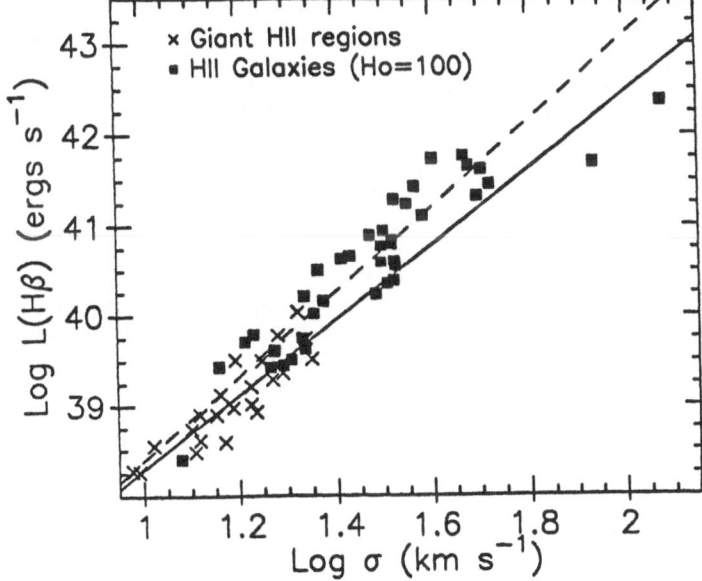

Figure 7 - Plot of the logarithmic velocity dispersion vs. the logarithmic H_β luminosity for giant HII regions and HII galaxies. A Hubble constant of 90 $km\,s^{-1}\,Mpc^{-1}$ was used to compute the luminosity of the HII galaxies.

2.3. A new distance estimator and the value of H_0.

Terlevich and Melnick (1981) analysed the relations between H_β luminosity, size, width of the emission lines and heavy element abundance of an heterogeneous sample of 25 giant HII regions and HII galaxies. They concluded that the relations:

$$Luminosity \quad \alpha \quad \left(line \quad width\right)^4, \quad Size \quad \alpha \quad \left(line \quad width\right)^2$$

valid for gravitationally bound stellar systems like elliptical galaxies, bulges of spirals and galactic globular clusters, are also valid for giant HII regions and HII galaxies. This result suggests that HII galaxies and Giant HII regions are also gravitationally bound systems in which the observed emission line widths represent the velocity dispersion of discrete gas clouds in the gravitational potential of the gas-star complex. Figure 7 shows the relation between velocity dispersion and luminosity for a new homogenous data set collected (Melnick et al. 1987), based on the SCHG. The small scatter of the relation is remarkable. This confirms Terlevich and Melnick (1981) findings and also implies that the line-width, a distance independent parameter, can be used to predict accurately the luminosity of HII galaxies and Giant HII regions and thus estimate their distances independently of redshift. We (Melnick et al. 1987; Melnick, Terlevich & Moles 1988) have recently used an improved version of this distance estimator to determine the value of the Hubble constant and find a value of 89 ± 10 km s^{-1} Mpc^{-1} compatible with some previous determinations and incompatible with others, notably that of Sandage and Tammann.

2.4. The Ly_α problem

Another surprising result of the early stages of our study of HII galaxies was the discovery that Ly_α was very weak or absent in their UV spectrum (Meier & Terlevich 1981). The usual explanation for the reduction of Ly_α emission in HII regions is absorption by dust in the ionized gas. Multiple scattering in the Ly_α line enhances the dust absorption (Meier & Terlevich 1981). But the observed lack of correlation of the Ly_α strength with the Balmer decrement does not support this simple picture. Even more strong constrains are provided by the lack of detection of Ly_α emission in objects that like POX 124, Mk 26 or Mk 12 have little evidence for significant visual or UV extinction (Deharveng, Joubert & Kunth 1986). This evidence together with the fact that Ly_α is usually detected in some extremely dusty objects like type 2 Seyfert nuclei suggests that may be some other "hidden" factor affecting the Ly_α emissivity. These results have important consequences for searches of primaeval galaxies. From strict empirical considerations we expect that young galaxies will have a strong UV continuum and weak emission in Ly_α. We do not expect to see any strong metal line, in particular CIV $\lambda1550$ is expected to be even weaker than Ly_α. Galaxies with strong Ly_α may be more related to AGNs than to young star-forming galaxies.

2.5. Morphology and environment

Using the Danish 1.5m telescope at La Silla we have recently obtained CCD pictures of 30 HII galaxies with the purpose of studying their morphology and environment. On the basis of these data HII galaxies can be divided into three morphological classes: 1 - Multiple systems: These galaxies are dominated by more than one giant HII region. The small radial velocity differences between the components indicate that they are probably not parts of interacting systems. Eight examples of this class are shown in plates 1 and 2.

2 - Interacting systems: Here the HII galaxy is part of an interacting system characterized by tidal tails and /or distortion. Examples of this class are shown in plate 3.

3 - Isolated objects: The HII galaxies belonging to this class are isolated and in most cases with stellar (unresolved) or semi-stellar images. Typical examples are shown in plate 4.

A fourth class of object which lies at the boundary of what we could call HII galaxies are amorphous galaxies with one or several giant HII regions in their central parts. Three examples of this class are shown in plate 5. The starburst component of NGC 1705 has no emission lines but is a massive, young ($\sim$ 10Myr) starburst cluster (Melnick, Moles and Terlevich 1985). The cluster has an absolute magnitude $M_b = -15.4$ that makes it the most luminous star cluster known.

In total, 20% of the objects surveyed appear to be parts of interacting systems, 30% have double or multiple HII region components and 50% are isolated with stellar or semi-stellar morphology. These results show that galaxy interactions are an important but by no means unique mechanism to induce starburst activity. In fact, the isolated semi-stellar objects which comprise half of the sample are probably the best candidates for being young galaxies.

CONCLUSIONS

The study of HII galaxies gives important clues about the intrinsic properties of young or unevolved galaxies. These clues may be relevant for searches of primaeval galaxies particularly if Starbursts are the dominant mode of galaxy evolution. A very interesting aspect is that a substantial fraction of low metallicity HII galaxies are isolated and stellar and may be truly young galaxies. I would like to finish this presentation with a summary of the properties of HII galaxies

M_B	-11 to -22
$L(H_\beta)$	10^{38} to 10^{43} erg s^{-1}
L_{Bol}	10^7 to 10^{12} L$_\odot$
L_{Ly_α}	up to 10 $L(H_\beta)$
U-V	-0.4 to -0.8
B-V	0.4 to 0.0
HII region size	a few hundred pc to few kpc
$W(H_\beta)$	$\geq$ 50 Å $<$ 400 Å
O/H	1/2 to 1/40 of the solar value
Y	0.23 to 0.26
M_t	up to a few 10^9 M$_\odot$
M_{gas}	up to a few 10^{10} M$_\odot$
FWHM	70 to 300 km s^{-1}
Environment	low density

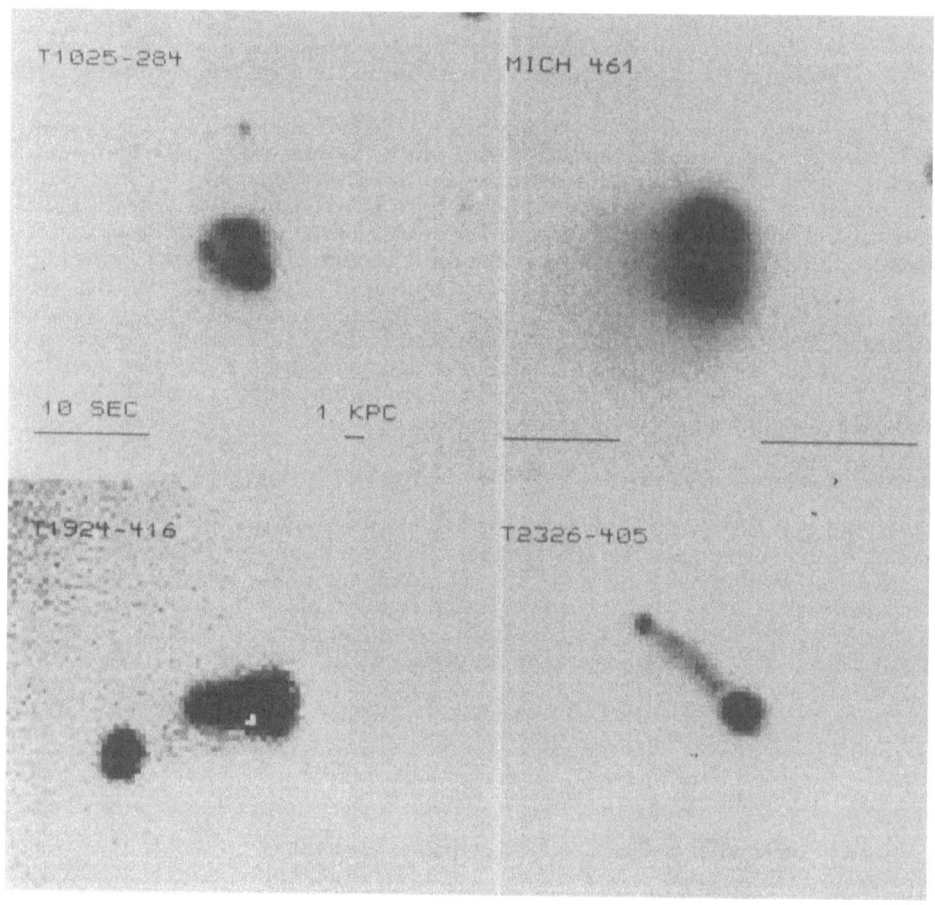

Plate 1 CCD images of four HII galaxies of multiple morphology. North is at the top, west is to the left. The linear and angular scales of each picture are also shown.

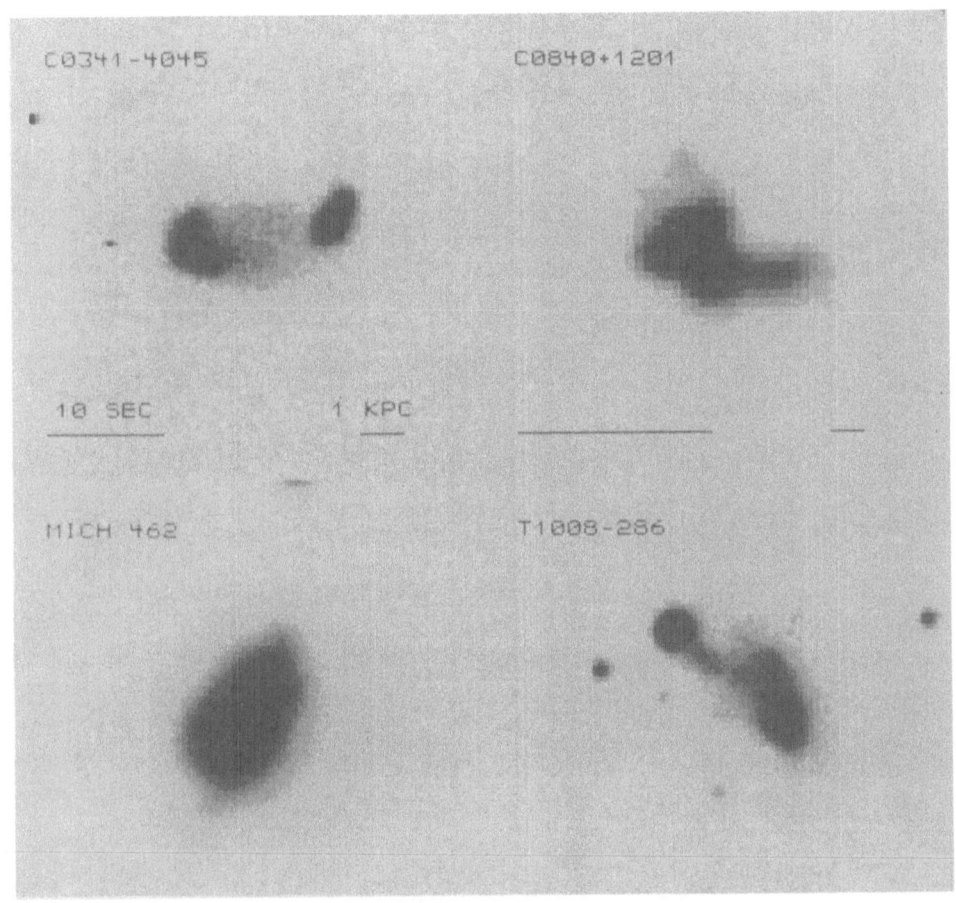

Plate 2 CCD images of four double or multiple HII galaxies. C0840+1201 is also known as Markarian 701.

80

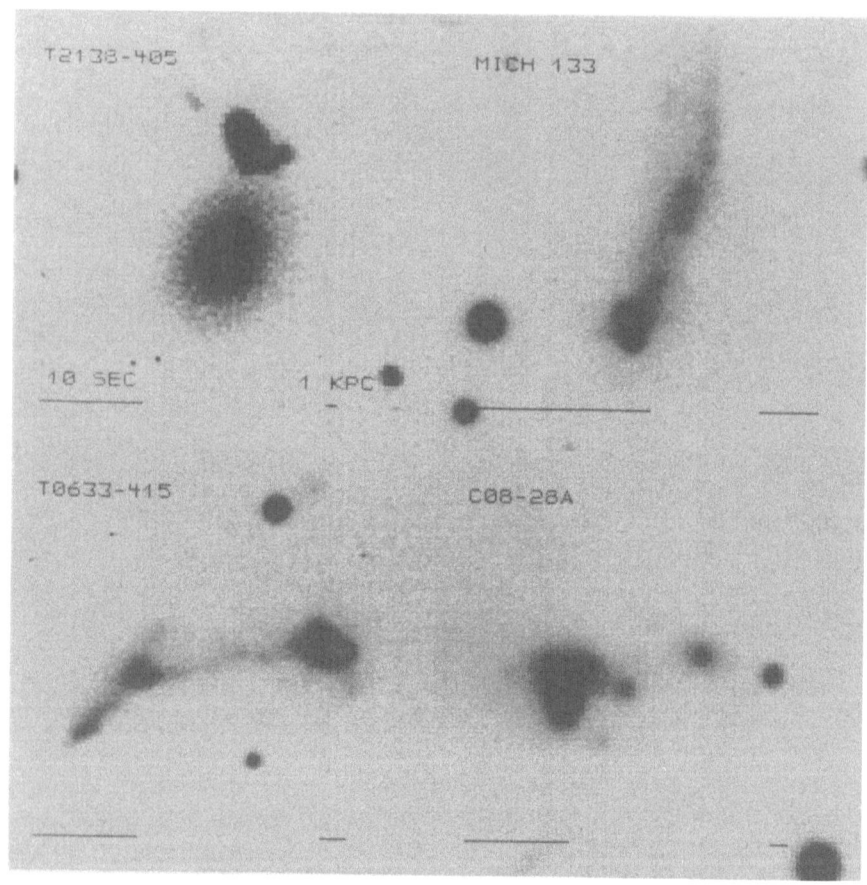

Plate 3 CCD images HII galaxies in interacting systems.

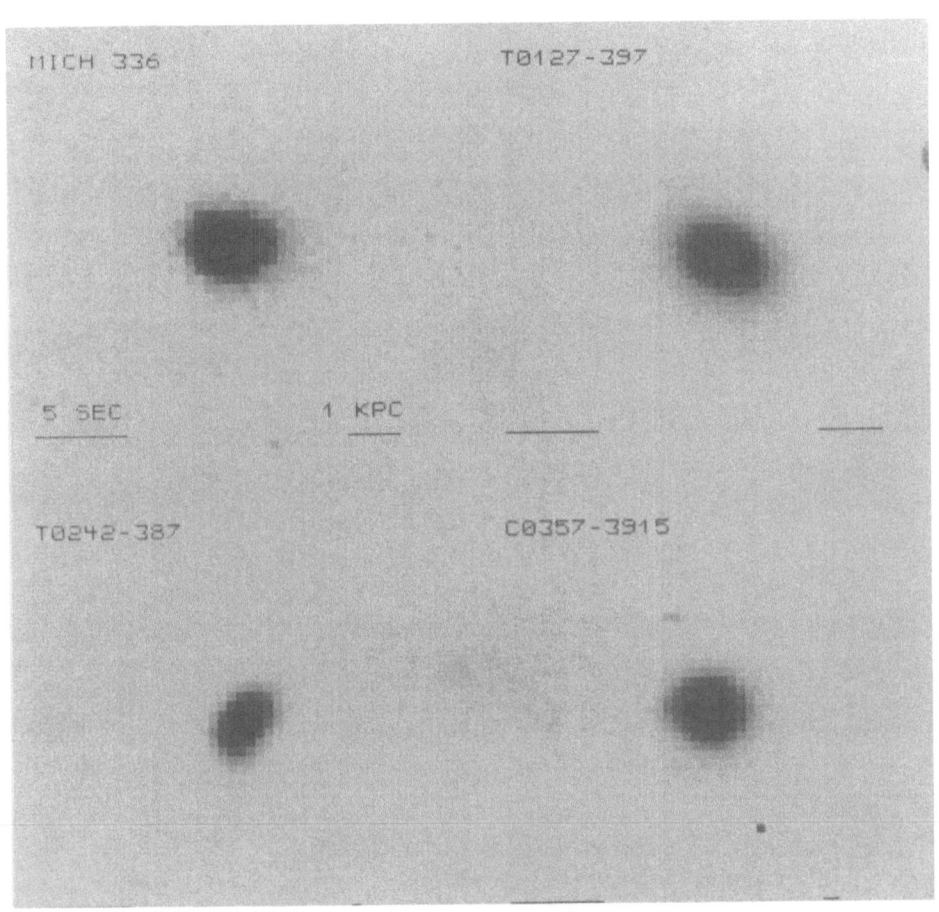

Plate 4 CCD images of typical compact and isolated HII galaxies.

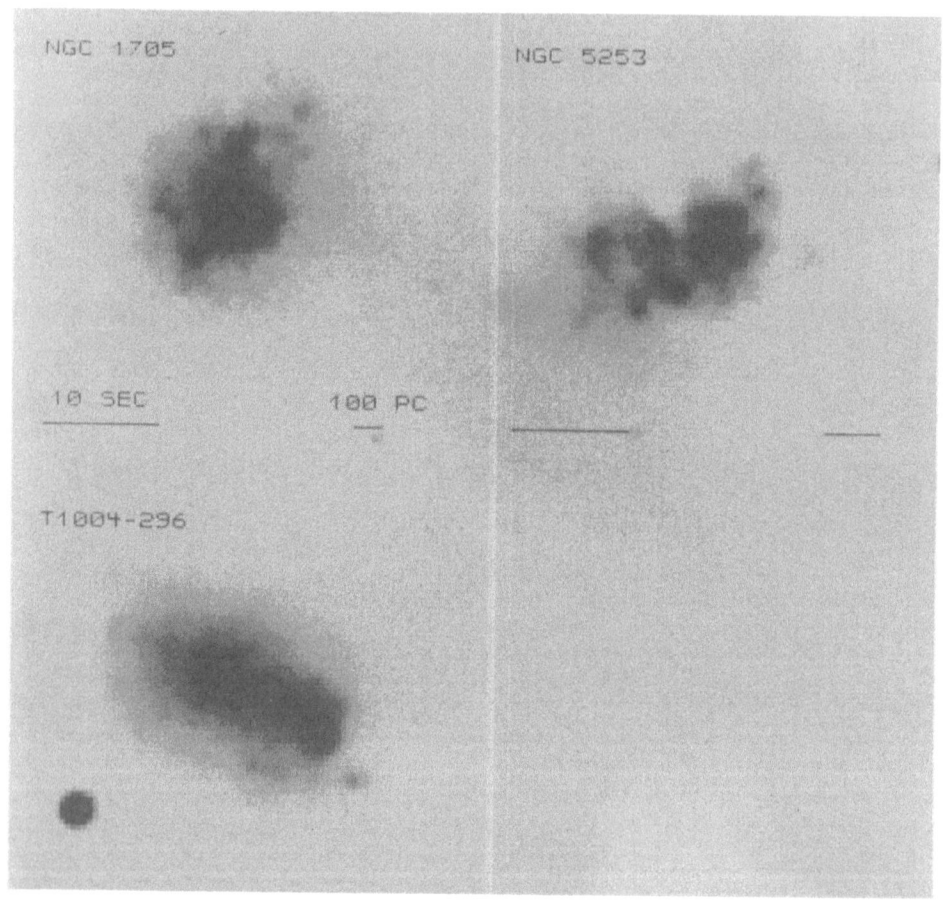

Plate 5 Amorphous galaxies with HII region spectral characteristics. The starburst component of NGC1705 is a massive young cluster about 10^7 years old.

REFERENCES

Baldwin, J.A., Phillips, M.M. and Terlevich, R., 1981. *Publ. astr. Soc. Pacif.* , **93**, 5 **(BPT)**

Bergeron, J., 1977. *Astrophys. J.* , **211**, 62

Campbell, A., Terlevich, R. and Melnick, J., 1986. *Mon. Not. R. astr. Soc.* , **223**, 811

Deharveng, J.M., Joubert, M. and Kunth, D., 1986. *First IAP workshop: "Star forming dwarf galaxies and related objects"*, eds. Kunth, D.and Thuan, T.X.

French, H.B., 1980. *Astrophys. J.* , **240**, 41

Gondhalekar, P.M., *et al.* , 1986. *Mon. Not. R. astr. Soc.* , **219**, 505

Lequeux, J. *et al.* , 1979. *Astr. Astrophys.* , **80**, 155

Meier, D. and Terlevich, R., 1981. *Astrophys. J.* , **246**, 409

Melnick, J., 1987. *XXIInd Reencontres de Moriond: "Starbursts and Galaxy Evolution"* ed. T.Monmerle and T.X.Thuan

Melnick, J., Terlevich, R. and Eggleton, P.P., 1985. *Mon. Not. R. astr. Soc.* , **216**, 255

Melnick, J. *et al.* , 1987. *Mon. Not. R. astr. Soc.* , **226**, 849

Melnick, J., Terlevich, R. and Moles, M., 1987. *Mon. Not. R. astr. Soc.* , in press

Sargent, W.L.W. and Searle, L., 1970. *Astrophys. J. Lett.* , **162**, L165

Skillman, E. *et al.* , 1988. *Astr. Astrophys.* , in press

Soifer, B.T, *et al.* , 1986. *Astrophys. J.* , **303**, L41

Soifer, B.T, *et al.* , 1987. *Astrophys. J.* , in press

Talent, D.L., 1980. *Ph.D. Thesis, Rice University*

Terlevich, R., 1984. *Frontiers of Astronomy and Astrophysics*, ed. Pallavicini, R., p. 123

Terlevich, R. and Melnick, J., 1981. *Mon. Not. R. astr. Soc.* , **195**, 830

Terlevich, R., Melnick, J., Masegosa, J. and Moles, M., 1988. in preparation, *(SCHG)*

Yang, J. *et al.* , 1984. *Astrophys. J.* , **281**, 493

THE REDSHIFT CUTOFF FOR STEEP-SPECTRUM RADIOGALAXIES

J E Baldwin
Cavendish Laboratory
Cambridge
England

This talk is about two related questions and attempts to solve them by David Rossitter, Stephen Dingley and myself. The first is

1) Is there a redshift cutoff in the radio galaxy population?

A cutoff has already been found in the optical QSO population at redshifts z ⩾ 3. There is still some dispute about selection effects and their influence on the completeness of samples at differing z. A cutoff has also been found for flat-spectrum radio sources, which are mainly quasars, (Peacock, 1985). The comoving space density shows a peak at redshifts of ~2 with a steep, or moderately steep fall at greater z. What has not been known is whether radio galaxies follow the same pattern. The answer bears on the question of whether quasars and radio galaxies are interchangeable aspects of the same phenomenon or not. If they are, then there should be a redshift cutoff.

2) Were the physical sizes of radiogalaxies the same or smaller in the past?

This is not a question about individual objects but the overall properties of the population. It is really a refinement of question 1): does the redshift cutoff for very large radiogalaxies occur at a smaller redshift than for small radiogalaxies? There has been some evidence affirming this (Eales 1985; Kapahi 1985).

We have new evidence indicating that the answer to the first question is yes and the second is both yes and no. Answering question 1) is straightforward in principle. Take radio samples selected by flux density and spectrum, identify the galaxies corresponding to the sources and measure their redshifts. There is no problem about the completeness of the samples but every problem about the next two stages. The radio galaxies are of course several magnitudes fainter than quasars and the line strengths weaker. Furthermore, the sources in the sample which are at high z constitute only a small fraction of the sample, because of the strong redshift factors. If in addition

N. Kaiser and A. N. Lasenby (eds.), The Post-Recombination Universe, 85–89.
© *1988 by Kluwer Academic Publishers.*

there is a cutoff in the comoving density of sources the fraction becomes minute. Indeed, but for the strong increase in space density out to z of 1-2, there would be no chance of detecting a cutoff beyond that.

One way round the requirement for massive amounts of spectroscopic time is to use alternative distance indicators. The m-z relation has been important for this since radiogalaxies are typically the most massive giant elliptials. The m_R-z relation is a good one up to z ~ 0.8. $m_R \approx 21$ but has a large scatter and a very flat slope beyond that redshift. There is some prospect that the m_k-z relation will be useful (Lilly et al, 1985) perhaps to redshifts as large as 2.5-3 since the corresponding rest wavelengths are still in part of the spectrum which is not too sensitive to the fraction of young stars in the stellar population. However, at present there are no redshifts for radio galaxies available beyond z ~ 1.8 to substantiate this belief.

We have therefore searched for another distance indicator. It has been known for several years that there are correlations of intrinsic radio luminosity with radio spectral index (Laing and Peacock, 1980) and with the so-called compactness, the fraction of the flux density arising in hotspots in double radio galaxies (Jenkins & McEllin, 1977). Each of these relations separately has considerable scatter for the bright source sample of Laing et al (LRL: 1983). However, the relation found by seeking the best linear fit between log P and the spectral index α, the compactness C and log D, where D is the overall physical size, is much tighter (Fig 1). The relation is

$$\log P = 4.08\ C + 2.86\ \alpha + 0.65\ \log(D/kpc) + 21.24$$

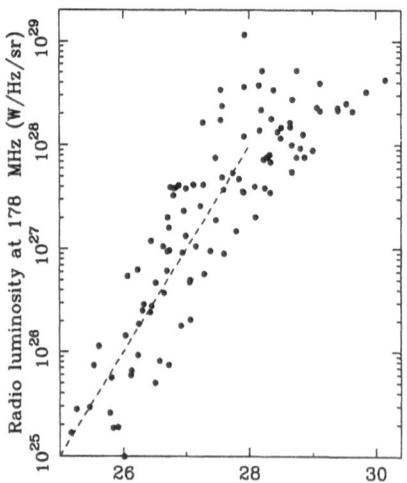

Fig 1. The relation between radio luminosity. P_{178} W H^{-1} sr^{-1}, for the LRL sample and the best-fitting linear combination of C, α and log (D/kpc), (4.08C + 2.86α + 0.65 log (D/kpc) + 21.24).

Assuming that this relationship is one depending on P only, independent of z, it can be applied to estimate the value of P for any source from radio evidence only. Given the flux density of the source, the redshift can be deduced. Three comments on the relation need to be made. Firstly, the compactness C is defined as the fraction of the total flux lying in the hotspots within regions 15 kpc in diameter. It might seem that both this term and that in log D already require a knowledge of the distance of the source. However, at the redshifts of interest in applying the relation, the variation of angular size with z is small. A first estimate of the redshift from the relation can be used to refine a further estimate. Secondly, whilst the relationship is entirely empirical and has no theoretical reasoning behind it, it is easy to think of explanations for the anticorrelation of C and α implied by it. The spectra of hotspots are typically flatter than those of bridges in double radio galaxies. If the beams illuminating the hotspots are currently strong then C will be large but α relatively low. If the beams are weak at the present time the reverse will be true. Thirdly the same relation applies to radiogalaxies and to quasars with double lobed structure.

Rossitter has tested whether the relation is independent of z by applying it to sources in the $2 < S_{151} < 4$ Jy sample of Eales (1985) and the sample of high z radio quasars of Barthel (1984). In each case the redshifts predicted from the relation agree with those that have been measured spectroscopically.

Three fainter samples have now been studied over the range 2 Jy to 50 mJy. The sources were selected from surveys made with the Cambridge Low Frequency Synthesis Telescope operating at 151 MHz in areas corresponding to several of the 5C surveys for which large amounts of data on spectra and identifications already existed. The samples were mapped with the VLA at 1465 MHz in the A array giving a resolution sufficient to just resolve the 15 kpc hotspot diameter adopted in the definition of compactness.

The results, when analysed using the relationship of Fig 1, indicate the expected increase in space density for FRII sources, ie double radio sources of high power ($P \geqslant 10^{27}$ WHz sr^{-1}) out to redshift of 2-2.5 with a decrease in space density at higher z(Fig 2). For those sources with $P < 10^{26}$ WHz^{-1} sr^{-1}, even the fainter samples do not reach beyond z = 1 so that no conclusions can be drawn about a possible cutoff. It is interesting that the cutoff found is at a similar redshift to those found for flat spectrum sources and QSOs to within the present limits of error.

A suprising feature of the results is that the distribution of physical sizes of the FRII sources in the faint samples is very similar to that found in LRL. The upper limit of size for the powerful sources is about 1 Mpc just as it is for nearby sources with $P > 10^{26}$ WHz^{-1} sr^{-1}. How does this fit in with the assertion by both Kapahi and by Eales that the median physical size of sources decreases with redshift?

The provisional answer seems to be that in fainter samples, whereas the doubles cover the same range in size, there is a larger fraction of very compact, unresolved sources. These are not susceptible to the same analysis using Fig 1.

The numbers of sources in these samples at the largest sizes are very small. In order to provide a more stringent test of the evolutionary behaviour of the largest sources, Dingley and I have selected a sample of 72 sources from 0.3 sr of the CLFST surveys with angular sizes of 1.3-3 arcmin and flux densities of $0.4 < S_{151} < 1.0$ Jy. Sources with $27.5 > \log P > 26.5$ will have redshifts in the range 0.9-2.3. The sample has been mapped with the VLA, initially with 15 arcsec resolution, from which candidate large radio galaxies can be selected.

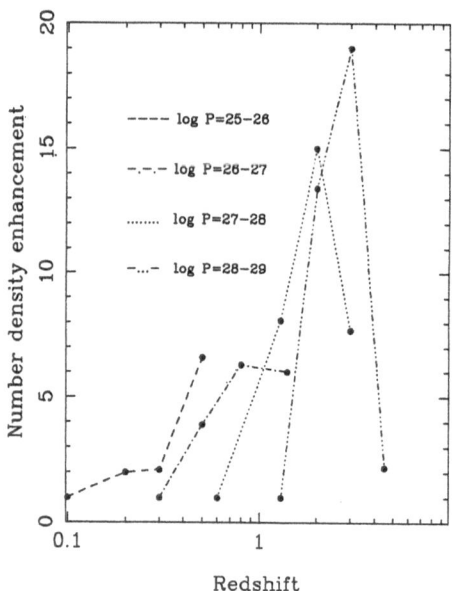

Fig 2. The evolution function of double radio galaxies as a function of radio luminosity. Since the 'local' density of radio galaxies of high power is too low to measure, the evoltion factor is normalised to unity in the highest flux density sample for each range of P.

Several objects, such as those in Fig 3, seem likely to be both large and distant. Further mapping is under way to establish the strength of their hotspots. At present it seems that conditions at large redshifts do allow radio galaxies to reach large sizes. Thus one cannot account simply for the cosmic evolution of the population by asserting that a denser medium surrounding galaxies in the past slows the expansion of sources and thereby makes them intrinsically more radio luminous. Whether that view survives the next set of observations remains to be seen.

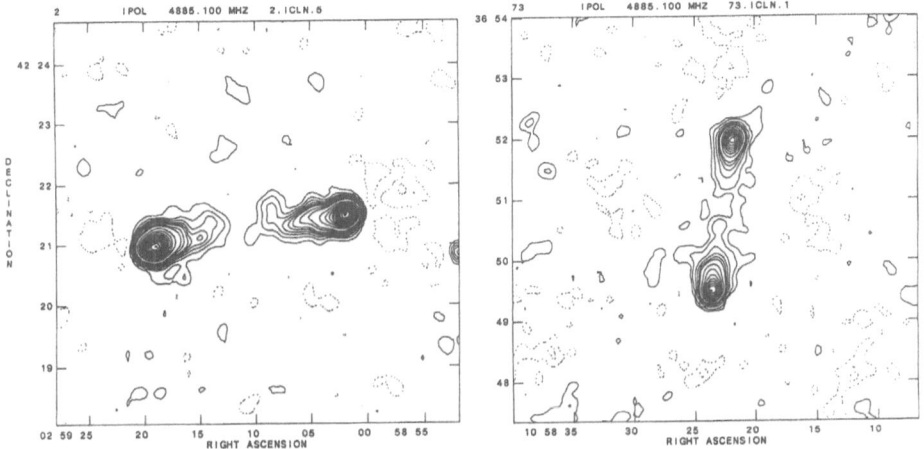

Fig 3. Examples of radio galaxies selected from the $0.4 < S_{151} < 1.0$ Jy sample which are likely to be sources with $P_{151} > 10^{26}$ WHz^{-1} sr^{-1} and sizes of ~1 Mpc at redshifts $\geqslant 1$.

References

Barthel, P., 1984. Ph.D. thesis. University of Leiden

Eales, S.A., 1985. Mon.Not.R.astr.Soc. 213, 899.

Jenkins, C.J. & McEllin, M., 1977. Mon.Not.R.astr.Soc. 180, 219.

Kapahi, V.K., 1985. Mon.Not.R.astr.Soc. 214, 19.

Laing, R.A. & Peacock, J.A., 1980. Mon.Not.R.astr.Soc. 190, 903.

Laing, R.A., Riley, J.M. & Longair, M.S., 1983. Mon.Not.R.astr.Soc. 204, 151.

Lilly, S.J., Longair, M.S., & Allington-Smith, J.R., 1985. Mon.Not.R.astr.Soc. 215, 37.

Peacock, J.A., 1985. Mon.Not.R.astr.Soc. 217, 601.

QUASARS CONTRIBUTING TO THE EXTRAGALACTIC X-RAY FLUX: THE DEEP X-RAY
SURVEY IN PAVO

R. E. Griffiths
Space Telescope Science Institute

I.R. Tuohy and R. J. V. Brissenden
Mt. Stromlo and Siding Spring Observatories
Australian National University

M. Ward
University of Washington

S. S. Murray and R. Burg
Center for Astrophysics

Abstract The nature of discrete x-ray sources contributing to the all-sky x-ray background has been investigated. X-ray selected quasars account directly for $\sim$ 30% of the all-sky x- ray background in the energy range of the Einstein Observatory. Quasars which were not detected in x-rays may account for a further 10–20% of the x-ray background. No examples have yet been found of any class of sources, other than quasars, which might make a comparable or substantial contribution to the extragalactic x-ray background.

1. Introduction

The only direct method of finding out the nature of the discrete sources contributing to the all-sky x-ray background is to resolve as many sources as possible in an x-ray deep survey, and to obtain optical spectroscopy of the corresponding optical counterparts.

The early results of the Einstein Observatory deep surveys in Draco and Eridanus were reported by Giacconi et al. (1979), who established that discrete sources contribute about a quarter of the extragalactic x-ray background in the energy range 1-3 kev, where such sources were detected at least five standard deviations above background. Giacconi et al. assumed that the spectrum of the x-ray background in the 1 to 3 kev range was a smooth extrapolation of the 3-40 kev background spectrum measured by Marshall et al. (1980) with HEAO A2. These discrete source detections were made partially with the Imaging Proportional Counter on the Einstein Observatory, with resulting x-ray astrometric errors of 60 arc seconds. The large 'error circles' and insufficient time for full optical spectroscopy of the possible candidates meant that the nature of most of the discrete sources remained unknown. Giacconi et al. identified only four of their 43 sources with quasars. In contrast, the Pavo deep survey was performed predominantly with the micro-channel plate High Resolution Imager (HRI), with source positional errors of 10 arc seconds or less. The initial optical spectroscopy of sources in the 0.55 square degree Pavo deep survey resulted in identifications with four quasars out of seventeen sources, of which only one redshift was definite, with a further seven objects having the broad-band colours of quasars but without the confirming spectroscopy because of the faintness of the optical counterparts (Griffiths et al. 1983, Paper I).

91

N. Kaiser and A. N. Lasenby (eds.), The Post-Recombination Universe, 91–97.
© 1988 by Kluwer Academic Publishers.

Large numbers of quasars have been identified as x-ray emitters as the result of Einstein Observatory and other satellite experiments which have examined optically and radio- selected samples. In addition, large numbers of x-ray selected quasars have been found in the Medium Sensitivity Survey performed with the Einstein Observatory (Gioia et al. 1987). Nevertheless, the sky surface densities of the objects found in these surveys are such that they do not contribute more than a few percent of the extragalactic x-ray background.

The purpose of this paper is to present a progress report on the optical spectroscopy of sources reported in Paper I, together with results on some of the objects found by reanalysis of the x-ray data. The Pavo field was also studied by a variety of methods in order to search for optically selected qso's or other emission line objects which were not detected in the deep x-ray survey. These observations have been used to estimate the contributions to the all-sky background of those quasars and emission-line objects which were not detected in the x-ray deep survey.

II. X-ray Observations

The x-ray observations of the Pavo deep-survey field have been described in Paper I, in which seventeen x-ray sources were reported, of which all but one were detected with the HRI, and were therefore positioned to accuracies of 10 arc seconds or better. Only one source was found to be extended (source #5), and associated with an elliptical galaxy at a redshift of 0.13, probably in a small group of galaxies. The other sixteen sources showed no evidence for extension, and were therefore considered to be likely to have compact optical counterparts. Further analysis of the x-ray data has confirmed the existence of all of the original sources, and added a further eight detections with the HRI, together with a further seven detections with the IPC. The latter were largely at the edge of the IPC field where there was no HRI coverage. Identification of these new sources is the subject of continuing optical spectroscopy. The purpose of the present paper is to comment mainly on the identifications of the original seventeen sources, together with a few of the new sources for which optical spectroscopy has already been carried out.

The significance of source detection in the HRI was lowered for the purpose of finding fainter sources than had hitherto been found. The detection significance is four standard deviations above background in some cases, so that flux determinations are not well known. Although some sources have fluxes of $\sim 7 \times 10^{-15}\,\mathrm{ergs\,cm^{-2}\,s^{-1}}$ ($\sim 1 - 3\,\mathrm{kev}$) the sample is not complete at this level. The purpose of the present work is not to define a complete sample for x-ray $\log(N)$ versus $\log(S)$ purposes (q.v. Giacconi et al. 1979) but to find and identify optical counterparts to the faintest sources detected in the Deep Survey. If there is a class of discrete objects other than quasars contributing significantly to the all-sky background, then examples of such a class might be found amongst the faintest Deep Survey sources.

III. The x-ray source couterparts—optical spectroscopy properties

Optical candidates for the x-ray selected sources were observed at the 3.9 metre Anglo-Australian Telescope on September 6 and 7 1986. The fibre-optically coupled aperture plate, FOCAP (Ellis et al. 1984; Gray and Sharples, 1985) was used to feed the RGO spectrograph (Robinson, 1985) and Image Photon Counting System (IPCS, see Lucey and Taylor, 1983) for the 3300 to 7300 Angstrom range. The output from another set of FOCAP fibers was fed to the Faint Object Red Spectrograph (FORS) for the 5500 to 9800 Angstrom range.

Fourteen of the x-ray selected objects were found to be quasars, for which the redshift distribution is shown in Figure 1, and compared with the redshift distribution of the x-ray selected quasars from the Einstein Observatory Medium Sensitivity Survey (MSS), taken from Gioia et al.

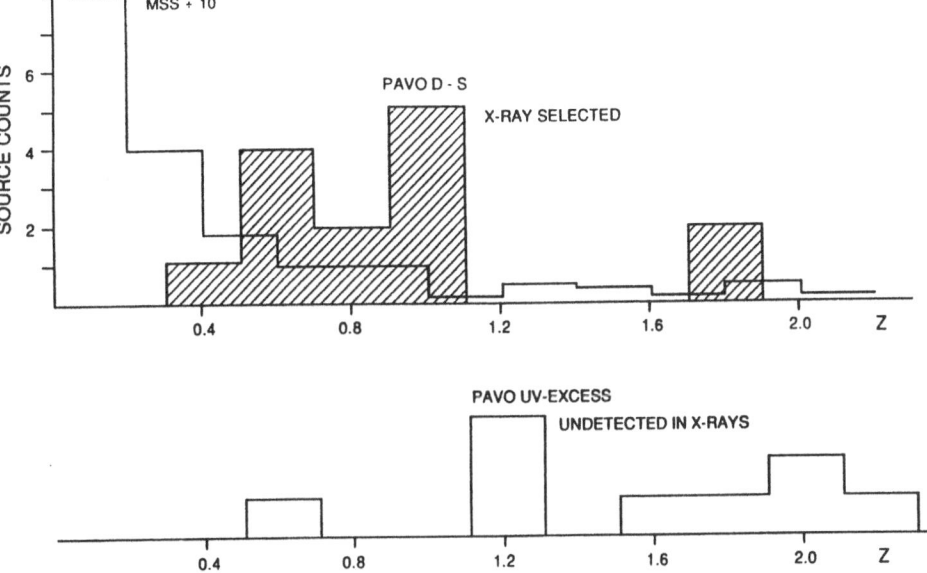

Figure 1. Redshift distribution of x-ray selected quasars in the Pavo deep survey, compared with that for the Medium-Sensitivity Survey (Gioia *et al.* 1987). The redshift distribution is also shown for the UV-excess quasars which were not detected in the deep x-ray survey.

(1987). The average redshift for the MSS objects is 0.4, compared with an average redshift of 0.9 for the Pavo Deep Survey quasars.

The present work shows that the deep survey selects more distant objects, rather than closer objects at the faint end of the x-ray luminosity function for Seyferts or quasars. This is illustrated in figure 2, by comparing the luminosity of the Pavo x-ray selected quasars with those in the MSS. The Pavo quasar luminosities fall within the envelope of the MSS qso's. At an average redshift of 0.9, the observed energy range of 0.5 to 3 kev corresponds to an emitted x-ray energy range of 1 to 6 kev, where x-ray absorption within the source is less important (c.f Elvis and Lawrence, 1985).

The nominal dividing line between "Seyfert galaxies' and quasars has been drawn at $M_b = -23$ by Schmidt and Green (1987). The deep-survey x-ray selected objects fall largely on the quasar side of such a division. This is in contrast to the predictions of Schmidt and Green (1986) and those of Khembavi and Fabian (1982), who predicted that the majority of the x-ray background is made up of low luminosity active galactic nuclei (LLAGN). In particular Schmidt and Green predicted that the contribution of LLAGN to the x-ray background is more than twice that of quasars, in direct contradiction with the results reported here.

The x-ray selected quasars occupy a different part of the $L_{opt} - z$ diagram from objects found in grism surveys or objects in the bright quasar survey of Schmidt and Green, for example. In particular, the x-ray deep survey quasars have typical values of $\log(l_{opt})$ between 29.5 and 30.5, at a median redshift of 0.84. For the same $\log(l_{opt})$, the bright quasar surveys find objects at much lower redshifts, typically 0.1. At similar redshift to the x-ray selected sample, optically selected quasars typically have greater $\log(l_{opt})$, between 30 and 32 (e.g. Avni and Tananbaum,1986).

The x-ray selected quasars had a mean $\langle \alpha_{0x} \rangle$ of 1.4, with a standard deviation of 0.13. α_{0x} shows a dependence on L_{opt} (fig. 3), as found for the optically selected samples (Avni and Tananbaum, 1986). A best fit regression analysis for this dependence shows that

$$\alpha_{0x} = 1.4 + 0.15(l_{opt} - 30.0)$$

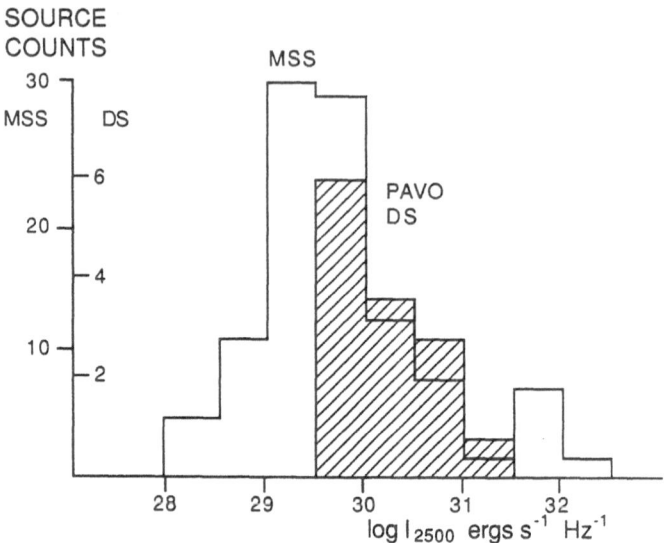

Figure 2. Luminosities of the x-ray selected quasars in the Pavo deep survey compared with the Medium Sensitivity Survey objects (Gioia *et al.* 1987).

The data suggest a steeper dependence of α_{0x} on l_{opt} than that found for optically selected objects (Avni and Tananbaum, 1986, Kriss and Canizares, 1985), but the statistical significance of this difference is small, and is the subject of continuing work. The residuals to this best fit are consistent with a symmetric Gaussian distribution with $\sigma = 0.1$. This value of σ on the α_{0x} residuals is significantly smaller than the value of 0.2 typically found for optically selected quasars (Avni and Tananbaum, 1986), and explains the discrepancy between x-ray source number counts and the counts predicted from optically selected samples (this possibility was discussed by Franceschini *et al.* 1987). There is marginal evidence that the sigma (α_{0x} residuals) found here is less than that for the Medium Survey quasars, viz. 0.14. The Pavo object luminosities are similar, but may be slightly higher to those of the MSS (i.e. $\langle M_v \rangle = -25.0$, c.f. -23.7 for the MSS).

Of the 14 x-ray selected quasars, half had (U-J) < -0.5 while the remaining half were not 'uv excess' objects under this criterion. A further 10 uv-excess quasars were not detected in x-rays, so that approximately 30% of the quasars would have been missed by
the uv-excess criterion alone, in rough agreement with estimates of Boyle *et al.* (1987) on the incompleteness of uv-excess surveys for quasars in the redshift range 0.5 to 0.9. This result would also imply that x-ray selection is capable of finding most of the remaining quasars in a magnitude-limited survey, given sufficient x-ray sensitivity.

IV. Selection of UV-excess quasars

Plates including the Pavo field were taken at the United Kingdom Schmidt Telescope (UKST) in 1983, in the U, J, R and I passbands. These plates were scanned and digitised using the SERC Automatic Plate Measuring facility (APM) at the Institute for Astronomy in Cambridge (Kibblewhite *et al.* 1983). Apparent magnitudes for all objects with stellar appearance were calculated by standard techniques (Bunclark and Irwin 1983).

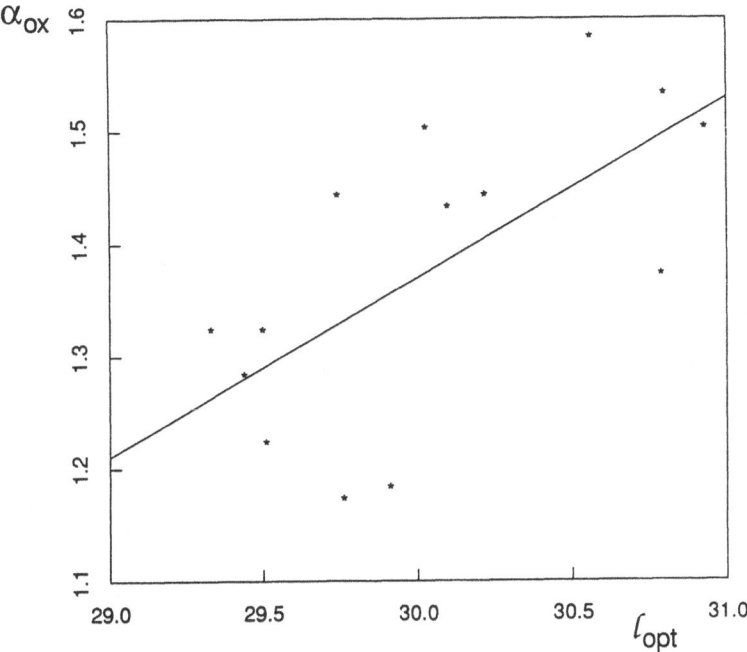

Figure 3. α_{0x} plotted against l_{opt} for the x-ray selected quasars in the Pavo field. The line is the best fit by regression.

UV-excess objects were selected based on the criterion U-J < -0.5. Of the objects selected in this way, it was found that 10 of the total of 35 had already been selected by virtue of x-ray emission, but there was no overlap between uv-excess selection and selection by prism or J-R colours.

V. Spectroscopy of UV-excess quasars

The uv-excess quasars which were undetected in the deep x-ray survey were observed spectroscopically at the AAT in the same manner to that described above for the x-ray selected objects.

The redshifts of the uv-excess quasars undetected in x-rays (UVNX objects) are compared in figure 1 with the x-ray selected objects. The redshift distribution of the UVNX objects is marginally different from that for the x-ray selected objects, with a median z of 0.84 for the x-ray selected quasars and a median z of 1.2 for the UVNX quasars. These UVNX objects have, on average, the same Jmag as the x-ray selected sample $\langle J \rangle = 20.2$, but a higher average optical luminosity $\langle M_v \rangle = -26.1$ compared with the $\langle M_v \rangle = -25$ of the x-ray selected quasars. The dependence of L_x on L_{opt} for optically selected qso's, viz. $L_x \sim (L_{opt})^{0.84}$ (Avni and Tananbaum, 1986) is in the sense that qso's with higher optical luminosity and greater redshifts have relatively lower x-ray luminosities.

VI. Contribution of observed x-ray selected quasars to the x-ray background

The HRI had no energy resolution over its operating range of 0.1 to 4.5 kev, cut off at the low energy end by the aluminum filter, and at high energy by the response of microchannel plates to x-rays, folded with the telescope mirror response function. An effective energy can be defined by the known instrument response function and the spectrum of incoming radiation. In order to convert source counts from the HRI into a fraction of the all-sky background, the assumption of spectral shape can reasonably be made in at least two ways:

1) from measurements of the spectral shape of the background made by other x-ray satellite experiments. Whereas the spectrum of the all-sky x-ray background has been relatively well determined in the energy range 3 to 40 kev (Marshall *et al.* 1980), the HEAO A-2 experiment indicated a turn-up in the spectrum below 3 kev, in agreement with the spectrum determined by the Low Energy Detectors on HEAO-A2 (Garmire and Nousek 1980). Extrapolation of the Marshall *et al.* spectrum is therefore inappropriate, and a photon power law spectral slope of ~ 2.0 is possibly indicated; or

2) to assume an average source spectrum which is the same as that of the qso's found in the MSS, or, better still, the average of qso's detected with the IPC in the Deep Surveys. This average spectrum for the MSS qso's has a hardness ratio (ratio of counts in the ~ 0.5 to 1.5 kev channel to counts in the ~ 1.5 to 4.5 kev channel) of 0.9, consistent with a photon power law of index ~ 2.0 (Maccacaro *et al.* 1987). The Pavo qso's, with an average redshift of 0.9, are observed in the emitted energy range from ~ 0.2 to ~ 9 kev ,so that the low energy excess found by Wilkes and Elvis (1987) in the lowest IPC energy channel (< 0.28 kev) is generally red-shifted into the lowest part of the detected energy range of the HRI. This may account, in fact, for the success of the HRI in detecting a large number of quasars in the deep survey fields.

Flux determinations for the individual x-ray sources reported here are generally subject to large errors. The individual sources have therefore been summed to find their total flux, for comparison with the total predicted flux from the HEAO-A2 LED extragalactic spectrum in the 0.1 to 4.5 kev range. For an energy spectral index of 1.0, the observed flux corresponds to $\sim 30\%$ of the extragalactic x-ray background.

VII. Contribution of UVNX quasars to the x-ray background

The UVNX quasars probably make a further contribution to the x-ray background, in addition to that of the x-ray selected objects. The overall distribution and limits on α_{0x} have been used to estimate that the UVNX quasars may contribute a further $\sim 10\%$ of the x-ray background, on the same assumptions as above.

VIII. Summary and conclusions

The discrete sources detected in the Einstein Observatory's deep survey in Pavo are predominantly quasars of redshift ~ 1, with an average absolute visual magnitude $M_v \sim -25$ ($L_{opt} \sim 10^{30}\,\mathrm{ergs\,s^{-1}}$ at 2500 Å). There is a distinct absence of low-luminosity AGN amongst the discrete sources.

It has been shown by Leiter and Boldt (1983), and Giacconi and Zamorani (1987) that the remaining contribution to the all-sky background must originate in discrete sources with Comptonized spectra. No evidence has yet been found for any unusual features in the optical spectra of the deep survey objects. The spectra of the x-ray selected quasars discussed here are likely to have power law index of about 1, by comparison with the Medium-survey objects discussed by Maccacaro *et al.* (1987). This may explain a turn-up in the background spectrum below 3 kev, but it is significantly different from the overall x-ray background spectrum above 3 kev. The present work does not necessarily address the 3-40 kev background, of which the origin is largely unknown.

References

Avni, Y. and Tananbaum, H., 1986, *Ap. J.*, **305**, 83.

Boyle, B. J., Fong, R., Shanks, T. and Peterson, B. A., 1987, *M.N.R.A.S.*, **227**, 717.

Ellis, R. S., Gray, P. M., Carter, D., and Godwin, J., 1984, *M.N.R.A.S.*, **206**, 285.

Bunclark, P. S., and Irwin, M. J., 1983, in Proc. Symp. "Statistical Methods in Astronomy", ESA SP-201, p. 195.

Elvis, M. and Lawrence, A., 1986, in "Astrophysics of Active Galaxies and Quasars", ed. J. Miller.

Franceschini, A., Gioia, I. M., and Maccacaro, T., 1987, *Ap. J.*, **301**, 124.

Garmire, G. and Nousek, J., 1980, private communication.

Giacconi, R., *et al.* 1979, *Ap. J.*, **234**, L1.

Giacconi, R. and Zamorani, G., 1987, *Ap. J.*, **313**, 20.

Gioia, I. M., Maccacaro, T., Schild, R. E., Wolter, A., Stocke, J. T., Morris, S. L., and Danziger, I. J., 1987, in Proc. IAU Symposium 121 on "Observational Evidence of Activity in Galaxies", eds. Khachikian, E. Ye., Fricke, K. J., and Melnick, J., p. 329.

Gray, P. and Sharples, R., 1985, AAO Fiber System Users Guide and Technical Manual.

Griffiths, R. E., *et al.* 1983, *Ap. J.*, **269**, 375.

Khembavi, A. K. and Fabian, A. C., 1982, *M.N.R.A.S.*, **198**, 921.

Kibblewhite, E. J., Bridgeland, M. T., Bunclark, P. S., and Irwin, M. J., 1983, in Proc. Conf. on "Astronomical Microdensitometry", NASA.

Kriss, G. A. and Canizares, C. R., 1985, *Ap. J.*, **297**, 177.

Leiter, D. and Boldt, E., 1982, *Ap. J.*, **260**, 1.

Lucey, J. and Taylor, K., 1983, IPCS User's Manual, AAO UM-10.

Maccacaro, T., Gioia, I. M., Wolter, A., Zamorani, G., and Stocke, J. T., 1988, *Ap. J.*, in press.

Marshall, F. E. *et al.* 1980, *Ap. J.*, **235**, 4.

Robinson, R. D., 1985, "The RGO Spectrograph", AAO UM-2.

Schmidt, M. and Green, R. F., 1986, *Ap. J.*, **305**, 68.

Wilkes, B. J., and Elvis, M., 1987, *Ap. J.*, **323**, 243.

Acknowledgements We would like to thank the Australian Telescope Allocation Committee for awarding us time at the Anglo-Australian Telescope, and also the staff at the AAT, especially Ray Sharples for observing assistance.

THE MILLI-JANSKY RADIO GALAXIES

H. van der Laan[*] and M.J.A. Oort
Sterrewacht
University of Leiden
The Netherlands

1. INTRODUCTION

The study of radio source populations is a long term programme at Leiden Observatory, based primarily on surveys with the Westerbork Synthesis Radio Telescope. The first two references[1] [2] and literature cited therein provide the overview of what was achieved since the early seventies. Oort's thesis[4] is the fourth one in this context and it enters new domains in terms of flux density range and angular size limits for statistically large samples.

2. THE LUMINOSITY FUNCTION OF RADIO GALAXY POPULATIONS

The radio galaxies everyone knows, at the 1 Jy level, number only a few thousands, all sky. At the 0.1 mJy level our surveys reach now, we have about twenty million sources in the sky and they have a very different distribution in the (P,z) plane than the apparently strong sources. Using the very deep surveys[3] [4] [9] [10] [11], correspondingly sensitive efforts at optical identifications[1] [4] and a concerted effort to acquire many of these sources' redshifts[4], the epoch-dependent radio luminosity function has been determined and interpreted in terms of distinct subpopulations' cosmic evolution[4] and a Danese & Franceschini (this volume).

3. BLUE RADIO GALAXIES

At flux levels below 1 mJy at 1.4 GHz, the mix of radio sources is dramatically different from that at the 1 Jy level. Blue radio galaxies now dominate and they differ markedly from the 'nearly standard candle' red radio galaxies famous in radio cosmology since the early sixties. The comparison is discussed extensively in Oort's thesis[4]; we summarize qualitatively: as their class names indicate the BRG and the RRG have very different colours; the BRG also have a greater fraction of compact ($\vartheta \leq 1.''5$) sources than the RRG. On the other hand the BRG are optically more extended, on average, than the

99

N. Kaiser and A. N. Lasenby (eds.), The Post-Recombination Universe, 99–100.

RRG. The radio compactness is reflected in the low frequency radio spectra: ~40% of the BRG show signs of opacity in the 21 to 92 cm wavelength range, compared to fewer than 5% if the RRG. The radio luminosity function of the BRG cannot possibly be explained in terms of normal spirals and Seyfferts, a conclusion confirmed by very deep identification contents[4].

4. EPOCH-DEPENDENT LINEAR SIZE DISTRIBUTIONS

Another major goal of the study reported here was the determination of the epoch dependence on red radio galaxies' linear size distributions. A radio- and optically complete set of sources, well distributed in the (P,z) plane, was formed and studied with the VLA. All the pitfalls of structure-dependent angular size definitions and of flux dependent selection risks were accounted for[4]. The VLA's combination of angular resolution and sensitivity reveals that $\vartheta(S_{1.4})$ does not flatten at 10" below $S_{1.4} \sim 20$ mJy, but continues to decrease. It was possible now[7) 6) 4)], to separate the power dependence and the redshift dependence of linear size. Remarkably, over three decades of power, the relative change of linear size with redshift has the same form. This result contains important clues for extended radio source physics[4].

PUBLICATIONS

1) van der Laan, H., Katgert, P., Oort, M.J.A.; 1986, Trieste Meeting "Structure and Evolution of Active Galactic Nuclei" (eds. Giuricin et al.), p.437-445 (Reidel, Dordrecht).
2) van der Laan, H., Katgert, P., Windhorst, R.A., Oort, M.J.A.; 1983, IAU symposium 104 "Early Evolution of the Universe and Its Present Structure" (eds. Abell, Chincarini), p. 73-79 (Reidel, Dordrecht).
3) Oort, M.J.A.; 1987, Astron. & Astrophys. Suppl. Series **71**, 221-243.
4) Oort, M.J.A.; 1987, Thesis, University of Leiden.
5) Oort, M.J.A.; 1988, Astron. & Astrophys. (in press).
6) Oort, M.J.A.; 1988, Astron. & Astrophys. (in press).
7) Oort, M.J.A., Katgert, P., Steaman, F.W.M., Windhorst, R.A.W.; 1987, Astron. & Astrophys. **179**, 41-59.
8) Oort, M.J.A., Katgert, P., Windhorst, R.A.; 1987, Nature **328**, 500-501.
9) Oort, M.J.A., van Langevelde, H.J.; 1987, Astron. & Astrophys. Suppl. Series **71**, 25-38.
10) Oort, M.J.A., Steemers, W.J.G., Windhorst, R.A.; 1988, Astron. & Astrophys. Suppl. Series (in press).
11) Oort, M.J.A., Windhorst, R.A.; 1985, Astron. & Astrophys. **145**, 405-424.

*New Address: ESO, Karl-Schwarzschild-Str. 2, 8046 Garching b. München

COSMOGONIC IMPLICATIONS OF THE FIRST QUASARS

Martin J. Rees
Institute of Astronomy
Madingley Road
Cambridge, CB3 OHA
England

ABSTRACT. Quasars offer clues to what the Universe was like at
redshifts z ~2-4. Before any quasars can turn on, some galaxies must
have evolved at least to the stage of harbouring enough concentrated
mass-energy to provide the observed luminosity. This paper summarises
some cosmogonic inferences that can already be drawn from studies of
high-z quasars.

1. INFALL FROM GALACTIC HALOS AT $Z \geq 2$.

The low densities, and dynamical timescales $\geq 10^9$ yrs, in the outer parts
of halos (>100 kpc) have the crucial implication - irrespective of what
these halos actually consist of - that galaxy formation (and, more
specifically, the formation of the outer parts of disc galaxies) was not
completed until the Universe was $>2 \times 10^9$ years old: i.e. not until
redshifts $z \simeq 2$, even if it started much earlier (Fall and Efstathiou
1980).
    For a total gas mass of $M_g = 10^{11}$ $M_\odot$ the characteristic mean
density in a protogalaxy is $\bar{n} \simeq 10^{-3}$ r_{100}^{-2} cm^{-3} where r_{100} is the radius
in units of 100 kpc. The free-fall time and the virial temperature
depend on the total mass of galaxy-plus-halo (which may not all be
baryonic). If the material at $r_{100} \simeq 1$ were at the virial temperature
$T \simeq 10$ K its luminosity, primarily in soft X-rays, would be only 10^{42}
erg s^{-1}. However, material falling within the radius where $t_{cool} < t_{infall}$
would develop a two-phase structure: gas would condense out as clouds,
or in sheets behind radiative shock fronts, cooling to 10^4 K (or a still
lower temperature if heavy elements were already present), and establish
pressure balance with the fraction of gas that remained at $T = T_{virial}$.
A fraction f(r) would condense out, the value of f adjusting itself so
that the cooling of the remainder (1 - f) was balanced by the available
heat input. The clouds would have densities of order $(T/T_{virial})^{-1}$
n(1-f). If there were no heating other than from adiabatic compression,
(1-f) would drop until the cooling and free-fall times were comparable
(cf. Fall and Rees 1985). The gas at 100 kpc radius would radiate a
luminosity of

N. Kaiser and A. N. Lasenby (eds.), The Post-Recombination Universe, 101–107.

$$\sim 10^{45} \ f(1 - f)^2 \left(\frac{M_g}{10^{11} \ M_\odot} \right)^2 \left(\frac{n_e}{n_{HI} + n_e} \right)^2 \ erg \ s^{-1} \qquad (1)$$

predominantly in H-recombination lines.

This luminosity could be very high if the clouds were fully ionized. Consequently, most of the gas would recombine (i.e. n_e / n_{HI} would become <<1) unless there were some powerful internal excitation mechanism capable of balancing these radiative losses. O and B stars could provide enough excitation if early star formation yielded a population with a sufficiently flat IMF, in which case all galaxies at this phase of their evolution would be conspicuous extended emission-line objects. York et al. (1986) make such estimates, on the assumption that a young galaxy is essentially an aggregate of gas-rich dwarfs. However, we do not know enough about the initial IMF and the early rate of star formation to quantify this.

But we do know that some galaxies at $z \geq 2$ possessed active nuclei, and the UV emission from a central quasar would contribute a still more potent input, capable of maintaining gas with the properties above in a fully ionized state, irrespective of the stellar content. Reprocessed or scattered emission of strength (1) would then be detected as narrow lines from a region extended by a few arc seconds. Quantifying this further depends on knowledge of uncertain geometry, covering factors, etc. However, even the existing limits on fuzz around high-z quasars discussed by Cowie in his contribution to this meeting may impose interesting constraints on galaxy formation.

Quasars serve as probes not only of conditions in their host galaxy, but also of diffuse gas in other galaxies along the line of sight. Some of the absorption lines are attributable to intervening galaxies: moreover, the fact that there are not even more systems with high column density places a significant constraint on the amount of infalling gas in cool pressure-confined clouds within a typical galaxy at $z \simeq 2$ (Hogan 1987, Rees 1987).

2. QUASARS WITH Z   4 AND THE EPOCH OF GALAXY FORMATION IN CDM COSMOGONY

The comoving density of quasars is well established to be much higher at $z \approx 2$ than at the present epoch. For a cosmological model with $\Omega = 1$ the number density of quasars at $z = 2$ brighter than $L = 10^{47} \ h_{50}^{-2}$ erg s^{-1} is

$$N_Q \approx 1.2 \times 10^{-7} h_{50}^{-3} \quad per \ comoving \ Mpc^3 \qquad (2)$$

(Schmidt and Green 1983; Hazard and McMahon 1985; Boyle et al. 1986; Weedman 1985; Hazard 1987; Koo and Kron 1987).

At redshifts ≥ 2.5, the quasar luminosity function is poorly known. There are strong indications that the prodigious rise in the comoving number density of luminous quasars seen at low redshifts does not continue beyond $z \geq 2.5$, but there is certainly no absolute redshift cut-off at $z \leq 4$ since some intrinsically very powerful quasars do exist at higher redshifts.

Some galaxies must therefore have formed, at least to the extent of having developed well-defined nuclei, by $z \simeq 4$. This is a constraint on the adiabatic "pancake" model, where no galaxies form until cluster-scale bound systems have collapsed. But high-z quasars also have important implications for the cold dark matter (CDM) model: George Efstathiou and I (1987) have addressed this problem recently, and I should like to summarise our conclusions here.

The central mass involved in a quasar depends on its luminosity, its lifetime t_Q, and the efficiency ϵ defined as the fraction of the central rest mass energy converted into radiation over its lifetime. The central mass associated with each $10^{47}h_{50}^{-2}$ erg s^{-1} quasar may therefore be written as

$$M_Q \approx 2 \times 10^9 h_{50}^{-2} \, t_{Q_8} \epsilon_{0.1}^{-1} \, M_\odot , \tag{3}$$

where t_{Q_8} denotes the lifetime in units of 10^8 years and $\epsilon_{0.1} = \epsilon/0.1$. There is no way of firmly estimating t_Q. Were it as long as $t_{4-3} \approx 5 \times 10^8 h_{50}^{-1}$ years, the elapsed time between $z = 4$ and $z = 3$, there need be only one generation of high redshift quasars. On the other hand, t_Q could be much shorter. There would then need to be many generations, the number density of hosts exceeding the observed number density of active quasars by $\approx (t_{4-3}/t_Q)$.

The density of quasars implied by (2) is of course very low compared to the present density of bright galaxies. However, the bulk of the matter in the CDM cosmogony condenses into galactic-mass objects at low redshifts. N-body simulations (e.g. Frenk et al. 1985, 1987; White et al. 1987) show that most galaxies with circular velocities ≥ 200 km s^{-1} would have formed at $z \leq 2$ in mergers of smaller systems; at $z > 4$, virialised systems would be almost exclusively of subgalactic mass. As we extrapolate back in time, galaxies could have formed only at progressively rarer upward fluctuations in the initial density field. Thus at high redshifts we must expect a precipitous drop in the number of galaxies capable of harbouring a central object as massive as required by (3). What then is the critical redshift z_{crit} above which the abundance of luminous quasars would have to fall below the value (2), and is this compatible with the observations?

The value of z_{crit} depends on how big a galaxy has to be in order to develop a central mass concentration of the requisite size (3). The ratio of the quasar's mass to the total galactic mass involves three factors:

(i) The fraction f_b of the matter in baryonic form, which is ≈ 0.1 if the $\Omega = 1$ CDM model is to be compatible with standard primordial nucleo-synthesis (Yang et al. 1984).

(ii) The fraction f_{ret} of the baryons originally associated with the dark halo which are retained within the galaxy rather than being expelled via a supernova-driven wind. This is likely to depend steeply on the virial velocity v_{vir} (Dekel and Silk 1986); it may well be ≤ 0.1 for $v_{vir} < 100$ km s^{-1}.

(iii) The fraction f_{hole} of the baryons retained in the galaxy which participate in the runaway gravitational collapse processes manifested in the quasar phenomena. This depends on the route whereby the central mass accumulates and evolves and on how efficiently gas in the outlying parts of the galaxy can lose angular momentum and sink towards the centre - two perennial uncertainties bedevelling all quantitative attempts to model quasars.

If for instance, one were to suppose $F = f_b f_{ret} f_{hole} = 0.01$ - likely to be a highly optimistic estimate (from the point of view of efficient formation of a central mass concentration) - then equation (3) implies a total halo masses of $\sim 10^{12} \epsilon_{0.1}^{-1} h_{50}^{-3} M_\odot$ if $t_Q = t_{4-3}$. More pessimistic, and probably more realistic, suppositions about the efficacy of black hole formation (i.e. values of $f_{ret} f_{hole}$ well below 0.1) would imply correspondingly larger galactic masses.

If quasar lifetimes were short, the accumulated central masses (equation (3)) need not be so great, and the host galaxies need not then be so massive. This suggests that very high-z quasars could more readily exist if t_Q were short: they could then be associated with the (much less rare) early-forming galaxies of lower mass.

There is, however, an astrophysical constraint on quasar masses. A quasar whose mass were below $7 \times 10^8 L_{47} M_\odot$ would be radiating at more than its Eddington luminosity. If accreted gas provides the power, then clearly (except in special geometry) there cannot be steady fuelling at a rate than generates $L > L_E$. Supercritical luminosities could nonetheless arise in at least two ways: tidally-disrupted stars could release fuel near the hole (though to generate quasar-level luminosities, the stars around the hole would have to be so densely-packed that stellar collisions would release still more gas (Frank, 1978)); electromagnetic energy extraction from a spinning black hole could also in principle yield a non-thermal power with $L > L_E$.

Unless the luminosity were indeed "super-Eddington", which would require one of the special models just mentioned, quasar lifetimes must be longer than $t \approx 4 \times 10^7 \epsilon_{0.1}$ years. Very short lifetimes may therefore lead to astrophysical difficulties; on the other hand the high central masses given by (3) imply host galaxies whose masses are too high to have formed at high z if $t_Q > 10^8$ years. The preferred hypothesis is therefore that quasars have lifetimes $t_Q \approx t_E < 4 \times 10^7$ years, and radiate with efficiency >0.1 at around the Eddington limit. Their masses would then be $\sim 10^9 M_\odot$, and they would reside in galaxies with masses $10^{11} - 10^{12} M_\odot$ (for $F_{0.01} \approx 1 - 0.1$) and virial velocities $v_{vir} < 200(1 + z)^{\frac{1}{2}}$ Km s^{-1}.

Although an abundance of luminous quasars at z = 4 similar to that at z = 2 is compatible with the CDM model, it requires that quasars form quickly and efficiently within the first galaxies. The quasar population should thin out exponentially at still larger redshifts. Even with optimistic values for and the fraction of baryons involved in a compact object, and for quasar lifetimes $\sim t_E$ a "cut-off" in the space density of luminous quasars is expected at $z \sim 5$, so an interesting test of the model may be within reach.

Since galaxy formation is so "recent" in the CDM model, there cannot be a long latency period between the virialization of a galaxy and the onset of activity in its nucleus (cf. Cavaliere and Szalay,

1986). But the rapid build-up of a $>10^8$ $M_\odot$ central object need not present a problem: unlike stars, black holes do not possess a "hard" surface; there is therefore no requirement that any specific amount of energy be emitted per unit mass accreted and so the timescale for growing a black hole could be much shorter than t_S. In effect the efficiency ϵ may adjust downwards so that the hole can accept mass at a high rate without the emergent luminosity exceeding L_E. As an extreme example, an entire supermassive star could collapse to a black hole in a free-fall time, due to post-Newtonian instability, without radiating any energy at all, although it could take several x 10^8 years for a dense star cluster to evolve to this point.

At high redshifts $z \geq 4$, we might expect to see strong differential evolution in the quasars numbers. This is because high luminosity quasars must be associated with massive proto-galaxies whereas low luminosity quasars could form in less massive and hence more abundant systems.

In the CDM model, luminous quasars at high redshifts would be associated with rare high peaks in the density field. They should therefore be correlated ab initio (see e.g. Bardeen et al. 1986) with a clustering amplitude comparable to that of local bright galaxies. One cannot, however, predict the z-dependence of quasar clustering (cf. Shaver's contribution) without a better understanding of what "triggers" the quasar phenomenon in a small fraction of galaxies at lower z.

All things considered, a high space density of luminous quasars at $z \geq 5$, could prove an embarrassment for CDM theory. It would require one or more of the following: (a) that an extremely high fraction of the baryons within a protogalaxy can collapse and from a compact object, (b) exotic models in which quasars radiate at super-Eddington luminosities, (c) a reappraisal of the normalization of the CDM spectrum, (d) a non-standard initial fluctuation spectrum.

3. THE PRE-QUASAR ERA

Even if the comoving density of quasars were as high back at $z \simeq 4$ as at $z = 2$, the resultant output of UV photons could not ionize a medium with $\Omega_B \sim 0.1$ to a sufficient degree to account for the lack of a Gunn-Peterson (1965) trough in any observed quasar (Shapiro and Giroux 1987; Donahue and Shull 1987; Steidel and Sargent 1987). This particular problem would be eased if there were many quasars at still larger redshifts, but as we have seen, this is difficult in the CDM cosmogony - at least for quasars with $L \geq 10^{47}$ erg s^{-1}. In the CDM model, the ionization of the intergalactic medium could naturally come about before any galactic mass systems (and a fortiori before any quasars) had condensed. The "first-light" in the CDM Universe would be provided by the capture and dissipative collapse of gas into ~10^5 $M_\odot$ microhalos (Bond and Szalay, 1983; Couchman and Rees 1986) starting at $z \geq 10$. The environmental effect of these first stars depends on their characteristic masses which cannot confidently be predicted. The radiation from metal-poor O and B stars, or from very massive objects (VMO's) would be mainly in the Lyman continuum. Conversion of hydrogen to helium yields about 7 MeV per proton, whereas the energy required to ionize an H atom is only 13.6 eV plus the kinetic energy of the dislodged electron; therefore even the conversion of only 10^{-4} of the matter into such stars could ionize the entire pregalactic medium (Carr, Bond and Arnett, 1984; Couchman and Rees 1986). If the IMF were steeper, a less trivial fraction of the primordial matter would be converted into stars before enough Lyman continuum was generated to ionize the entire medium. (The total pregalactic background light would then be much stronger. This would be in the near infrared unless re-processed by dust in which case it would now contribute a sub-millimetre background).

The material would, after photoionization, be on a higher adiabat, with a consequent rise in the Jeans mass to $10^8 - 10^9$ $M_\odot$. There would, as a result, be a lull in the cosmogonic process until objects of this mass-scale turned around - it is these that could confine gravitationally-bound clouds of ionized matter, and perhaps account for the multiple Lyman absorption features in quasar spectra (Rees 1986).

I am grateful for collaboration with George Efstathiou, and thank Avishai Dekel, Cyril Hazard, and Joel Primack for helpful discussions.

REFERENCES

Bardeen, J.M., Bond, J.R., Kaiser, N. and Szalay, A.S., 1986, Astrophys. J., **304**, 15.
Bond, J.R. and Szalay, A.S., 1983, Astrophys. J., **277**, 443.
Boyle, B.J., Fong, R., Shanks, T. and Peterson, B.A., 1987, Mon. Not. R. astr. Soc., **277**, 717.
Carr, B.J., Bond, J.R. and Arnett, W.D., 1984, Astrophys. J., **277**, 445.
Cavaliere, A. and Szalay, A.S., 1986, Astrophys. J., **311**, 589.
Couchman, H.M.P. and Rees, M.J., 1986, Mon. Not. R. astr. Soc., 221, 53.
Dekel, A. and Silk, J., 1986, Astrophys. J., **303**, 39.
Donahue, M.E. and Shull, M.J., 1987, preprint.
Efstathiou, G. and Rees, M.J., 1987, Mon. Not. R. astr. Soc. (submitted)
Fall, S.M. and Efstathiou, G., 1980, Mon. Not. R. astr. Soc., **193**, 189.
Fall, S.M. and Rees, M.J., 1985, Astrophys. J., **298**, 18.
Frank, J., 1978, Mon. Not. R. astr. Soc., **184**, 87.
Frenk, C.S., White, S.D.M., Efstathiou, G. and Davis, M., 1985, Nature, **317**, 595.
Frenk, C.S., White, S.D.M., Davis, M. and Efstathiou, G., 1987, Astrophys. J., in press.
Gunn, J.E. and Peterson, B.A., 1965, Astrophys. J., **142**, 1633.
Hazard, C., 1987, in "Modern Instrumentation and its Influence on Astronomy" (Hanbury Brown Birthday Symposium), ed. J.V.W. Wall (in press).
Hazard, C. and McMahon, R., 1986, Nature, **314**, 238.
Hogan, C.J., 1987, Astrophys. J. (Lett.), **316**, L59.
Koo, D.C. and Kron, R.G., 1988, Astrophys. J., in press.
Rees, M.J., 1986, Mon. Not. R. astr. Soc., **218**, 25P.
Rees, M.J., 1987, in "Quasar Absorption Lines", ed. C. Blades and D.

Turnshek (CUP), in press.
Schmidt, M. and Green, R.F., 1983, Astrophys. J., **269**, 352.
Shapiro, P.R. and Giroux, M.L., 1987, Astrophys. J., **321**, L107.
Steidel, C.C. and Sargent, W.L.W., 1987, Astrophys. J., **318**, L11.
Weedman, D.W., 1985, Astrophys. J. Suppl., **57**, 523.
White, S.D.M., Frenk, C.S., Davis, M. and Efstathiou, G., 1987,
    Astrophys. J., **313**, 505.
Yang, J., Turner, M.S., Steigman, G., Schramm, D.N. and Olive, K.A.,
    1984, Astrophys. J., **281**, 493.
York, D.G., Dopita, M., Green, R. and Bechtold, J., 1986, Astrophys. J.,
    311, 610.

MEASUREMENTS OF CBR ANISOTROPIES
AT INTERMEDIATE ANGULAR SCALES

Paolo de Bernardis , Silvia Masi , Francesco Melchiorri

Dipartimento di Fisica , Universita' "La Sapienza"

P.le A.Moro 2

I00185 Roma , ITALY

Bianca Melchiorri

Istituto di Fisica dell' Atmosfera del CNR

P.le Sturzo 31

I00144 Roma , ITALY

Nicola Vittorio

Dipartimento di Fisica Universita' de l'Aquila

P.le dell' Annunziata 1

I67100 L' Aquila , ITALY

ABSTRACT. We present a review of current CBR anisotropy theories and experiments at intermediate angular scales. The data from four recent measurements are compared using the maximum likelihood method.

1. INTRODUCTION

"Instead of the steady state theory being wrong because it cannot produce the microwave background, it is the hot big bang that is wrong because it cannot make galaxies".

This statement by Fred Hoyle in 1953 appears to be prophetic now. We observe galaxies, but we have been unable, up to now, of detecting the seeds responsible for galaxy formation at $z = 1000$. In more quantitative words, Big Bang theory allows

N. Kaiser and A. N. Lasenby (eds.), The Post-Recombination Universe, 109–123.

us to extrapolate the present density contrast $\delta\rho/\rho \simeq 1 \sim 5$ (where $\delta\rho$ is the excess density in galaxies with respect to the average density ρ in the universe) down to $z = 1000$ simply scaling by a factor $1 + z$.

Therefore, one has to explain the difference between a density contrast $\delta\rho/\rho \simeq 10^{-3}$ and the upper limits already obtained on the brightness contrast $\delta I/I \simeq \delta T/T \lesssim 10^{-5}$.

Unfortunately, big bang theory anticipates a tight coupling between matter and radiation: the excess of 10^9 photons per barion is explained as the result of the anihilation between matter and antimatter, so that any primordial fluctuation in matter-antimatter density is reproduced in the cosmic radiation bath. If we accept this " adiabatic view " suggested by big bang theory we face with the paradox of the existence of the galaxies versus the astonishing isotropy of cosmic background radiation. Two different kinds of solutions to this paradox have been suggested.

The first one is related to the ionization history of the universe. Late reheating of the intergalactic medium would result in an angular smearing transverse to the line of sight on angular scales from $\alpha_c \simeq 8'\Omega_o^{1/2}$ (thickness of the last scattering surface), up to several degrees (reionization at $z < 40$ is irrelevant, being the density of the universe too low). Therefore, any upper limit at angular scales less than $1^o \sim 5^o$ can be reconciled with big bang theory if we introduce an appropriate amount of ionized matter.

The second solution makes use of non-barionic matter to build up galaxies. This is formally equivalent to reduce the coupling between matter and radiation. It is reasonable to assume the same density contrast for barionic and non-barionic matter at the time when the perturbations first formed (Planck era ? After Inflation ?). Therefore the difference in the two density contrasts $(\delta\rho/\rho)_{barionic}$ v/s $(\delta\rho/\rho)_{non-barionic}$ at $z = 1000$ depends on the history of the perturbations, i.e. on the type of non-barionic matter involved.

If we take in mind the problems and the possible solutions we arrive at the conclusion that the angular scale of the order of a few degrees is the best choice for the search for CBR anisotropies. Larger angular scales are significantly affected by the emission from our galaxy, while at smaller scales the primordial fluctuations could be completely erased by a possible strong reionization.

The CBR spectrum has its maximum at a wavelength of about 1 mm , between the lower edge of the infrared and the higher edge of the microwaves . For these reasons both the experimental techniques can be used , but technologies at the state of the art are required : bolometers cooled at 0.1 or 0.3 K for the infrared ; SIS mixers in the microwaves .

The small thermodynamic temperature fluctuations $\Delta T/T$ of the CBR are converted in flux fluctuations $\Delta I/I$ with an higher efficiency in the millimetric region (fig.1) ; moreover the far infrared bolometers are sensitive to a wide band of wavelengths , thus providing the higher possible sensitivity when observing a continuum radiation spectrum.

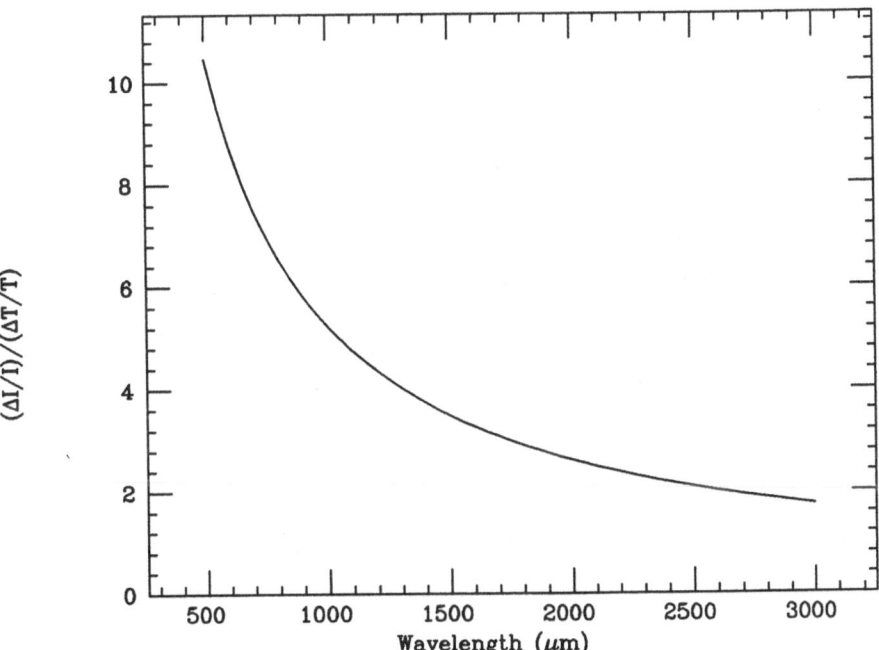

fig.1 : The ratio between the flux fluctuations $\Delta I/I$ and the thermodynamic temperature fluctuations $\Delta T/T$ is plotted for a blackbody with $T = 2.75°K$

These two facts seem to favour the far infrared approach ; on the other hand the small ($\Delta T/T \lesssim 10^{-4}$) CBR fluctuations are embedded and hidden in the local backgrounds , and one of these , the atmospheric emission , is higher and noisier at shorter wavelengths . For this reason CBR experiments at wavelengths shorter than 2 mm are carried out at ballon or rocket altitude .

Patchy galactic dust emission , increasing at short wavelengths , can contaminate the faint CBR signal in the far infrared , while synchrotron and free-free emission from galactic electrons can be dangerous in the radio band : as a result , high galactic latitude regions must be observed and the ' cosmological window ' at wavelengths between 2 and 4 mm must be selected for this kind of measurements.

In the following section we briefly review the theoretical background ; in section 3 we describe four recent experiments on the intermediate scale anisotropies ; in section 4 we analyze the data from these experiments using a common procedure .

2. THEORETICAL BACKGROUND

On intermediate angular scales ($1^{\circ} < \alpha < 10^{\circ}$) CMB anisotropies are expected because of the gravitational potential fluctuations on the last scattering surface (Sachs and Wolfe, 1967). In fact , photons are gravitationally redshifted streaming through the gravitational potential gradients between the last scattering and the present epoch . As an order of magnitude , $\delta T/T \sim \delta\phi(R)/c^2$: here $\delta\phi$ is the gravitational potential associated with a density perturbation of size R.

In the gravitational instability scenario the density fluctuation field is completely described by its power spectrum $|\delta_k|^2 = Ak^n$, under the usual random phase approximation . To the Sachs and Wolfe effect contribute large scale perturbations ($> 100h^{-1}Mpc$), which experience uninterrupted growth. In this range of scales, the initial slope of the power spectrum is conserved. As $\delta\phi \sim k^3|\delta\phi_k|^2 \propto k^{n-1}$, for $n = 1$ the potential fluctuations are independent on the perturbation scale. The inflationary scenario provides with a physical mechanism for producing a Harrison-Zeldovich spectrum (i.e., n=1) through quantum fluctuations of the Higgs field. The observations on the intermediate and large scale anisotropy should at least in principle constrain not only the amplitude of the density fluctuations, but also the initial spectrum.

Given the initial density fluctuation power spectrum, under the assumption of gaussian distributed, adiabatic fluctuations, it is possible to calculate the large scale pattern of the CBR: in particular, the number of hot spots in the microwave sky (regions with temperature fluctuations greater than ν times the rms value) and their angular dimensions. Beeing the result of a pattern analysis, the predictions are independent of the actual value of the temperature fluctuations and do not discriminate among models which share the same initial conditions.

The information on the hot spot number and angular diameter can at least guide in defining the best observational strategy for a positive detection of CBR temperature fluctuations. For example, as the observations are differential, it would be desiderable to beam switch at an angular scale larger than the typical hot spot angular diameter, for not looking to same hot spot with both the antenna hornes. If the primordial fluctuations have a Zel'dovich spectrum, the number of hot spots expected in all the sky is (Vittorio and Juszkiewicz (1987)) :

$$N_{>\nu} = \frac{650}{\sigma^2} \frac{\nu exp(-\nu^2/2)}{[-ln\ 2\sigma + 3.78]} \tag{6}$$

while their angular dimension is

$$D = \frac{5.6}{\nu}\sigma[-ln\ 2\sigma + 3.78] \tag{7}$$

In the above formulae, the antenna beam has been modeled with a gaussian profile of dispersion σ, here measured in degrees. As expected for a scale invariant process, the number and the angular dimension of the hot spot is determined mainly by the smearing of the antenna beam. For $\nu \sim 2$ and $\sigma \lesssim 3°.5$, $D \sim 20°$. A more quantitative assessment of the best experimental strategy can be obtained by simulating CBR observations of the theoretical sky, taking into account all the important experimental parameters (modulation geometry, beam pattern, sky coverage, etc.) (Vittorio et al. (1988)). As we said before , the above calculations depend only on the initial density fluctuation power spectrum. Of course the amplitude of these fluctuations depends upon the specific model. Attention has been drawn in the last years to a very specific model for the formation and evolution of the large scale structure in the Universe (see Blumenthal et al. (1984), for a review). According to inflation , the universe is assumed to be flat,(see e.g. Turner (1987)),and today dominated by massive , weakly interacting , 'cold' relics of the early universe, for consistentcy with primordial nucleosynthesis bounds on the baryonic content. This model seems to be the most efficient at our disposal for forming galaxies without perturbing the observed isotropy of the CBR. For matching the model predictions with the small scale non linear clustering of galaxies , it is necessary to require that galaxies are biased to form in the rare and dense peaks of the density field. This implies that the galaxy distribution is more clustered than the underlying mass density field , and in particular that the variance in the counts of bright galaxies isgreater than the variance in the mass fluctuation by a biasing factor $b > 1$.

The rms value $\Delta T/T$ of the CBR temperature anisotropies expected in a cold dark matter scenario (Scaramella and Vittorio (1988)) are shown in Fig.2. In Fig.2a $\Delta T/T$, as measured in a single subtraction experiment, is plotted vs. the beam switching angle α. Here an antenna beam size $\sigma = 2^o.2$ has been assumed. If n=1, the predicted anisotropy for $\alpha = 6^o$ is $\Delta T/T = 7.4 \cdot 10^{-6}/b$. In Fig.2b, $\Delta T/T$, as measured in a double subtraction experiment, is plotted for $\sigma = 3^o.5$. If n=1, at $\alpha = 8^o$, $\Delta T/T = 4.3 \cdot 10^{-6}/b$.

The CDM scenario predicts a rms temperature fluctuation a factor ~ 6 below the Melchiorri et al. upper limit and a factor ~ 10 below the Davies et al. detection.

fig.2 : Rms value $\Delta T/T$ of CBR anisotropies expected in the cold dark matter scenario. In A a $2^o.2$ beamsize and a single subtraction were assumed. In B a $3^o.5$ beamsize and a double subtraction were assumed. Three different spectral indexes were assumed: $n = 0$ (dotted lines), $n = 1$ (solid lines), $n = 2$ (dashed lines).

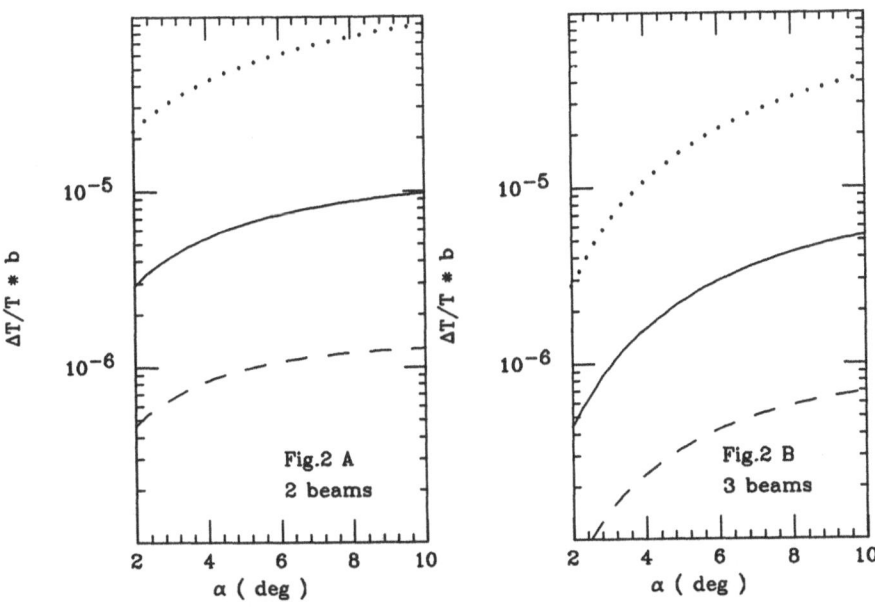

3. INTERMEDIATE SCALE ANISOTROPY EXPERIMENTS

The first experiment was carried out at balloon altitude (Fabbri et al (1979), Melchiorri et al (1981)). Beam switching was performed between two fields of view at the same elevation on the horizon ; a slow gondola rotation allowed the scansion of a wide sky strip . Beam switching was performed between two fields of view with a separation $\alpha = 6^o$; the beam size was 5° FWHM ($\sigma = 2^o.2$). The modulator was a plane wobbling mirror and the detector was a criogenic composite bolometer , sensitive to radiation with wavelengths larger than $700\mu m$). The experiment is sketched in fig. 3 . An high galactic latitude region was scanned for several hours , and a system noise as low as 0.02 mK was achieved in each independent field of view. The data from these scans are shown in fig. 4a . The variance of the data is significantly larger than system noise fluctuations : this means that a non-istrumental anisotropy at a level of about $\Delta T/T \sim 4 \cdot 10^{-5}$ has been detected ; this value is usually referred to as an upper limit to CBR anisotropies since at FIR wavelengths galactic dust patchy emission can represent a significant fraction of the detected signal . Melchiorri et al (1981) have used the 21 cm emission as a dust indicator and estimated that the dust emission contribution should be negligible in the observed region ; a more direct comparison using IRAS data is now in progress.

The second experiment was ground based and was performed using a radio receiver at $\lambda = 3cm$ (Mandolesi et al (1986)). The sky circle at $\delta = 80°$ was scanned by a differential radiometer : the radiation collected by two corrugated horns pointed in two directions with a separation $\alpha = 10°$ was continuously compared using Dicke switching ; the beamsize was $2°.8$ FWHM ($\sigma = 1°.2$). The observations were carried out both at Medicina (near Bologna , Italy) and near Tromso (Norway) . No correlation between the two data sets is evident from the published data. The averaged data are reported in fig. 4b ; the large scatter of the points is due to atmospheric fluctuations . The result of this experiment is an upper limit to CBR anisotropies at a level $\Delta T/T \sim 3 \cdot 10^{-4}$.

The third experiment was carried out in Terra Nova Bay , Antarctica (Dall'Oglio and de Bernardis (1988)) using a 1m parabolic flux collector and a ^3He cooled bolometer as the detector , at $\lambda = 2mm$. A sketch of the experimental setup is shown in fig. 5. Beam switching ($\alpha = 1°.3$) was accoplished for by wobbling the flux collector at a frequency of 10 Hz . The beamsize was $1°.3$ FWHM ($\sigma = 0°.53$). Part of the sky circle at $\delta = -75°$ was continuosly scanned. The data collected at high galactic latitudes are visible in fig. 4c and are dominated by atmospheric fluctuations ; an upper limit $\Delta T/T \sim 2.5 \cdot 10^{-4}$ to CBR anisotropies was found .

The fourth experiment was carried out in Izana , Tenerife (2300 m osl) (Davies et al.(1987)). A Dicke switching , two horns radio receiver at 10.4 GHz was coupled to a wobbling plane mirror in order to obtain a three beams sky modulation ; this technique greatly reduces atmospheric emission contributions to the signal, because

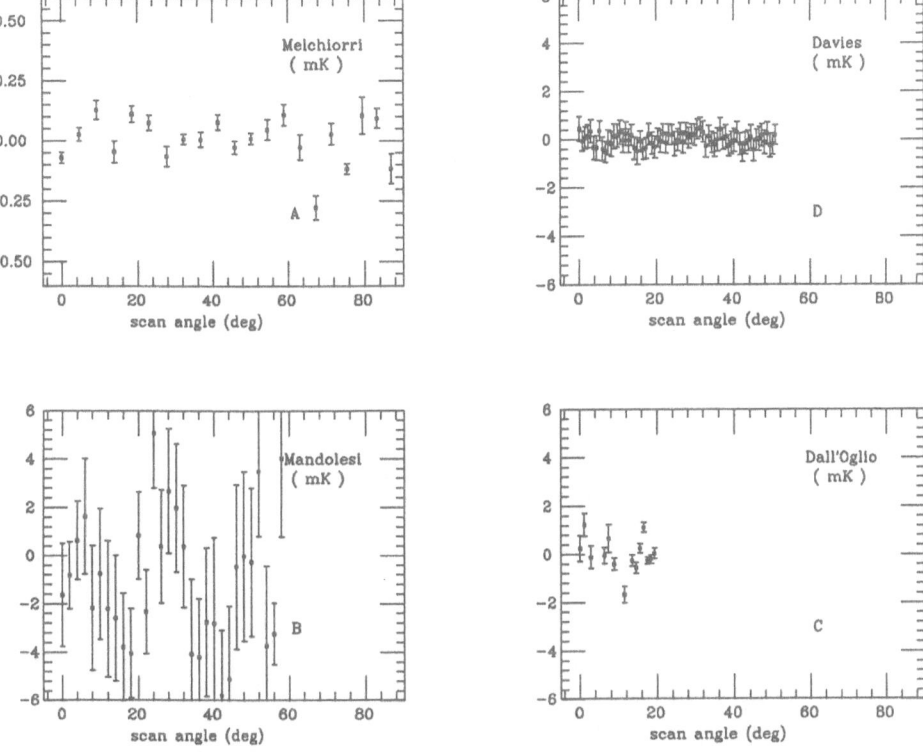

fig.4 : Data from recent CBR intermediate scale anisotropy measurements: in order to have a quick comparison between different experiments, the measured temperature differences in mK have been plotted vs the line of sight position , measured in degrees along the scan direction. Note that the temperature scale is amplified a factor 10 in the Melchiorri et al. (1981) experiment.

Tab. 1 : Four recent intermediate scale anisotropy experiments

Reference	Melchiorri et al. (1981)	Mandolesi et al. (1986)	Dall'Oglio et al. (1987)	Davies et al. (1987)
experiment	balloon	ground	Antarctica	ground
detector	bolometer	radioreceiver	bolometer	radioreceiver
beams	2	2	2	3
beamthrough α	$6°$	$10°$	$1°.3$	$8°.2$
beamsize σ	$2°.2$	$1°.2$	$0°.53$	$3°.5$
sampling angle θ_o	$5°.5$	$2°$	$1°.1$	$0°.77$
standard deviation (mK)	0.10	2.79	0.69	0.26
ΔT (mK , N.P.lemma 95%)	0.11	2.55	0.69	0.20
$C_o^{1/2}(mK), LR \geq 10, n = 1$	0.12 ± 0.06	≤ 3.6	$2.6^{+2.9}_{-1.5}$	$0.16^{+0.24}_{-0.16}$

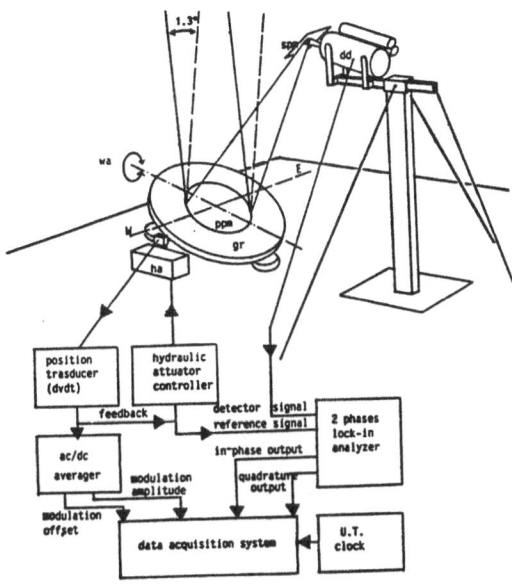

fig.5 : The ground based isotropometer used by Dall'Oglio and de Bernardis (1988) in Antarctica . It consists of a wobbling off axis paraboloid 1 meter in diameter (ppm) and an ^3He cooled bolometric detector (dd) at 2 mm of wavelength .

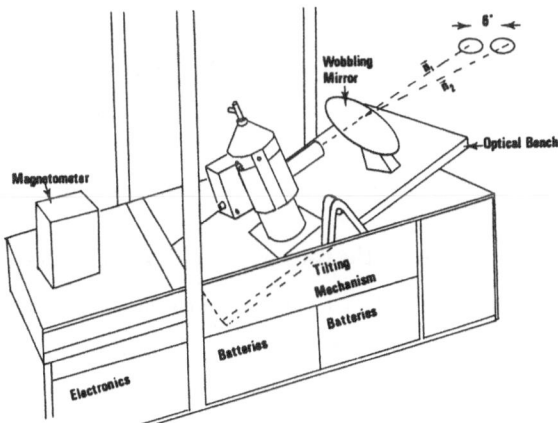

fig.3 : The balloon borne isotropometer used by Fabbri et al. (1979).

is insensitive to linear intensity gradients . The beams were separated by $\alpha = 8^\circ.2$ and the beamsize was $8^\circ.3$ FWHM ($\sigma = 3^\circ.5$). A sky strip at $\delta = 40^\circ$ was scanned . The data in the region far from the galactic plane are shown in fig. 4d ; at first sight they seem to be dominated by detector noise ; however evidence for a non-zero sky anisotropy has been found using a likelihood analysis , at a level $\Delta T/T \sim 3.7 \cdot 10^{-5}$. Also in this experiment a certain fraction of galactic contamination is expected (synchrotron and free-free from HII regions).

4. DATA ANALYSIS

When one compares the anisotropy estimates coming from different experiments, differences in the data reduction and analysis , statistical methods used , modulation geometry , observation strategy and assumed anisotropy pattern must be taken into account . As a result , a direct comparison between experiments with different fields of view or sky coverage is usually very difficult .

This kind of problem can now be solved using a likelihood analysis and an analytical evaluation of the CBR anisotropy autocorrelation function for the expected CBR temperature field (Vittorio et al. (1988)). The smoothed CBR anisotropy autocorrelation function $C(\alpha, \sigma)$ in the case of adiabatic , gaussian distributed, scale free density perturbations has been calculated (Scaramella and Vittorio, 1988). It is then straightforward to calculate the autocorrelation matrix C_{ij} for the data ΔT_i which are assumed to be the sum of system noise N_i and true sky anisotropy Δ_i :

$$C_{ij}(\theta(\vec{d}_i, \vec{d}_j)) = \langle \Delta_i \Delta_j \rangle + \delta_{ij} N_i^2$$

Here $\theta(\vec{d}_i, \vec{d}_j)$ is the angle between the two directions $\vec{d}_i$ and $\vec{d}_j$; $\langle \Delta_i \Delta_j \rangle$ is in general a linear combination $\sum a_k C(\theta_k, \sigma)$ with coefficients a_k and angles θ_k depending only on the experiment modulation geometry and sky coverage.

When C_{ij} is known , the observed anisotropies ΔT_i can be analyzed by searching for the best value of the autocorrelation function amplitude C_{mo} (i.e. the variance of the sky fluctuations) , by maximizing the Likelihood function $L(C_o)$. An idea of the statistical significance of the hypothesis $C_{mo} > 0$ is given by the likelihood ratio

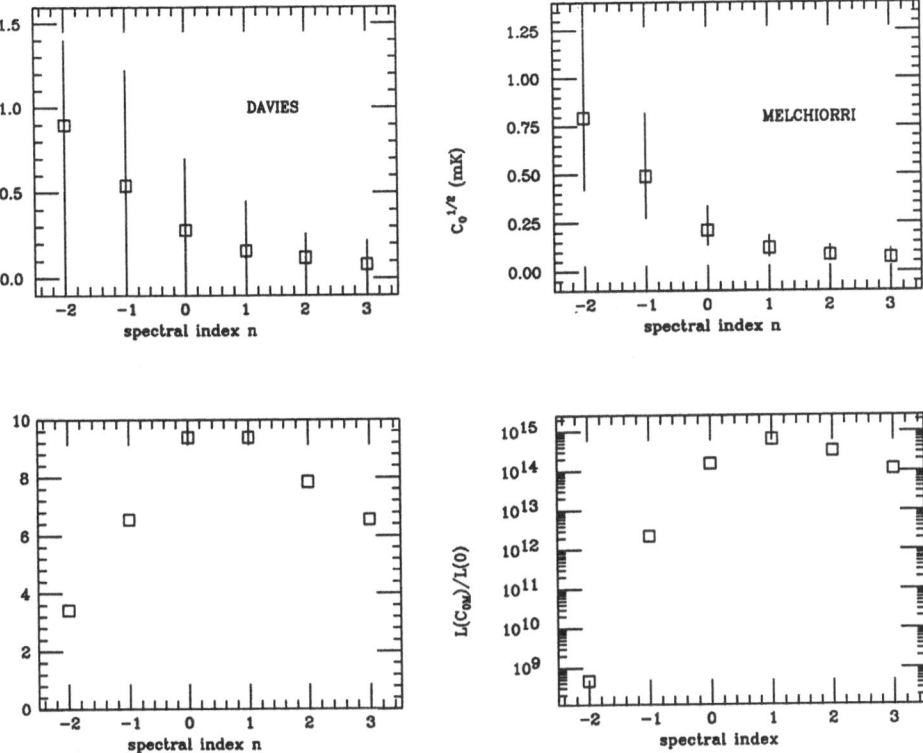

fig.6 : Results of the maximum likelihood analysis of the Davies et al. (1987) and Melchiorri et al. (1981) data . In the upper panels the rms value of the sky temperature fluctuations $C_o^{1/2}$ is plotted as a function of the spectral index n of the density perturbation spectrum. In the lower panels the values of the likelihood ratios are plotted. It is interesting to note that for both the experiments values of n between 0 and 1 are preferred.

$LR = L(C_{mo})/L(0)$; a likelihood ratio larger than 10 or 20 is usually assumed to be the threshold for accepting such hypothesis ; however the problem of stating a confidence level on the likelyhood results is quite controversial (Lasenby and Kaiser (1987)).

The above described analysis has been carried out for the 4 experiments described before, in the case of a cosmological fluctuation spectrum with $n = 1$. The main features of the experiments and the C_{mo} values which maximize the likelihood are reported in table 1 , together with the results of the standard Neyman-Pearson lemma analysis (however the latter does not take into account the correlation between different beams (Boynton and Partridge (1973)). As a first step , the error in the C_{mo} value can be estimated in a simple way by searching for the $C_{m\pm}$ values which satisfy the relation $L(C_{mo}) = 10L(C_{m\pm})$. It can be seen that the Melchiorri et al. result is a significant detection of sky anisotropies ; the Davies et al. result is marginally significant and the other two results can be considered only upper limits .

In fig. 6 are summarized the likelihood analysis results , in the case of Melchiorri and Davies experiments , when the spectral index n is varied between -2 and 3 . This is a wide range of values , but shows the dependence of the results on different assumptions. In both the experiments spectral indexes between 0 and 1 are preferred (i.e. have larger likelihood ratios) ; the $C_{om}^{1/2}$ values increase for smaller spectral indexes , as expected for differential experiments : for $n \to -3$ adjacent beams are expected to be very correlated ; so , in order to "fit" a fixed level of detected sky fluctuations , larger values of the amplitude C_o are preferred.

5. CONCLUSIONS

It is impressive the fact that two completely different experiments (Davies et al. (1987) and Melchiorri et al. (1981)) at similar angular scales ($6^o \sim 8^o$) report evidence for the same level of sky anisotropy . Moreover if one assumes adiabatic density perturbations and temperature fluctuations generated by the Sachs and Wolfe effect, a Likelihood analysis shows that scale invariant or white noise density fluctuation spectra are preferred. However we must stress the fact that the extragalactic origin of the detected anisotropies has not been proved . Since radio and infrared emission by our galaxy are not well known at the required level of sensitivity , we believe that only several observations carried out at well spaced wavelengths can allow to disentangle the cosmological contribution from the local one. On this respect the discovery of a good correlation between the data of the two experiments should be an important step in favour of the cosmological origin.

ACKNOWLEDGMENTS

This work has been supported by Piano Spaziale Nazionale of CNR and by Ministero della Pubblica Istruzione.

REFERENCES

Blumenthal G., Faber S., Primack J., Rees M., *Nature*, **301**, 584, (1984)

Boynton P.E. and Partridge R. , *Astrophys.J.* , **181** , 243 (1973)

Dall' Oglio G., de Bernardis P., *Observations of CBR anisotropies from Antarctica* , *Astrophys.J.* accepted for publication , August 1988 issue.

Davies R.D., Watson R., Daintree E.J., Hopkins J., Lasenby A.N. Beckman J., Sanchez-Almeida J., Rebolo R. *Nature* , **326** , 6112 (1987)

Fabbri R., Guidi I., Melchiorri F., Natale V.; *Proc. of the second Marcel Grossmann meeting on General Relativity* , 889 , R. Ruffini ed. , North Holland (1982)

Lasenby A.N., Kaiser N., *preprint* (1987)

Mandolesi N., Calzolari P., Cortiglioni S., Delpino F., Sironi G., Inzani P., De Amici G., Solheim J.E., Berger U., Partridge R.B., Martenis P.L., Sangree P.H., Harvey R.C. *Nature* , **319** , 751 (1986)

Melchiorri F., Olivo Melchiorri B., Ceccarelli C., Pietranera L., *Astrophys. J. Lett.* , **250** , L1 (1981)

Scaramella R. and Vittorio N. , On the large scale anisotropy of the Cosmic Background Radiation , *Ap.J.Lett.*, submitted (1987)

Vittorio N., de Bernardis P., Masi S., Scaramella R., *Monte Carlo simulations of CBR anisotropy experiments* , *Nature* , submitted (1987)

Vittorio N., Juszkiewicz R., *Ap.J.Lett.*, **314**, L29, (1987)

COSMOLOGICAL IMPLICATIONS OF THE INFRARED BACKGROUND RADIATION

Michael Rowan-Robinson and Bernard Carr
Astronomy Unit,
Queen Mary College,
Mile End Road,
London. E1 4NS.

ABSTRACT. We review some exciting new observational developments in the infrared background (1 - 1000 μm). The Nagoya-Berkeley collaboration have made accurate measurements of the infrared background intensity between 100 and 1000 μm and have detected excess radiation, compared to a 2.74 K blackbody, peaking at 700 μm. New detailed studies of the IRAS background data show the clear possibility of a cosmological background at 100 μm, though there is still not a concensus on the correct model for the foreground zodiacal emission. Finally, the Nagoya group claim a significant background intensity at 2 μm. Possible interpretations of the far infrared background as radiation from starburst galaxies undergoing strong cosmological evolution are discussed. The spectrum of the 700 μm excess does not agree well with such a model. More promising are models involving a pregalactic generation of stars which make dust and light. The expected infrared background from a variety of pregalactic sources (primeval galaxies, Population III stars, accreting black holes, large-scale structure formation, decaying elementary particles) are discussed in detail. These backgrounds would generally appear in the near infrared unless they have been absorbed by galactic or intergalactic dust, in which case they should have been reprocessed to the far infrared.

1. REVIEW OF OBSERVATIONS

1.1 Nagoya-Berkeley Collaboration

Microwave background measurements have been reviewed recently by Smoot et al. (1987). All recent measurements at $\lambda > 1$ mm are consistent with a blackbody spectrum with T = 2.74 ± 0.02 K. The significant excess claimed by Woody & Richards (1979) at 1-2 mm has not been confirmed in subsequent measurements.

A very important recent development in the infrared background has been the successful flight of a rocket-borne, helium cooled, multi-band radiometer experiment by a collaboration between Nagoya

N. Kaiser and A. N. Lasenby (eds.), The Post-Recombination Universe, 125–140.
© *1988 by Kluwer Academic Publishers.*

University and Berkeley (Matsumoto et al. 1987). This experiment, which involves in-flight calibration with a blackbody source, gives the first accurate measurements of the background intensity between 100 and 700μm. For each of the six wavelength channels there was a significant portion of the flight free from contamination by earthshine, atmospheric emission and emission from rocket-borne contaminants. Table 1 summarises the infrared background temperature and intensity, together with the excess over a 2.74 K blackbody.

Table 1: Matsumoto et al.(1987) results

$\lambda(\mu m)$	T_{CBR}	νI_ν	νI_ν(2.74K b.b.)	difference
		$(10^{-11}\ Wcm^{-2}sr^{-1})$		
1160	2.795 ± 0.018	7.87 ± 0.20	7.08 ± 0.31	0.79 ± 0.37
709	2.963 ± 0.017	5.00 ± 0.20	2.83 ± 0.15	2.17 ± 0.25
481	3.146 ± 0.022	1.65 ± 0.11	0.40 ± 0.035	1.25 ± 0.12
262	–	< 0.75	0.0005	< 0.75

The uncertainty quoted in the difference includes the uncertainty of 0.02 K in the microwave background temperature (2.74 K). Fig. 1, taken from Matsumoto et al. (1987), shows the total background spectrum from 60 μm to 1.6 cm. They have compared their 100-260 μm data with a $\nu^2 B_\nu(T_d)$ curve with T_d = 20 K, to represent emission from interstellar dust. Fig. 2, also taken from Matsumoto et al (1987), shows the blackbody temperature as a function of wavelength together with the predictions of two simple models: (i) a Comptonization model with T_0 = 2.795 K, y = 0.02 (cf Guilbert & Fabian 1986), (ii) redshifted dust emission in the early universe (Rowan-Robinson et al.1979, Negroponte et al.1981, Hayakawa 1984, Bond et al. 1986, Negroponte 1986, McDowell 1986) with T_d = 3.55(1 + z) K and an excess radiation energy density of $5 \times 10^{-14}\ (1 + z)^4$ erg cm^{-3}, where z is the redshift of the emitting dust.

1.2 IRAS Background

First results on the IRAS background radiation were presented by Hauser et al. (1984), who gave the total intensity at 12, 25, 60 and 100 μm at the ecliptic poles and in the ecliptic plane. In the ecliptic plane emission from interplanetary dust, peaking at 25 μm, dominates the extended emission. On the basis of a detailed model of far infrared emission from interstellar dust and a preliminary zodiacal emission model,

Figure 1. The spectrum of the far infrared background (Matsumoto et al. 1987). The fluxes obtained by Matsumoto et al. are shown by ◖ with the vertical error bars and the horizontal bars for the effective bandwidths. The results of other measurements are shown for comparison by O (Peterson et al. 1985), ☐ (Meyer and Jura 1985), ■ (Crane et al. 1986), X (Smoot et al. 1985, 1987) and Δ (Johnson and Wilkinson 1987). IRAS data at 60 and 100 μm, are shown by ◆ .

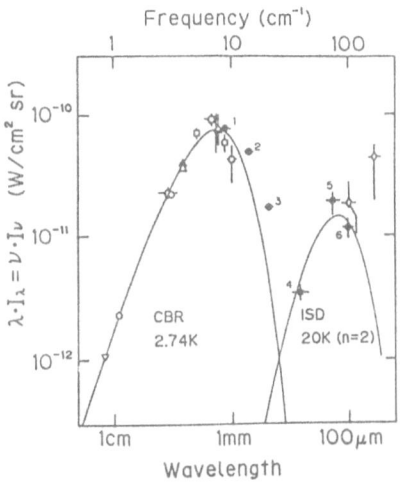

Figure 2. The equivalent blackbody temperature versus wavelength (Matsumoto et al 1987). The results of other measurements are shown for comparison by O (Peterson et al. 1985), ☐ (Meyer and Jura 1985), ■ (Crane et al. 1986), Δ (Mandolesi et al. 1986), X (Smoot et al. 1985, 1987) and ▲ (Johnson and Wilkinson 1987).

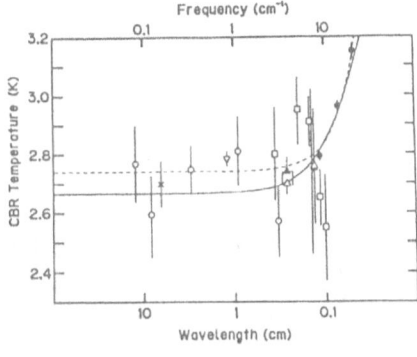

Figure 3. The spectrum of the observed surface brightness at the galactic pole region at 1 – 10 μm (Matsumoto 1987). SL and ZL mean star light and zodiacal light, respectively. The square represents IPD emission at 12 μm towards the observed sky based on the data by IRAS (Hauser et al. 1984). The emission component around 5 μm is fitted by three types of blackbody emission.

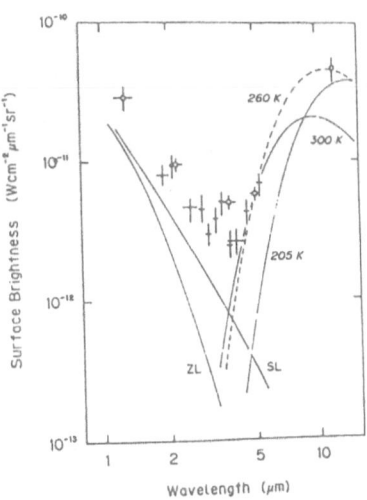

Rowan-Robinson (1986) concluded that there was an unexplained isotropic component at 100 μm of ~ 5 MJy/sr.

At the 3rd IRAS Conference (Q.M.C., London, July 1987) models for zodiacal emission were presented by no less than 6 independent groups. The diversity of the models used shows that there is so far no concensus on the correct density law for the zodiacal dust. However the two most detailed solutions, those by Good (1987) and Rowan-Robinson (1987a) agree quite well on the existence and magnitude of constant offset terms (i.e. possible isotropic background components) at 12, 25, 60 and 100 μm (see Table 2).

Table 2: Constant offset terms (possible isotropic background components) in IRAS zodiacal emission solutions (MJy/Sr).

	λ = 12 μm	25 μm	60 μm	100 μm
Good (1987)	2.6	9.1	1	3
Rowan-Robinson (1987a)	3.5	7.7	1.1	3.8
Zero-point uncertainty (Beichman et al. 1984)	1.6	3.6	1.0	1.6

Since we cannot at present rule out the possibility that these offset terms are simply zero-point errors, the safest statement on the IRAS background radiation is:

$$I_\nu \ (100 \ \mu m) = 3.5 \ ^{+1.6}_{-3.5} \ MJy/Sr$$

$$I_\nu \ (60 \ \mu m) \ = 1 \pm 1 \ MJy/Sr.$$

The data of Matsumoto et al. (1987) at 100 μm will, when fully analysed, provide a check on the IRAS 100 μm zero-point. The Cosmic Background Explorer (COBE) Mission should settle the issue when it is launched in 1989. Further refinement of the zodiacal emission model may modify the figures of Table 2.

1.3 Japanese Near-infrared Data

Matsumoto et al. (1987) have made rocket-borne measurements of background radiation in the wavelength range 1 - 6 μm. The bulk of the emission they see can be attributed to zodiacal emission (3 - 6 μm) or starlight + zodiacal scattered light (1 - 2 μm). Figure 3 shows their data together with model curves for the foreground radiations. The peak at 2.57 μm remains unexplained at present. If it were due to cosmological background radiation, it would correspond to a rather sharply peaked spectrum.

2. PREDICTED 30 - 1000 μm BACKGROUND FROM INFRARED GALAXIES

It is inevitable that at some level there will be background radiation from galaxies in the far infrared. It is therefore important to consider carefully whether the observed far infrared background could be due to galaxies.

The infrared spectra of galaxies can be modelled in terms of 4 components (Rowan-Robinson and Crawford 1987):

(i)   emission from the disc component of normal spiral galaxies due to reradiation of starlight by interstellar dust (the infrared 'cirrus' in our Galaxy),

(ii)   emission from bursts of star formation heavily shrouded in dust, mainly concentrated to galactic nuclei ('starbursts'),

(iii)   emission from warm dust in the narrow-line region of Seyfert nuclei, peaking at 25 μm,

(iv)   emission with a power-law spectrum, presumably of non-thermal origin, from quasars.

The latter two components can be shown to make a negligible contribution to the 30 - 1000 μm background, even with strong cosmological evolution at the rate derived from optical and radio source-counts.

To estimate the contribution of normal and starburst galaxies, we note that the background intensity of frequency ν is

$$I_\nu = \overline{nL_\nu} \ ct_0 \int_0^1 \frac{L(\nu Z)}{L(\nu)} \ f(Z) \ d(t/t_0) \qquad (1)$$

where $\left\{ \begin{array}{l} \overline{nL_\nu} \text{ is the luminosity density at } \nu \\ L(\nu) \text{ is the source spectrum} \\ Z = 1 + z \\ f(Z) \text{ is the evolutionary factor, assumed to be separable from} \\ \text{the luminosity density and spectrum (i.e. we assume that the} \\ \text{evolution consists of some combination of pure luminosity and} \\ \text{pure density evolution and that the source spectrum does not} \\ \text{change with redshift).} \end{array} \right.$

We take the luminosity density from the luminosity function of Lawrence et al (1986) and the form of $L(\nu)$ for starburst and normal disc components from the models of Rowan-Robinson and Crawford (1987). For simplicity we use the $\Omega_0 = 1$, $\Lambda = 0$, cosmological model : for $\Lambda = 0$, $0 < \Omega_0 < 1$, the effect of varying the cosmological model is small. For the evolutionary function $f(Z)$ we use the simple exponential form

$$f(Z) = \exp Q(1 - 1/Z) \qquad (2)$$

where the IRAS source counts limit the value of Q. Figure 4 compares IRAS 60 μm source-counts from Rowan-Robinson et al. (1986) and Hacking and Houck (1987) with models with no evolution and with pure luminosity evolution with Q = 5 and 8. For each evolutionary case the luminosity function of Lawrence et al. (1986) has been recalculated to take evolution into account. Luminosity evolution has been chosen

(i)   because it gives a good approximation to radio counts of active galaxies and quasars (Rowan-Robinson 1970, Condon 1984, Rowan-Robinson 1987b) and to optical counts of quasars (Cheney & Rowan-Robinson 1981, Schmidt & Green 1983),

(ii)   because most spirals are far infrared sources at the present epoch, so that density evolution is not an option.

From Figure 4 we see that Q < 8 and that either the rate of evolution of the infrared galaxy population is more rapid than that for active galaxies and quasars or the deep counts of Hacking and Houck (1987) are subject to confusion. To see whether a steeper evolutionary rate is reasonable, we need to consider the history of massive star formation in galaxies. Most massive star formation in spiral galaxies today takes place in giant molecular clouds where $A_v \sim 30 - 100$. The heavy elements providing this dust opacity were mostly generated during the first 10^9 years of the galaxy's life, as also was most of the galaxy's luminosity. During only a few percent of this time would there have been naked ($A_v < 1$) star-formation. Thus much of the stellar radiation associated with this nucleosynthesis emerged in the far infrared. We may therefore expect to see strong cosmological evolution in the luminosity of the starburst component in spiral galaxies. There is no a priori reason why the rate of evolution should not be more rapid than that for active nuclei for which the gas supply to a massive black hole is probably the controlling factor.

Assuming that a proportion of the light from regions of massive star formation escapes to illuminate interstellar dust (at least 50% does today in our Galaxy), we will also see strong evolution in the luminosity of the cooler disc component. In practice the spectrum of both starburst and disc components may evolve as the luminosity in starlight and dust content changes, but we have not taken this into account in the present calculation.

In evaluating eqn (1) we have assigned 50% of the total 60 μm luminosity density to the starburst component and 50% to the disc component. Refinement of these proportions would require a far more detailed picture of the generation of far infrared radiation in normal and starburst galaxies than is available at present.

Figure 5 shows the predicted background intensity from 30 - 1200 μm for the starburst and cirrus components for different rates of evolution, compared with the excess observed by Matsumoto et al. (1987) and that derived from IRAS data by Rowan-Robinson (1987a). With the maximum permitted rate of evolution, Q = 8, background intensities comparable to those observed are achieved. From the point of view of the energetics of the far infrared background, an origin from infrared galaxies is plausible. However there is a rather poor fit to the spectrum of the observed excess in the range 240 - 1100 μm. If the uncertanties in the observations and in the microwave background

Figure 4. Differential source counts at 60 μm. Data are from Rowan-Robinson et al. (1986) and Hacking and Houck (1987). Solid curve: predicted counts assuming no evolution. Broken and dotted curves: luminosity evolution with Q = 5, 8.

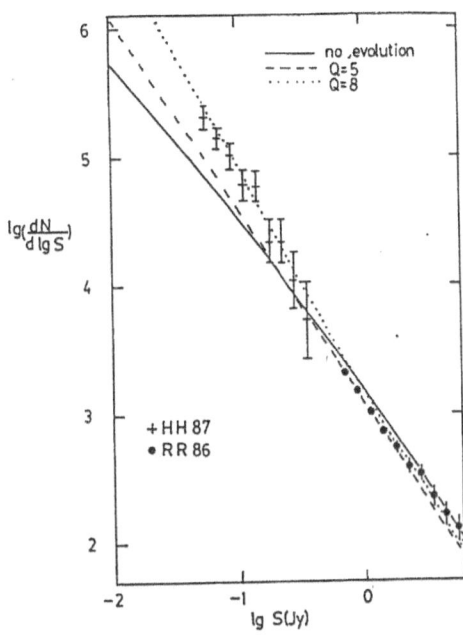

Figure 5. Far infrared and submillimetre background. Data points labelled M are the Matsumoto et al. (1987) excess taken from Table 1. Data points labelled R are the IRAS isotropic component (Rowan-Robinson 1987a). Solid curves are predicted backgrounds for starburst component, broken curves for normal disc component, with luminosity evolution.

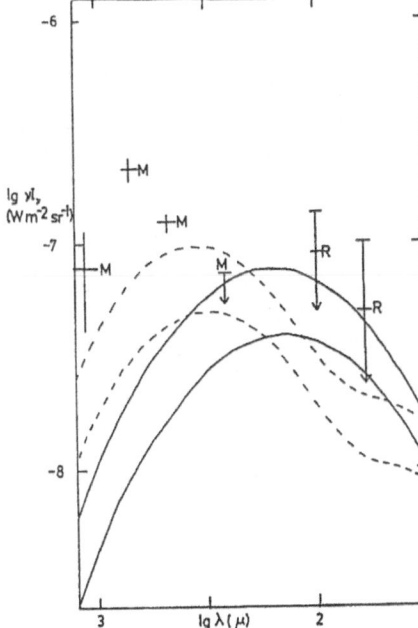

temperature are correctly stated, then it seems unlikely that the background from infrared galaxies can produce a spectrum peaking so sharply at 700 μm. Population III dust models of the kind studied by Rowan-Robinson et al. (1979) and Negroponte et al. (1981), with starlight and dust formation taking place at $z_f \sim 70$, in order to shift the 10 μm silicate feature to 700 μm, would be a much more promising way to generate this type of spectrum (see Section 3).

3. PREGALACTIC SOURCES

There could be several kinds of astrophysical generators of IR in the period between $z=10$ and $z=10^3$: for example, primeval galaxies, Population III stars, accreting black holes, large-scale explosions, or decaying particles. If the radiation from these sources propagated to us unimpeded (unaffected by dust), it would presently reside in the near-IR to UV range. We now indicate the total energy density Ω_R and peak wavelength λ_{peak} associated with each background; here

$$\Omega_R(\lambda) = 2 \times 10^5 \ h^{-2} (\lambda I_\lambda / Wcm^{-2} sr^{-1})$$

is the density in units of the critical density and h is the Hubble constant in units of 100 km/s/Mpc. The associated spectra are summarized in Figure 6. Most of them have a dilute black-body form, although this would require modification if there was enough neutral hydrogen in the background Universe to absorb photons shortward of the Lyman cut-off. In this case, most of the radiation would come out as recombination lines. If the backgrounds are reprocessed by dust, λ_{peak} will change but Ω_R will be roughly the same (as discussed in Section 4). More detailed calculations can be found in Bond et al. (1986).

Primeval Galaxies. Several arguments suggest that galaxy formation was accompanied by an initial burst of massive star formation which generated the first metals (Truran & Cameron 1971). The stars must have generated a metallicity Z of order 10^{-3} at a redshift z_G in the range 3-10 (depending on the epoch of galaxy formation) and must have had a mass M in the metal-producing range 10-100 M$_\odot$. Since these stars must also generate light, one can predict a minimum integrated background radiation density (Peebles and Partridge 1967). The curve "PG" in Figure 6 corresponds to a model with M=25M$_\odot$ and z_G=9. An analogous argument shows that the background associated with the stars which produce the solar metallicity is given by the curve "SM" in Figure 6. Since the appropriate normalizations are now $Z \approx 10^{-2}$ and $z \approx 1$, λ_{peak} is reduced by a factor of 10 and Ω_R is increased by a factor of 100. This is already in conflict with the background UV constraint, which suggests that the radiation must have been reprocessed by dust.

Population III Stars. It has been proposed that the dark matter in galactic halos is baryonic (Ashman & Carr 1987), in which case a large fraction of the Universe must have been processed through a generation of "Population III" stars which left dark remnants. Background light and nucleosynthetic constraints imply that the objects must be either jupiters or the black hole remnants of "Very Massive Objects" (VMOs) in the mass range above $M_c \approx 200$ $M_\odot$ (Carr *et al.* 1984). Since VMOs have a surface temperature of 10^5K (independent of M) and radiate at the Eddington limit, one can show that the background light they generate will be given by the curve "VMO" in Figure 6. This assumes that the stars burn at a redshift $z_* \approx 100$ and that they have the density $\Omega_* \approx 0.1$, in units of the critical density, required to explain galactic halos. The background from jupiters would have a much smaller density but it would peak in the range 10–100μ and extend over a wider waveband. The "J" curve in Figure 6 assumes a jupiter mass of 0.085 $M_\odot$ and is based on the calculation of Karimabadi & Blitz (1984). Note that this background is non-thermal because it is associated with the *formation* of the jupiters (viz. a Hayashi phase, followed by a degenerate cooling phase) rather than their emission at the present epoch. The prospects of detecting individual jupiters directly are remote because they are so dim.

Black Hole Accretion. In order to explain quasars and active galactic nuclei, it is commonly supposed that some galaxies have giant black holes in their nuclei with a mass M of order $10^8 M_\odot$ (Rees 1978a). If the holes radiated at the Eddington limit for a "mass-doubling" time t_E, then they would generate the background indicated by the curve "AGN" in Figure 6. This assumes that the efficiency with which the accreted material generates radiation is 10% and that most of the radiation is generated at $z_E \approx 10$, the M-independent redshift corresponding to the epoch when the Hubble time is t_E. The wavelength estimate assumes that one has an optically thick accretion torus at a temperature of 2×10^4K, as suggested by the models of Begelman (1984). If galactic halos are the black hole remnants of VMOs, they would also accrete but the accretion would generally be sub-Eddington. In the pregalactic era, one can usually assume that they accrete at the Bondi rate from a medium with the cosmological gas density and a temperature $T \approx 10^4$K. In this case, they should produce the background given by the curve "HBH" in Figure 6. This assumes that the holes have a mass $10^6 M_\odot$, since there may be dynamical evidence for halo black holes of this mass (Lacey & Ostriker 1986), and that they generate radiation with 10% efficiency at a redshift $z_* \approx 10$. The choice of z_* corresponds to the redshift at which most of the radiation is generated (Carr et al. 1983).

Pregalactic Explosions. It has been proposed that some features of the large-scale cosmic structure can be explained by pregalactic explosions (Ostriker & Cowie 1981, Ikeuchi 1981). One envisages each explosive seed (a star or a cluster of stars) generating a shock which sweeps up a shell of gas. In order to explain the existence of giant voids and the form of the galaxy correlation function, the shells must eventually overlap with a characteristic radius of order 10 Mpc (Saarinen *et al.* 1986). However, the stars which generate the explosive energy will also generate light, so one can predict a minimum background radiation density as follows. To generate density fluctuations of order 1 on a comoving scale of 10 Mpc corresponds to a minimum kinetic energy density (Hogan 1984). If we identify this kinetic energy density with the explosive energy density and assume that each star generate about 100 times as much radiation energy as explosive energy then the background radiation density must be given by the curve "ES" in Figure 6. This density is already in conflict with the optical to UV limits unless the radiation is reprocessed by dust.

Decaying Particles. Elementary particle relics of the Big Bang would be expected to pervade the Universe and, if their mass is sufficiently large, they could have an appreciable cosmological density. In certain models, these particles would be expected to decay radiatively on some timescale τ_d. For $\tau_d < 50y$, they would contribute to the CBR, while for $50y < \tau_d < 3 \times 10^4 y$ they would distort its spectrum (Silk & Stebbins 1983).

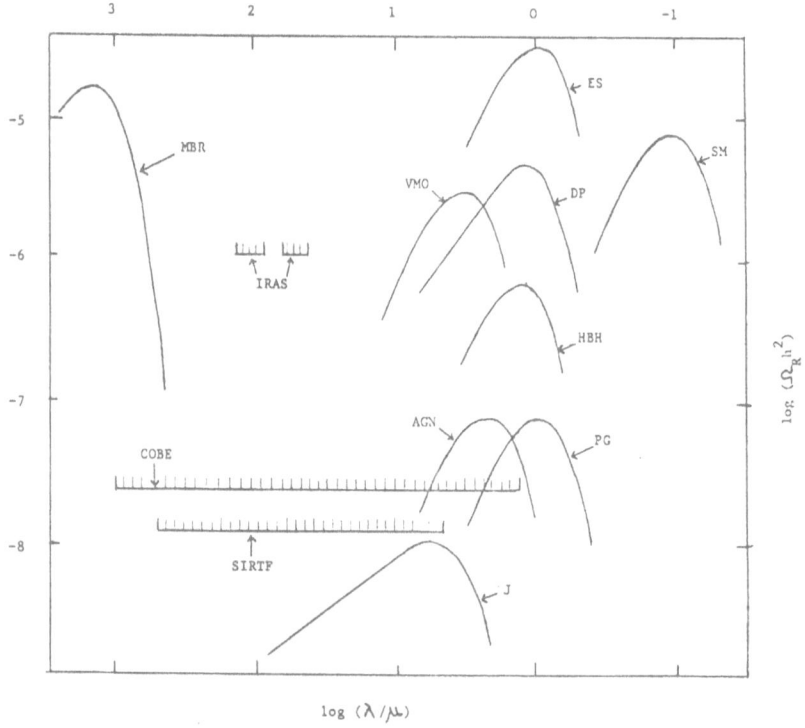

Fig.6. This summarizes the theoretical backgrounds which may have been generated by pregalactic and protogalactic events and compares them to the sensitivity of various space experiments. If the backgrounds are reprocessed by dust, Ω_R remains the same but λ moves into the far-IR.

However, for $\tau_d > 3 \times 10^4 y$, they would just generate an IR background. The curve "DP" in Figure 6 corresponds to $\Omega_X = 0.01$, $m_X = 1 \text{keV}$ and $z_d = 10^3$; here Ω_X is the density parameter which would be associated with the particles had they not decayed, m_X is the particle mass, and z_d is the decay redshift. Note that the spectrum deviates somewhat from the black-body form

4. REPROCESSING OF PREGALACTIC RADIATION BY DUST

The predicted spectra of Figure 6 apply only if the radiation propagates freely between us and the source. However, most of the backgrounds discussed above are originally in the optical or UV and one might expect such backgrounds to be absorbed by intervening dust. In this case, they would be re-emitted at a longer wavelength. The dust could either be confined to galaxies (if galaxies cover the sky) or it could be uniformly spread throughout the Universe. We first determine the condition for dust absorption and then calculate the characteristics of the re-emitted radiation.

For simplicity we will assume that each grain has a radius r_d, that its absorption cross-section is πr_d^2 for wavelengths $\lambda < 2\pi r_d$, and that it falls off as λ^{-1} for $\lambda > 2\pi r_d$. Then we can write

$$\sigma \approx \pi\, r_d^2 \left[1 + \left[\frac{\lambda}{2\pi r_d} \right] \right]^{-1} \tag{3}$$

In fact, the exponent of λ must increase to 2 at very long wavelengths (Draine & Lee 1984); a more general analysis is given by Bond $et\ al.$ (1986). If the grains are uniformly distributed throughout the Universe and have a density Ω_d in units of the critical density, then the optical depth back to a redshift z for photons with present wavelength λ can be expressed as

$$\tau(\lambda, z) \approx 1.3 \left[\frac{\Omega_d}{10^{-5}} \right] \left[\frac{\rho_{id}}{2} \right] \left[\frac{r_d}{0.1\mu} \right]^{-1} \left[\frac{1+z}{10} \right]^{3/2} \left[1 + \frac{5\lambda}{6\pi r_d(1+z)} \right]^{-1} h\Omega^{-1/2} \tag{4}$$

Here ρ_{id} is the internal grain density in g/cm^3 and we have normalized the grain density Ω_d to the dust density associated with galaxies. Thus the optical depth at short wavelengths reaches unity at a redshift

$$1+z \approx 8 \left[\frac{\Omega_d}{10^{-5}} \right]^{-2/3} \left[\frac{r_d}{0.1\mu} \right]^{2/3} \left[\frac{\rho_{id}}{2} \right]^{2/3} \Omega^{1/3} h^{-2/3} \tag{5}$$

In practice, of course, Ω_d is itself a function of z, so it is useful to express the opaqueness condition in terms of the (Ω_d, z) space of Figure 7. This figure assumes the wavelength of the source radiation is less than $2\pi r_d$ at the time when it encounters the dust, as is usually the case.

If one thinks of the mean cosmological dust density as following a trajectory $\Omega_d(z)$ in Figure 7, then photons from pregalactic sources will be absorbed by intervening dust providing there is some redshift between their emission and now at which the trajectory penetrates the shaded region of Figure 7. It is not clear whether this is the case. Ω_d clearly starts off above the shaded region since $\Omega_d=0$ initially; observations of distant quasars also imply that a uniform dust distribution must have $\Omega_d < 6 \times 10^{-5} h^{-1}$ back to $z=2$ (Wright 1981), so one certainly has $\tau < 1$ at the present epoch. However, one could still have $\tau > 1$ in some intermediate redshift range. For example, one would only need $\Omega_d > 10^{-7}$ at $z=300$, the earliest epoch at which VMOs could complete their nuclear burning. Of course, one could only expect an intergalactic grain abundance like this if there was some pregalactic star formation.

Even if there is no intergalactic grain abundance, the dust *within* galaxies could still absorb any pregalactic radiation providing two conditions are satisfied. Firstly, the *mean* dust density (i.e. the density which would be obtained if the dust was spread uniformly throughout the Universe instead of being confined to galaxies) must be large enough for τ to exceed 1 at the redshift of galaxy formation (z_G). The contribution of galactic dust to Ω_d can be written as

$$ \Omega_d \simeq 10^{-5} \left[\frac{f_d}{0.01} \right] \left[\frac{f_g}{0.1} \right] \left[\frac{\Omega_G}{0.01} \right] \tag{6} $$

where f_g is the fraction of the galaxy's mass in gas, f_d is the fraction of the gas in dust, and Ω_G is the density parameter associated with galaxies. This just corresponds to the normalization in eqn (5). Secondly, we need the galaxies to cover the sky at z_G. Otherwise, most of the background photons would be unaffected. This requires

$$ 1 + z_G > 11 \left[\frac{R_G}{10\text{kpc}} \right]^{2/3} \left[\frac{\rho_{iG}}{10^{-24}} \right]^{2/3} \left[\frac{\Omega_G}{0.1} \right]^{-2/3} \Omega^{1/3} h^{-2/3} \tag{7} $$

where R_G is the radius of the dust-containing part of the galaxy and ρ_{iG} is the density within a galaxy in g/cm^3. Since the redshift of galaxy formation is in the range 3 to 10, it is not clear whether these two conditions are satisfied. It is certainly possible, and Ostriker and Heisler (1984) have even proposed this as the explanation for why quasars cut off at a redshift of 4, but it is not necessarily the case. Note that we have normalized Ω_G, R_G and ρ_{iG} to values appropriate for our galaxy. In practice, these parameters will span a range of values; one would generally expect the smallest galaxies to contribute most to the covering factor.

If the pregalactic radiation, whatever its source, is absorbed by dust, one can readily calculate how its spectrum is modified. If the radiation density is Ω_R, then thermal balance implies that the dust temperature T_d evolves according to

$$T_d(z) = T_c(z) \left[1 + \left[\frac{\Omega_R}{\Omega_c} \right] \left[\frac{r_d}{0.1\mu} \right]^{-1} \left[\frac{1+z}{10^4} \right]^{-1} \right]^{1/5} \qquad (8)$$

Here $T_c(z)$ is the temperature of the CBR photons, Ω_c is the CBR density, and we have assumed $T_d \ll r_d^{-1}$ and $T_c \ll r_d^{-1}$ in appropriate units. Thus if the radiation density is less than

$$\Omega_{crit} \simeq 2 \times 10^{-7} h^{-2} \left[\frac{r_d}{0.1\mu} \right] \left[\frac{1+z}{100} \right] \qquad (9)$$

the dust temperature will just be the CBR temperature (the CBR heating alone ensuring that it never drops below this). However, if Ω_R exceeds the value given by eqn (9), the dust temperature will be somewhat larger than T_c. In this case, one expects a far-IR background with a spectrum peaking at a present wavelength

$$\lambda_{peak} \simeq 700 \left[\frac{\Omega_R h^2}{10^{-6}} \right]^{-1/5} \left[\frac{r_d}{0.1\mu} \right]^{1/5} \left[\frac{1+z}{10} \right]^{1/5} \mu \qquad (10)$$

The spectrum will not be exactly black-body: $\Omega_R(\lambda)$ will scale as λ^{-4} longward of the peak rather than λ^{-3}. There will also be spectral features associated with resonance effects, although these will tend to be smeared out by cosmological redshift effects.

What is the appropriate value of z to use in eqn (10)? Strictly speaking, one is dealing with a range of values since the reprocessed radiation comes from a shell: the outer edge of the shell corresponds to the redshift z_d at which the dust or radiation is generated (whichever is smaller) and the thickness of the shell is determined by the condition that the optical depth be unity in the waveband of the source radiation. However, the thickness is generally small, so the effective redshift is just z_d. The striking feature of eqn (10) is that λ_{peak} depends only weakly on r_d, Ω_R and z. Thus one expects all the reprocessed radiation from pregalactic sources to pile up at roughly the same wavelength. In a sense, this is unfortunate since it means that the spectrum itself contains little information about the origin of the radiation. On the other hand, it is interesting because it means that one can *predict* that a far-IR background with these characteristics ought to exist.

These considerations show that, if $\Omega_R > \Omega_{crit}$, one expects the total background spectrum to have three parts: the CBR component (peaking at 1400μ), the far-IR dust component (peaking at λ_{peak}), and the residual source component (peaking in the optical or near-IR). If $\Omega_R < \Omega_{crit}$, the dust radiation will be superposed on the CBR, so the overall spectrum will have only two peaks. However, since the dust radiation does not have a black-body spectrum, it will distort the CBR spectrum unless it is itself absorbed (Rowan-Robinson *et al.* 1979, Negroponte *et al.* 1981, Puget & Heyvaerts 1980). If self-absorption occurs, the radiation can be thermalized, in which case the dust is actually generating part of the CBR. The condition for this is that the

138

value of τ given by eqn (4) should exceed 1 at the wavelength given by eqn (10). This requires

$$1 + z > 65 \left[\left[\frac{\Omega_d}{10^{-5}} \right] \left[\frac{\rho_{id}}{2} \right] h \; \Omega^{-1/2} \right]^{-0.4} \left[\frac{r_d}{0.1\mu} \right]^{0.1} \left[\frac{\Omega_R}{\Omega_C} \right]^{-0.1} \quad (11)$$

which corresponds to the heavily shaded part of Figure 7. In principle, provided the $\Omega_d(z)$ trajectory passes through this double-shaded region, one could hypothesize that the *entire* microwave background derives from grains (Layzer & Hively 1973, Rees 1978b, Wright 1982.) However, for a reasonable grain density, this requires the radiation to be produced at a very high redshift and the grains to be rather exotic.

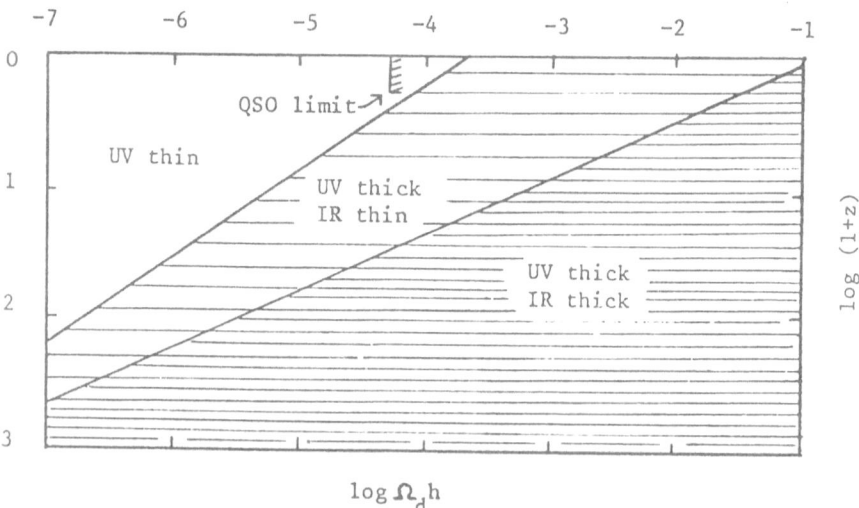

Fig. 7. This shows the dust density required to absorb UV radiation at a redsdift z, thereby reprocessing it into the far-IR. In the heavily shaded region the far-IR radiation is itself absorbed, leading to thermalization. We assume $r_d = 0.1\mu$.

5. DISCUSSION

The question now arises of whether the theoretical predictions in Figure 6 could be relevant to any of the claimed detections of an IR background. The sort near-IR background reported by Matsumoto (1987) could be generated by several of the models, provided one does not have too much dust absorption. Indeed, if the spectrum is cut off beyond the Lyman limit by neutral hydrogen absorption, Ly-α emission might even produce a narrow line feature (Carr et al. 1983, Matsumoto 1987). If one wants to produce a far-IR background, one must clearly invoke dust. McDowell (1986) and Negroponte (1986) have already tried to explain Rowan-Robinson's 100μ background using dust and pregalactic stars and they find values of λ_{peak} which are in good agreement with the simple analytical estimate of eqn (10). Matsumoto et al. (1987) and McDowell (1987) discuss whether their 700μ background can be explained in a similar way. Even without a detailed analysis, however, it is clear that one can explain both backgrounds providing one imposes suitable constraints on the grain size and formation redshift. Indeed the energy density involved is comparable to that expected in several of the scenarios discussed in Section 3 and it is striking that the value for λ_{peak} predicted by eqn (10) is almost exactly what is required. If the claims to have detected an IR background are not confirmed, comparison with the background levels which will be accessible to COBE and SIRTF in Figure 6, shows that most of the predicted backgrounds are potentially observable. Only the background associated with jupiters in galactic halos appears to be below the detectability threshold.

REFERENCES

Ashman, K.M. and Carr, B.J., MNRAS (submitted).
Beichman, C.A., Neugebauer, G., Habing, H.J., Clegg, P.E. & Chester, T.J., 1984, IRAS Introductory Supplement.
Begelman, M.C., 1984. In Astrophysics of Active Galaxies and Quasi-Stellar Objects, ed. J.S. Miller (Mill Valley: University Science Books), p.411.
Bond, J.R., Carr, B.J. and Hogan, C.J., 1986, Astrophys.J., 306, 428.
Carr, B.J., Bond, J.R. and Arnett, W.D., 1984. Astrophys.J., 277,445.
Carr, B.J., McDowell, J.C. and Sato, H., 1983, Nature, 306, 666.
Cheney, J.E. and Rowan-Robinson, M., 1981, MNRAS, 185, 497.
Condon, J.J., 1984, Astrophys.J., 287, 461.
Draine, B.T. and Lee, H.M., 1984, Astrophys.J., 285, 89.
Good, J., 1987, reported by M. Hauser, 'Comets to Cosmology', ed. A. Lawrence (Springer-Verlag).
Guilbert, P.W. and Fabian, A.C., 1986, MNRAS, 220, 439.
Hacking, P. and Houck, J.R., 1987, Astrophys.J.Supp., 63, 311.
Hauser, M.G. et al., 1984, Astrophys.J., 278, L15.
Hayakawa, S., 1984, Adv. Space Res., 3, 449.
Hogan, C.J., 1984, Astrophys.J., 284, L1.
Ikeuchi, S., 1981, Pub.astr.Soc.Japan, 33, 211.
Karimabadi, H. and Blitz, L., 1984, Astrophys.J., 283, 169.

140

Lacey, C.G. and Ostriker, J.P., 1985. Astrophys.J., 299, 633.
Lawrence, A., Walker, D., Rowan-Robinson, M., Leech, K.J., Penston, M.V., 1986, MNRAS, 219, 687.
McDowell, J.C., 1986, MNRAS, 223, 763.
Matsumoto, T., 1987, 'Comets to Cosmology', ed. A. Lawrence (Springer-Verlag).
Matsumoto, T., Akiba, M. and Murakami, H., 1984, Adv.Space Res., 3, 469.
Matsumoto, T., Akiba, M. and Murakami, H., 1987, Preprint.
Matsumoto, T., Hayakawa, H., Matsuo, H., Musakami, H., Sato, S., Lange, A.E., and Richards, P.L., 1987, Astrophys.J. (in press).
Negroponte, J., 1986, MNRAS, 222, 19.
Ostriker, J.P. and Cowie, L.L., 1981, Astrophys.J., 273, L127
Ostriker, J.P. and Heisler, J., 1984, Astrophys.J., 278, 1.
Peebles, P.J.E. and Partridge, R.B., 1967, Astrophys.J., 148, 377.
Puget, J.L. and Heyvaerts, J., 1980, Astr.Astrophys., 83, L10.
Rees, M.J., 1978a, Phys.Scripta, 17, 193.
Rees, M.J., 1978b, Nature, 275, 35.
Richards, P.L., 1987, 'Comets to Cosmology', ed. A. Lawrence (Springer-Verlag).
Rowan-Robinson, M., 1970, MNRAS, 149, 365.
Rowan-Robinson, M., 1986, Negroponte, J., and Silk, J., 1979, Nature, 281, 635.
Rowan-Robinson, M. 1986, MNRAS, 219, 737.
Rowan-Robinson, M., 1987a, in preparation.
Rowan-Robinson, M., 1987b, Vatican Conference, 'Theory and Observational Limits in Cosmology'.
Rowan-Robinson, M., and Crawford, J., 1987, MNRAS (submitted).
Saarinen, S., Dekel, A. and Carr, B.J., 1987, Nature, 325, 598.
Silk, J. and Stebbins, A., 1983, Astrophys.J., 269, 1.
Schmidt, M., and Green, R.F., 1983, Astrophys.J., 269, 352.
Smoot, G. et al., 1987, Astrophys.J., 317, L45.
Truran, J.W. and Cameron, A.G.W., 1971, Astrophys.Space Sci., 14, 179.
Wright, E.L., 1981. Astrophys.J., 250, 1.
Wright, E.L., 1982, Astrophys.J., 255, 401.
Woody, D.P. and Richards, P.L., 1979, Phys.Rev.Letters, 42, 925.

ORIGIN AND ANISOTROPY OF THE COSMIC SUBMILLIMETER BACKGROUND

C.J. Hogan
Steward Observatory
University of Arizona
Tucson, AZ 85721 USA

J.R. Bond
Canadian Institute for Theoretical Astrophysics
McLennan Physical Laboratories
University of Toronto
Toronto, ON M5S IA1, Canada

ABSTRACT. Some cosmological implications of the recently discovered submillimeter background (SMB) are discussed. Opacity and isotropy arguments show that if the background is due to dust emission it must originate at a high redshift $z \gtrsim 10$. Isotropy measurements are used to constrain the amplitude of dust optical-depth fluctuations and hence the scale of dust inhomogeneity. It is shown that the comoving scale of $(\delta\rho/\rho) \simeq 1$ fluctuations in dust was likely no larger than $\sim 3\,h^{-1}\,\mathrm{Mpc}$ when the background was generated. The observed flux, the inferred high z and high opacity imply that on small scales the mock gravitational instability driven by radiation pressure may have dominated gravitational effects.

1. ORIGIN OF THE SUBMILLIMETER BACKGROUND

Bond, Carr and Hogan (1986; hereafter BCH1) surveyed a wide ranging sample of models of the "post recombination universe," and found that a generic outcome of many scenarios was the production of cosmic background radiation in the very far infrared, between 100 and 1000 μm, due to emission from dust heated by sources of energy such as pregalactic stars (see Carr and Rowan-Robinson's contribution to these proceedings). Matsumoto et $al.$ (1987) recently announced the discovery of just such excess isotropic radiation in the submillimeter region peaking at about 700μm. They state that the spectrum of this radiation is not consistent with Comptonization but seems to arise from optically thin dust emission. Thus one is immediately tempted to invert the arguments of BCH1. In this paper we will assume that this background is due to dust and deduce several interesting properties of the radiating dust, including constraints on its redshift, its scale of inhomogeneity, and the importance of radiation pressure in determining its distribution. A more detailed discussion is given in Bond, Carr and Hogan (1987, BCH2).

141

N. Kaiser and A. N. Lasenby (eds.), The Post-Recombination Universe, 141–149.
© *1988 by Kluwer Academic Publishers.*

Matsumoto *et al.* observe a flux in "channel 2" (centered at $14.1\,\text{cm}^{-1} = 4.2 \times 10^{11}\,\text{Hz} \rightarrow 710\,\mu\text{m}$) of

$$\nu I_\nu = 5 \times 10^{-11}\,\text{W}\,\text{cm}^{-2}\,\text{sr}^{-1} \tag{1}$$

For a $T = 2.74$ black body, νB_ν at this frequency is 2.7×10^{-11}; therefore, a new background is present with $\nu I_\nu \simeq 2 \times 10^{-11}\,\text{W}\,\text{cm}^{-2}\,\text{sr}^{-1}$. They state that channels 1, 2, 3 are best fit using the CBR plus a λ^{-2} emissivity graybody, such as dust grains, with $T = 3.55\,\text{K}$. From the flux and spectrum, the optical depth of the dust can be estimated:

$$\tau_{dust}(710\,\mu\text{m}) \simeq I_\nu / B_\nu (T = 3.55)$$

$$\nu B_\nu (T = 3.55, \lambda = 710\,\mu\text{m}) = 1.6 \times 10^{-10} \quad \text{so} \quad \tau_{dust} \simeq .15!! \tag{2}$$

Although this result is indicative, eq.(2) ignores dust absorption and stimulated emission which can significantly alter the spectrum. In BCH2, we find that a λ^{-1} dust absorption cross section gives as good a fit to the data as a λ^{-2} cross section does. A redshifted dust temperature of 3.7K and a depth of 0.13 across the region over which the emission takes place give the best fits. It is quite possible that the true depth at $700\mu\text{m}$ is significantly higher if the emission region is narrow. See §2.

In any case, remarkably high optical depths for such a long wavelength are required for any dust model. This seems to imply a high redshift of dust formation as we now show. Suppose that the dust has an emissivity going like $\lambda^{-\alpha}$ for $\lambda < 700\,\mu\text{m}$. Suppose further that the dust absorption cross section per hydrogen atom is the same as Galactic dust, $10^{-21}\,\text{cm}^2$ at 1000 Å. Each hydrogen atom at redshift $z \gg 1$ would have a cross section at wavelength $710\,\mu\text{m}/z$ smaller than this by a factor of $z^\alpha 7100^{-\alpha}$. Thus, the optical depth produced at an observed $\lambda = 710\,\mu\text{m}$ out to redshift z by gas with Ω_g of the critical density and fraction f_d of galactic absorption is

$$\tau = \int_0^z dz'(dt/dz')n(z')\sigma(\lambda(z'))$$
$$= \begin{cases} 6 \times 10^{-3} f_d \Omega_g h_{100} z^{5/2} & \alpha = 1 \\ 6 \times 10^{-7} f_d \Omega_g h_{100} z^{7/2} & \alpha = 2 \end{cases} \tag{3}$$

To achieve the observed optical depth requires a redshift

$$z \gtrsim \begin{cases} 3(\Omega_g h_{100} f_d \delta/0.1)^{-2/5} & \alpha = 1 \\ 30(\Omega_g h_{100} f_d \delta/0.1)^{-2/7} & \alpha = 2 \end{cases} \tag{4}$$

where $\delta = (\Delta z/z)$ is the fractional redshift interval over which the radiation is emitted (assumed to be small based on τ_{dust}). The parameter f_d is expected to be of order 10^{-2}, since it represents either the fraction of the baryons in the universe which belong to collapsed objects of Population I abundance and are not in stars or (at most) the ratio of Population II to Population I metallicity if the dust is distributed. The redshifts required are then quite high, especially for the $\alpha = 2$ case.

What about making the background local? To avoid unacceptable extinction at $z \lesssim 4$, at $\lambda \lesssim 1\,\mu m$, one requires a nearly flat emissivity. A flat $(\alpha = 0)$ extinction could be produced by conducting needles or by *very* large $(\simeq 1\,mm)$ grains. The latter would have the problem of implying an enormous column density in heavy elements. However, the strongest argument against a local or a low-redshift extragalactic origin for the radiation comes from its extreme isotropy; it is very hard to reconcile the low anisotropy limits below with any plausible galactic (e.g. cirrus) distribution or even a distribution as anisotropic as galaxies within the local Hubble volume.

2. LIMITS ON ANISOTROPY

Our aim is to use observations of anisotropy of the microwave background, made at a variety of wavelengths, to measure the anisotropy of the SMB and therefore obtain information about the distribution of the emitting dust. The first step is to estimate the change in flux I_ν caused by the dust using the following simple model. Assume that the dust is at constant temperature T_d and that we send a beam of $T = T_0$ blackbody radiation through an optical depth τ of dust. Then the emerging beam has (*e.g.*, Spitzer 1978)

$$I_\nu = B_\nu(T_0)e^{-\tau_\nu} + B_\nu(T_d)\{1 - e^{-\tau_\nu}\}. \qquad (5)$$

where B_ν is the Planck function. This solution also holds in a cosmological setting provided the ratio T_d/T_0 remains constant throughout the evolution or at least over the period of dominant dust emission. This is expected in many classes of models for the evolution of the source radiation (BCH1, BCH2).

For $\tau \ll 1$, we have a spectral perturbation from the original blackbody

$$\left(\frac{\delta I}{I}\right)_\nu \equiv \frac{I_\nu - B_\nu(T_0)}{B_\nu(T_0)} = \tau_\nu F(T_d, t_0) \qquad (6)$$

where $F = [B_\nu(T_d)/B(T_0)] - 1$. Setting $T_0 = 2.7$, $T_d = 3.55$, and $\tau(\lambda = 700\,\mu m) = 0.2$ we have the perturbations listed in table 1. We have evaluated eq. (6) for three different assumptions about the λ-dependence of the dust absorption optical depth τ_ν to illustrate the importance of CMB absorption and stimulated emission in ensuring that the deviations in the Rayleigh Jeans part of the spectrum are small. First, we have taken $\tau_\nu \propto \lambda^{-\alpha}$ with $\alpha = 1$ and $\alpha = 2$, as is conventionally done in cosmic dust studies. But we have also evaluated (6) with

$$\tau_\nu \propto \lambda^{-1}\left(1 - e^{-h\nu/kT_d}\right) \propto \tau_{\nu,s} \qquad (7)$$

which corresponds to dust which intrinsically has $\tau_\nu \propto \lambda^{-\alpha}$ with $\alpha = 1$ at zero temperature, but adds a multiplicative factor to correctly account for the effect of stimulated emission. This modification has the effect of asymptotically producing '$\alpha = 2$' behaviour at long wavelengths, as is found empirically in Matsumoto *et al.*'s data on galactic dust emission. In BCH2 we demonstrate that the $\alpha = 1$ cross section fits to the the Matsumoto *et al.* $1160\mu m$ and $480\mu m$ data points are adequate, and at least as good as the $\alpha = 2$ fits.

λ	F	$(\delta I/I)_{\nu,\alpha=1}$	$(\delta I/I)_{\nu,\alpha=2}$	$(\delta I/I)_{\nu,s}$
$700\,\mu\mathrm{m}$	5.	1.0	1.0	1.0
$1\,\mathrm{mm}$	2.6	0.34	0.25	0.33
$3\,\mathrm{mm}$	0.71	3.3×10^{-2}	7.7×10^{-3}	2.4×10^{-2}
$1\,\mathrm{cm}$	0.41	5.7×10^{-3}	4.0×10^{-4}	1.9×10^{-3}
$1.25\,\mathrm{cm}$	0.39	4.5×10^{-3}	2.5×10^{-4}	1.2×10^{-3}
$1.5\,\mathrm{cm}$	0.37	3.4×10^{-3}	1.6×10^{-4}	8.0×10^{-4}
$2\,\mathrm{cm}$	0.36	2.6×10^{-3}	9×10^{-5}	4.7×10^{-4}

Table 1. Spectral perturbations at various wavelengths in the single-redshift cosmic dust model with various models of the dust emissivity.

Next we estimate the fluctuation in observed flux, $(\delta I/I)_{observed}$, caused by a fractional fluctuation in dust optical depth $\delta\tau/\tau$. (In general there will also be fluctuations in T_d; depending upon the specific model these may enhance or decrease the values of $(\delta I/I)_{observed}$ in the models below. The decreases are never larger than a factor of two for any of the models treated in BCH2 in the Rayleigh Jeans regime.) From (5) we find, again for $\tau \ll 1$,

$$\delta I_\nu = \delta\tau_\nu \left(B_\nu(T_d) - B_\nu(T_0) \right) \tag{8}$$

so that

$$\frac{\delta\tau}{\tau} = (\delta I/I)_{observed} \left(\frac{I_\nu}{I_\nu - B(T_0)} \right). \tag{9}$$

Anisotropy measurements in the Rayleigh-Jeans part of the spectrum give $(\delta T/T)_{observed} = (\delta I/I)_{observed}$, and hence $\delta\tau/\tau \cong (\delta I/I)_\nu^{-1}(\delta T/T)_{observed}$. The best limits (or observations) at present are listed in table 2. (The quoted numbers for balloon observations of 0.11 mK anisotropy by the Florence group which are not in the Rayleigh Jeans part of the spectrum, reflect ambiguity in how to interpret the calibration of their published results.)

Name	θ_{beam}	λ	$(\delta I/I)_{observed}$	$(\delta\tau/\tau)_{\alpha=2}$	$(\delta\tau/\tau)_{\alpha=1}$
VLA	$6''$	$2\,\mathrm{cm}$	$< 6 \times 10^{-4}$	< 6	< 0.2
BP	$80''$	$3.5\,\mathrm{mm}$	$< 1.6 \times 10^{-3}$	< 0.23	$< 4.9 \times 10^{-2}$
OVRO	$1.8'$	$1.5\,\mathrm{cm}$	$< 2 \times 10^{-5}$	$\lesssim 0.12$	$< 5.6 \times 10^{-3}$
Balloon	$6°$	$720\,\mu\mathrm{m}$	$4 - 10 \times 10^{-5}$	$8 - 20 \times 10^{-5}$	$8 - 20 \times 10^{-5}$
(VLA)	$(4'')$	$(1.25\,\mathrm{cm})$	(3×10^{-4})	(1)	(7×10^{-2})

Table 2. Current (and projected) observations of background anisotropy and the corresponding fractional anisotropy in dust optical depth.

The techniques for using the VLA to measure fluctuations are described by e.g., Kellerman et al. (1984) and Martin and Partridge (1987). The figures quoted for the 2 cm VLA sensitivity are estimates of 95% confidence upper limits made from recent data obtained by R. B. Partridge and collaborators based on 18 hours of integration on a single, nearly source-free field. Interferometric data can also be treated by "tapering", or discarding the largest baselines along with their noise;

this results in improved sensitivity, but degrades the resolution and dynamic range. Future VLA observations at 1.25 cm are expected to have a similar $T_{sys} \simeq 120\,\mathrm{K}$ to the 2 cm results (and the same 50 MHz bandwidth per bank of correlators) but better (4″) angular resolution. For the same integration time one can expect a factor of 2 to 3 improvement in $(\delta\tau/\tau)$ sensitivity, so with a long (~50 hr) integration one may set limits at the $(\delta\tau/\tau) \lesssim 1$ level. In other words, if the SMB is resolved into sources at 4″ resolution, anisotropy should be detected, even if $\alpha = 2$.

The Owens Valley ("OVRO") result represents the current state-of-the-art in single-dish anisotropy measurements, as reported at this Institute by Lawrence and Readhead, again representing a high-confidence upper limit. It is interesting that the Boynton and Partridge (1973) limit, even with a much less sensitive measurement, sets comparable constraints on $\delta\tau/\tau$ because of the shorter wavelength. Finally, we have included the balloon measurements of Melchiorri *et al.* They quote a measured anisotropy of 0.11 mK at 700 μm. However, as this is not in the Rayleigh-Jeans limit more information is needed to convert to a $(\delta I/I)_{observed}$. We adopt two possible interpretations of this result: (1) Assume that $(0.11\,\mathrm{mK}/2.7\,\mathrm{K}) = 4 \times 10^{-5}$ gives $\delta\tau$; thus using $\tau = 0.2$ as derived previously we have $\delta\tau/\tau \simeq 2 \times 10^{-4}$. (2) Assume that $(0.11\,\mathrm{mK}/2.7\,\mathrm{K})$ gives $(\delta I/I)_{observed}$, as it would for large λ. Then using eq. (9) above with $I_\nu \cong 2(I_\nu - B_\nu(T_0))$ at 700 μm we have $\delta\tau/\tau \simeq 8 \times 10^{-5}$. Other interpretations are possible also so these numbers should be regarded as estimates.

What about the new generation of ground-based submillimeter telescopes and receivers? The ones on good sites are able to work well in the 345 GHz (870 μm) atmospheric window and occasionally in the (less good) windows at shorter wavelengths. When the atmosphere has $\tau < 1$, its thermal noise contribution is comparable to or less than the reciever, and an optimistic estimate is $T_{noise} \simeq T_{sys} + \tau T_{atmos} \simeq 400\,\mathrm{K}$. If there is no additional noise source (such as variable anisotropy in τT_{atmos}, etc.) then the brightness detectable in a time t_{int} is

$$(\nu I_\nu)_{noise}(t_{int}\Delta\nu)^{-1/2} \simeq 10^{-12}\left(\frac{t_{int}}{10^3\,\mathrm{s}}\frac{\Delta\nu}{10^9\,\mathrm{Hz}}\right)^{-1/2}\left(\frac{T_n}{400}\right)\,\mathrm{W\,cm^{-2}\,sr^{-1}} \quad (10)$$

evaluated at 710 μm, assuming a bandwidth $\Delta\nu \simeq 1\,\mathrm{GHz}$ typical of SIS receivers (larger bandwidth ~50 GHz may be attained with bolometers but at the cost of allowing more nonthermal τT_{atmos} noise into the window). Thus, such a telescope can detect a 3% difference in SMB flux between two patches of sky in $10^3\,\mathrm{s}$ at 700 μ, and perhaps a 6% difference at 900 μ. This sensitivity is shown in figure 1 as a horizontal line. The angular scale depends on the beam of the telescope being used; the diffraction limit is $\gtrsim 20″$ for 10-m class telescope. With very favorable conditions these telescopes may set lower limits than radio techniques, and without as much ambiguity from spectral modelling.

3. CONSTRAINTS ON MODELS

The fact that the spectrum of the SMB is now known allows these anisotropy measurements to be used to place stringent observational constraints on simple BCH-type models of the dust distribution. It is important to note that for scales much smaller than the horizon, the fractional anisotropy of emission $\delta\tau/\tau$ is much larger than the classical $\delta T/T$ produced by gravitational effects or by the Sunyaev-Zeldovich effect; therefore, interesting constraints can be placed on the dust distribution once it is known that it is producing a background even though the $\delta\tau/\tau$

limits are not as good as those on $\delta T/T$. We use two simplified models here for illustration.

Discrete Source Shot Noise Model. Randomly distributed objects with angular separation θ_{sep} lead to

$$(\delta\tau/\tau) \simeq (\theta/\theta_{sep})^{-1}, \tag{11}$$

shown as curve (2) in fig. 1 for $\theta_{sep} \simeq 6''$, characteristic of Tyson's (1987) high-redshift galaxies (see Cowie's contribution in these proceedings).

Dust Blob Model. Suppose the universe is divided into cubic cells of side r_c, and half of these, at random, are filled with dust blobs. The mean density of dust in a cell is thus twice the mean. Within each cell the blobs have a power-law correlation function $\xi \propto r^{-\gamma}$, falling abruptly to zero above r_c. The "blob model" is thus a simple scheme encapsulating continuum correlation properties similar to those of galaxies, and ignoring their discreteness.

Suppose this dust radiates the observed background during a redshift interval Δz centered on z. The correlation angle subtended by r_c is $\theta_c = r_c z H_0/2c$ in the high z limit. Using the results of BCH we can calculate the rms fluctuations in τ in square pixels on the sky of size θ. On large angular scales $\theta > \theta_c$, shot noise in the number of overlapping blobs gives

$$\begin{aligned}(\delta\tau/\tau) &= (\theta/\theta_c)^{-1}(r_c H/c\Delta z)^{-1/2} \\ &= \sqrt{2}(\theta/\theta_c)^{-1}\theta_c^{1/2} z^{-1/4}\delta^{-1/2}\end{aligned} \tag{12}$$

and for $\theta < \theta_c$, their correlation gives

$$(\delta\tau/\tau) \propto \theta^{(1-\gamma)/2}. \tag{13}$$

In figure 1, curves (1a,b) plot $\delta\tau/\tau$ for this model with parameters chosen to be characteristic of the modern galaxy distribution, i.e. $z = 1, \Delta z = z, r_c = 15\,h^{-1}$ or $30\,h^{-1}$ Mpc, $\theta_c = 8'$ or $17', \gamma = 1.8$. These curves when added to the discreteness contribution above should bracket a reasonable estimate of fluctuations $\delta N/N$ in $z \simeq 1$ galaxy counts. The arbitrariness of choosing r_c between $\simeq 15\,h^{-1}$ Mpc and $\simeq 30\,h^{-1}$ Mpc reflects uncertainty about large-scale clustering and the simplifications of the model.

The double line in figure 1 shows the values of $(\delta\tau/\tau)_{\theta=\theta_c}$ evaluated for other values of θ_c and shows how $\delta\tau/\tau$ would vary for smaller blob correlation scales. The utility of this plot is to place upper limits on θ_c from experiments. Figure 1 shows the experimental measurements, upper bounds, and theoretical sensitivites discussed above. We see immediately that at present the tightest constraints come from the balloon measurements. Combined with our current knowledge of the SMB spectrum these observations limit the structure in the emitting dust to $\theta_c \lesssim 1.5$ arc min, or $r_c \lesssim 3\,h^{-1}$ Mpc. This is a considerably smoother distribution than the galaxy distribution today, and appears to preclude any scenario, such as mock gravity or explosions, in which the SMB is a concomitant result of creating large-scale structure.

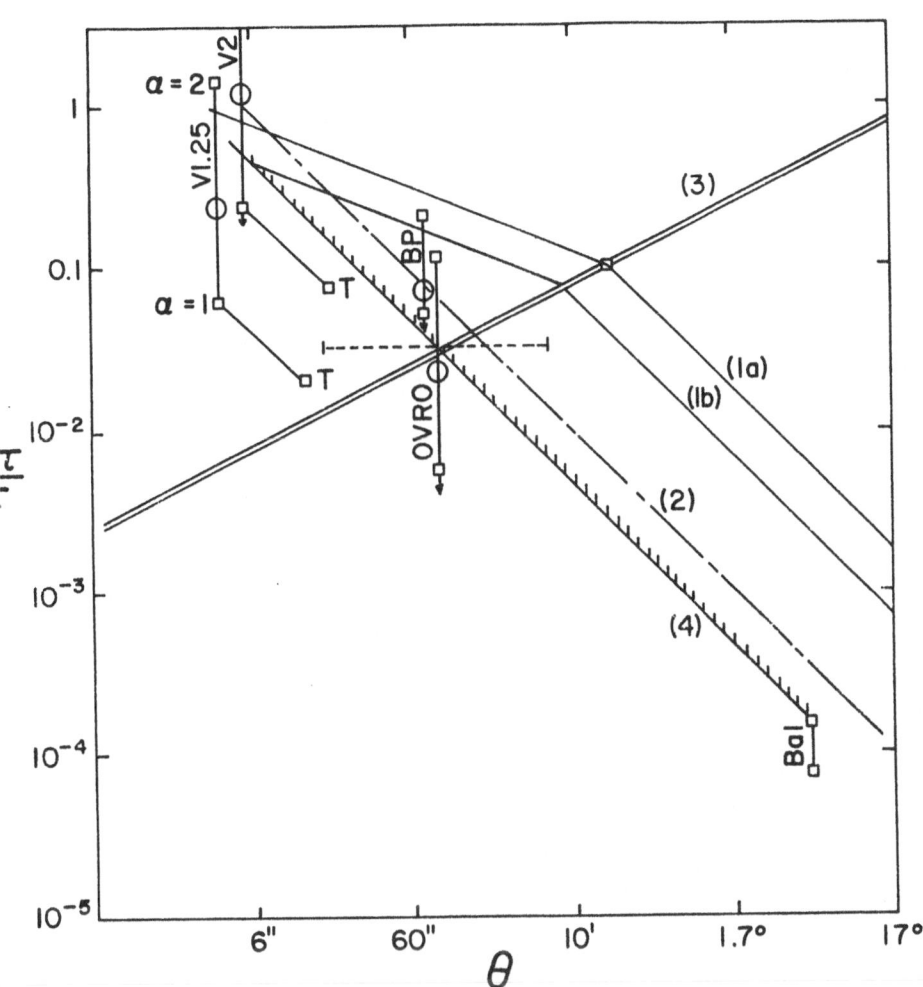

Figure 1. Anisotropy of optical depth as a function of angular scale. Squares indicate observations or upper limits; the ranges indicate uncertainties due to spectral index α or to calibration ambiguities, and circles indicate $(\delta I/I)_{\nu,s}$ model with stimulated emission included; "T" indicates effect of tapering VLA data. (1a) and (1b) show anisotropy in the "dust blob model" with $r_c = 15\,h^{-1}$ and $30\,h^{-1}\,\mathrm{Mpc}$; (2) shows anisotropy of discrete objects with $\delta N/N = 1$ on 6 arc sec scale; (3) shows $(\delta\tau/\tau)_{\theta_c}$ as a function of θ_c; and (4) shows the constraint imposed by the balloon data of Melchiorri et al. , leading to a limit of $r_c < 3\,h^{-1}\,\mathrm{Mpc}$ as described in text. For $\alpha = 1$ dust, the OVRO limit gives $r_c < 1\,h^{-1}\,\mathrm{Mpc}$. The horizontal dashed line is the approximate sensitivity level expected for ground-based submillimeter telescopes.

This small r_c result also provides an independent argument that the background must originate at high z, when the universe was more uniform than today. The data are barely consistent with structures on the scale of the mean separation of galaxies, which implies that the SMB is either pregalactic or associated with the initial collapse of unclustered galaxies. Figure 1 shows that VLA observations at 1.25 cm might confirm this through detection of very small-scale fluctuations.

4. IS MOCK GRAVITY CONFIRMED?

If the SMB is indeed coming from high-redshift dust, it greatly strengthens the case for a "mock gravity" instability in the early universe caused by the mutual shadowing of dust grains from the intense background radiation pressure. Hogan and White (1986) show that the mock gravitational growth rate of perturbations exceeds H/δ (as required for significant growth in a time δ/H) provided $A_0 \gtrsim 1$, where

$$A_0 = 4\ell\sigma\delta/9mcH^2 \tag{14}$$

Here ℓ is the luminosity density, and σ/m is the absorption cross section per unit mass. As above we adopt $\sigma/m = f_d \cdot 10^{-21}\,\mathrm{cm^2}/m_p \simeq 600 f_d\,\mathrm{cm^2 g^{-1}}$. If ϵ_0 is the present energy density and the radiation is released between z and $(1+\delta)z$,

$$\ell \cong z^4 H \delta^{-1} \epsilon_0 \tag{15}$$

Thus, using the observed value of $\epsilon_0 = 5 \times 10^{-14}\,\mathrm{erg\ cm^{-3}}$ we obtain

$$A_0 = \frac{4}{9} z^{5/2} \epsilon_0 H_0^{-1} \sigma/mc \cong 3 \times 10^{-4} z^{5/2} f_d h_{100}^{-1} \tag{16}$$

So, if the background originates at $z \gtrsim 30$ the instability can occur. Inserting our lower limits to z derived above we have

$$A_0 \gtrsim \begin{cases} 5 \times 10^{-3} f_d^0 \delta^{-1} (\Omega_g h_{100}^2/0.1)^{-1} & \alpha = 1 \\ 1.5 f_d^{2/7} \delta^{-5/7} (\Omega_g/0.1)^{-5/7} h_{100}^{-12/7} & \alpha = 2 \end{cases} \tag{17}$$

As discussed by Field (1971, 1987), imperfect coupling between gas and dust results in an even more rapid instability in the dust component, increasing A_0 by a factor of ten or more. As it is likely that $h_{100} < 1$, $\Omega_g < 0.1$ and $\delta < 1$, mock gravity on some scale is quite likely to be important.

Since the optical depth across the emission length $(\delta c/H)$ is inferred we may also derive the scale for which $\tau = 1$, using as above

$$\tau_{uv} = 0.2(7100/z)^\alpha \tag{18}$$

The scale with $\tau_{uv} = 1$ is given by

$$L_1 = \tau_{uv}^{-1}(c/H)\delta = 5(c/H)\delta(z/7100)^\alpha \tag{19}$$

which translates into a present-day comoving scale

$$\Lambda_1 = zL_1 = \delta \times \begin{cases} 7\,\mathrm{h^{-1}Mpc}(\frac{z}{10})^{1/2} & \alpha = 1 \\ 0.3\,\mathrm{h^{-1}Mpc}(\frac{z}{100})^{3/2} & \alpha = 2 \end{cases} \tag{20}$$

Up to this scale a rapid instability driven by radiation pressure is a likely consequence of the dust model of the SMB and is particularly important for modifying the dust distribution. Since this instability is very rapidly growing for $A_0 > 1$ it is also natural to credit it, rather than the pure gravitational instability, with influencing the formation of galaxies, particularly for Λ_1 comparable with their mean separation, $\sim 4\,h^{-1}$Mpc.

A related effect is that radiation pressure can convert linear luminosity-density fluctuations into (amplified) linear mass density fluctuations on large scales which later collapse gravitationally (see Hogan and Kaiser 1983). If Λ_1 approaches $20\,h^{-1}$Mpc this possibility would enable the radiation pressure to produce large-scale structure as well as galaxies and not conflict with the constraints on r_c derived above.

Certainly the detection of the predicted radiation background has rendered such schemes more plausible than before, and offers the possibility of additional observational verification. It seems likely that we are glimpsing our first view of the elusive "pregalactic era" and its far-reaching effects on the present day universe.

We wish to thank B. Carr, R. Martin, C. Walker, J. McDowell, and R. B. Partridge for useful conversations. C.J.H. was supported by NASA grant NAGW-763, and J.R.B was supported by a Canadian Institute for Advanced Research Fellowship and the NSERC of Canada. Both of us gratefully acknowledge Alfred P. Sloan Foundation Fellowships.

REFERENCES

Bond, J. R., Carr, B. J. and Hogan, C. J., 1986. *Astroph. J.*, **306**, 428. [BCH1]
Bond, J. R., Carr, B. J. and Hogan, C. J., 1987. Preprint. [BCH2]
Boynton, P. E. and Partridge, R. B., 1973. *Astroph. J.*, **181**, 243. (BP)
Field, G. B., 1971. *Astroph. J.*, **165**, 29.
Field, G. B., 1987. Harvard Preprint.
Fomalont, E. B., Kellermann, K. I. and Wall, J. V., 1984. *Astroph. J. (Letters)*, **277**, L23.
Hogan, C. J. and Kaiser, N., 1983. *Astrophy. J.*, **274**, 7.
Hogan, C. J. and White, S. D. M., 1986. *Nature*, **321**, 575.
Martin, H. M. and Partridge, R. B., 1987. *Astroph. J. Letters*, in press.
Matsumoto, T., Hayakawa, S., Matsuo, H., Murakami, H., Sato, S., Lange, A. E. and Richards, P. L., 1987. IRAS Conference, London (Nagoya preprint).
Melchiorri, F., Melchiorri, B., Ceccarelli, C. and Pietranera, L., 1981. *Astroph. J. Letters*, **250**, L1. (Balloon)
Readhead, A. C. S. and Lawrence, C., 1988. This meeting (OVRO).
Spitzer, L., 1978. *Physical Processes in the Interstellar Medium*, (New York: Wiley).
Tyson, J. A., 1987. Preprint, Bell Labs.

NON-GAUSSIAN STATISTICS AND THE MICROWAVE BACKGROUND

P. Coles
Astronomy Centre
University of Sussex
Falmer
Brighton BN1 9QH
UK

ABSTRACT. We describe some simple geometric properties of microwave background temperature anisotropies whose expectation values can be derived in cases where the random field of temperature fluctuations is not necessarily a gaussian random field. We explain how these properties may be useful for discriminating between gaussian and non-gaussian fluctuations.

1. INTRODUCTION

Fine scale anisotropies in the sky temperature of the Microwave Background provide one of the few direct observational tests of theories of galaxy formation. In models where galaxies are formed by the growth and collapse of small primordial density perturbations, the statistical properties of the Microwave Background fluctuations should mirror the statistical properties of the primordial density fluctuations. In most models of galaxy formation the initial perturbations are assumed to constitute a gaussian random field in three dimensions in which case the small angle temperature fluctuations on the sky will form a two dimensional gaussian random field. The gaussian assumption makes it possible to calculate a number of interesting statistical properties of such fields both in two and three dimensions [1,2,3,4] such as the number and size of high level regions as well as more detailed topological properties [7].

Although there is some physical motivation for the gaussian assumption (notably from inflationary theories), it is posssible that for some reason the initial fluctuations were not gaussian (e.g. in Cosmic String models). It is also true that temperature fluctuations produced by nonlinear phenomena at low redshifts (such as the Sunyaev-Zeldovich effect) would be non-gaussian. It is therefore interesting to study non-gaussian fields and show how, in may cases, it is possible to study the statistical properties of such fields to see which properties are very different from the gaussian case and which are similar. The details of this work can be found in references [5,6] so here we merely

N. Kaiser and A. N. Lasenby (eds.), The Post-Recombination Universe, 151–153.
© *1988 by Kluwer Academic Publishers.*

show the method used to treat non-gaussian fields, we describe some of the properties that can be obtained in this way and suggest that these properties may provide useful statistical methods for discriminating between gaussian and non-gaussian fluctuations.

2. NON-GAUSSIAN RANDOM FIELDS

The definition of a gaussian random field $X(\underline{r})$ is that the probability distribution of X is a gaussian and furthermore that all the finite dimensional joint distributions of $X(\underline{r})$ and $X(\underline{s})$ are multivariate gaussian. Statistical properties of gaussian random fields are therefore, at least in principle, simple to study. The simplest way to study corresponding properties of non-gaussian fields is to consider fields that are obtained via some transformation of a gaussian field or combination of gaussian fields. For example, if $X(\underline{r})$ is a gaussian random field then $Y = \exp(X(\underline{r}))$ is a non-gaussian field possessing the lognormal distribution. By finding appropriate transformations we have studied in [5,6] a number of random fields obtained from the gaussian in this way. As one can calculate the statistical properties of the underlying gaussian fields, it is possible thus to calculate the corresponding properties of the transformed gaussians. Cases we have studied so far include lognormal, Rayleigh, Maxwell, Chi-squared, rectangular and Gumbel-I distributed fields although many other fields could, in principle, be studied in this way. These fields range from those which are very similar in structure to the gaussian to those which are very different. In [5], we applied this method to determine the mean number density and mean area of "hotspots" -i.e. regions where the field exceeds a certain level - of the above mentioned non-gaussian fields. This analysis is considerably simplified by the fact that one can reduce the calculation to a one dimensional one by noting that the mean area of a hotspot for a gaussian field is merely the square of the mean length of an excursion above the level along a line [8].

3. TESTS FOR NORMALITY BASED ON PATTERN GEOMETRY

Although the above calculations give a useful insight into the different pattern properties that can be obtained if one relaxes the gaussian assumption, the *mean* properties of hotspots are not very useful discriminators between gaussian and non-gaussian fluctuations, as they contain very little information about the texture of the field. Fortunately it is possible to calculate a more sensitive pattern statistic namely the *Euler-Poincaré* characteristic of the contour set above a certain level. This is, roughly speaking, the number of disjoint contour regions minus the number of holes and in 3-D is related to the *genus* of the contour surfaces (see [9] and references therein). This is an extremely useful statistic to use as not only is it possible to calculate its expectation value for both the gaussian case and the non-gaussian cases given above [6] but also it is very simple to devise a computer algorithm to approximate the characteristic on a regular two dimensional grid [7]. Such an algorithm is useful because it allows one to evaluate the usefulness of the Euler-Poincaré characteristic as a discriminator between gaussian and non-gaussian fields using Monte Carlo simulations. Such simulations have already been performed for gaussian fluctuations in [4] by generating realisations of the power spectrum appropriate to the model under consideration with random phases and then Fourier transforming to recover the sky pattern. Because of the gaussian fields underlying our non-gaussian fields, we can generate non-gaussian sky maps with identical covariance functions by combining gaussian maps with appropriately chosen power spectra. For a further discussion of the use of these simulations, see [6]. At first thought, one would expect that all one would have to do to test whether the sky fluctuations were gaussian would be to perform a goodness-of-fit test between the distribution of the recorded temperatures and a gaussian. If one bears in mind the definition of a gaussian field given above, however, then it is clear that one would have to test all of the finite-dimensional joint distributions for multivariate normality – clearly an impossible task. Added to this, the full distribution of fluctuations may not be available if the signal is contaminated by noise. The standard method for testing a random field for normality involves the *bispectrum* or, equivalently, the three-point covariance function. There are problems with this method, however, as the bispectrum would be difficult to estimate from a patchy, censored map as is likely to be obtained from any experiment. For these reasons we are investigating the use of simple geometric measures of pattern such as those mentioned above, with a view to incorporating them in robust and powerful statistical tests of normality.

Acknowledgements. I thank John Barrow for constant advice and encouragement and SERC for the receipt of a postgraduate studentship.

REFERENCES

1. Bardeen, J.M., Bond, J.R., Kaiser,N. and Szalay, A., 1986. Astrophys. J., **304**, 15.
2. Peacock, J.A. and Heavens, A.F., 1985. Mon. Not. R. astr. Soc., **217**, 805.
3. Couchman, H.M.P., 1987. Mon. Not. R. astr. Soc., **221**, 53.
4. Bond, J.R. and Efstathiou, G., 1987. Mon. Not. R. astr. Soc., **226**, 655.
5. Coles, P. and Barrow, J.D., 1987. Mon. Not. R. astr. Soc., in press.
6. Coles, P. 'The Statistical Geometry of the Microwave Background', in prep.
7. Adler, R.J., 1981. *The Geometry of Random Fields*, John Wiley, New York.
8. Vanmarcke, E.H., 1983. *Random Fields : Analysis and Synthesis*, MIT Press, Cambridge, Massachusetts.
9. Hamilton, A.J.S., Gott, J.R. and Weinberg, D. Astrophys. J., **309**,1.

COSMOLOGICAL GRAVITATIONAL WAVES

ERIC V. LINDER
Department of Physics and Center for Space Science and Astrophysics,
Stanford University, Stanford, California USA

A cosmological background of gravitational waves would alter the propagation of radiation, inducing redshift fluctuations, apparent source position deflections, and luminosity variations. By comparing these astrophysical effects with observations we can deduce upper limits on the energy density present in gravitational waves, Ω_{GW}. We concentrate on microwave background anisotropy from the redshift deviations and galaxy clustering correlation functions from the angular deviations. Many of the gravitational wave effects are shown to be generalizations of gravitational lensing formalism.

1. INTRODUCTION

We examine some astrophysical effects of a cosmological background of stochastic gravitational waves, and use these effects to impose constraints on their energy density Ω_{GW}. As metric perturbations the waves alter the propagation of radiation and matter, giving rise to fluctuations in redshift, angular and depth position, and luminosities. Note that the waves are only changing our *observations* of the universe and not the actual constituents – they fool our perceptions in the same way as gravitational lensing does with multiple imaging, say.

2. FLUCTUATION FORMALISM

Both gravitational wave and gravitational lensing theory can be viewed as special cases of the general problem of light propagation through gravitationally perturbed spacetimes. Although the calculations are carried out in a perturbed geometry, we can aid our physical intuition by viewing this as an unperturbed cosmology filled with a refractive, turbulent medium. This effective index of refraction paradigm can be justified by Fermat's principle (see Linder 1987).

We concentrate on cosmological scale gravitational waves, but wavelengths $\lambda > 1$ kpc imply periods in excess of a thousand years, much greater than the characteristic time scale of an astronomer. Therefore we will not observe the fluctuations but only a freeze frame picture, which retains the correlations of the shifts induced by the gravitational waves. This key point will be utilized in sections 3 and 4.

The fluctuations themselves are derived by solving the photon geodesic equation under the assumptions of a flat Friedmann universe with $\lambda_{\text{photon}} \ll \lambda_{GW} \ll H_0^{-1}$. The redshift result is

$$z = (a_t^{-1} - 1) - \frac{1}{2} a_t^{-1} \int_t^{t_o} dt\, h_{11,0}, \qquad (1)$$

N. Kaiser and A. N. Lasenby (eds.), The Post-Recombination Universe, 155–158.
© 1988 by Kluwer Academic Publishers.

where the first term is the usual expansion factor and the second is a Sachs-Wolfe (1967) term. Angular position deviations are given by

$$\theta_j = \frac{dx^j}{d\bar{x}^1} = \theta_j(0) - h_{1j}|_t^{t_0} + \frac{1}{2}\int_t^{t_0} dt\, a^{-1} h_{11,j}, \tag{2}$$

and the *rms* fluctuations are

$$\delta z_{rms} \approx 5 \times 10^{-6} \lambda_0\, \Omega_{GW4}^{1/2}(1+z), \tag{3a}$$

$$\delta\theta_{rms} \approx 3''\lambda_0\, \Omega_{GW4}^{1/2}(1+z), \tag{3b}$$

where λ_0 is the present day gravitational wave wavelength in units of $1\,Mpc/(H_0/\,100\,kms^{-1}Mpc^{-1})$ and Ω_{GW4} is $\Omega_{GW}/10^{-4}$.

3. GALAXY CLUSTERING CORRELATIONS

Concentrating on the angular two-point galaxy-galaxy correlation function

$$W(\Phi) = \langle\delta\rho(\phi)\,\delta\rho(\phi')\rangle/\langle\rho\rangle^2, \tag{4}$$

we can write the solid angle number density ρ in terms of the gravitational wave distorted solid angle $\omega(z)$. This distortion is, as in gravitational lensing, just the jacobian of the mapping (eq. [2]) between source and image position. Evaluating equation (4) then yields

$$W_{GW}(\Phi) \approx \frac{9}{35}(H_0 L)^2 \Omega_{GW}(1 - \frac{3}{8}\Phi^2). \tag{5}$$

Contrasting both the magnitude and survey depth L scaling with the observed correlation functions (see, *e.g.* Peebles 1980) gives the upper limit quoted in section 5,

$$\text{clustering correlations} \Longrightarrow \Omega_{GW} < 10^{-3}. \tag{6}$$

4. MICROWAVE BACKGROUND ANISOTROPY

We can use a similar analysis to that of section 3 on the temperature correlation function of the microwave background,

$$C(\Phi) = \langle\frac{\Delta T}{T}(\phi,t)\,\frac{\Delta T}{T}(\phi',t')\rangle, \tag{7}$$

with the added complication of including the structure of the last scattering surface. Temperature fluctuations correspond to redshift fluctuations, and are composed of two parts — variations in "emission" times of the photons due to the blurriness of the last scattering surface, and the gravitational wave effect of section 2.

We find anisotropies are diluted due to blurring by a factor

$$d \approx (60/\lambda_0)(\frac{\Delta z}{z}/0.1), \tag{8}$$

and have a coherence angle

$$\Phi_c = [-C''0)/C(0)]^{-1/2} \approx 10''\lambda_0. \tag{9}$$

Fig. 1. Upper limits on the energy density of a cosmological gravitational wave background plotted versus wavelength.

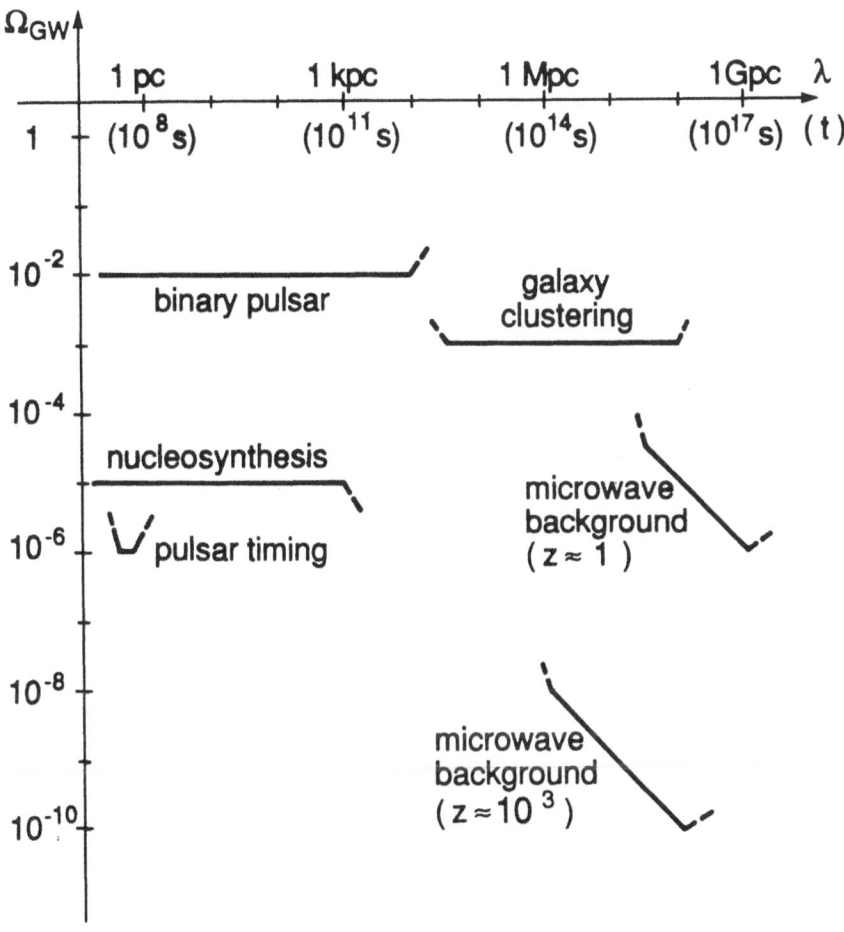

158

Choosing a gravitational wave spectrum yields the anisotropy form, or alternately, $\Delta T/T$ limits impose a spectrum-dependent constraint on $\Omega_{GW}(\lambda) = d\ln\Omega_{GW}/d\lambda$ of approximately

$$\Omega_{GW}(\lambda) \leq 10^{-9}\langle(\frac{\Delta T}{T}/10^{-5})^2\rangle. \tag{10}$$

5. CONCLUSION

Figure 1 graphically presents the upper limits on the gravitational wave energy density discussed here, as well as some shorter wavelength limits. These are the *only* constraints below one percent of closure density at cosmological wavelengths. For the pulsar timing curve see Taylor (1987), for the binary pulsar see Rees (1983), and for the nucleosynthesis constraints see Carr (1980). The microwave background curve labelled with $z = 1$ gives the limits of equation (9) if the gravitational wave background was generated as recently as $z = 1$.

I would like to thank Nick Kaiser for arranging my attendance at the workshop at the last minute and I gratefully acknowledge an NSF travel grant. This work was supported in part by grants from NASA (NAGW-299) and NSF (PHY-86-03273).

REFERENCES

Carr, B.J. 1980, *Astr. Ap.* **89**, 6.
Linder, E.V. 1987, submitted to *Ap. J.*
Peebles, P.J.E. 1980, *Large-Scale Structure of the Universe*
    (Princeton: Princeton University Press).
Rees, M.J. 1983, in *Gravitational Radiation*, ed. N. Deruelle and T. Piran
    (Amsterdam: North-Holland), p. 297.
Sachs, R.K. and Wolfe, A.M. 1967, *Ap. J.* **147**, 73.
Taylor, J.H. 1987, in *Proc. 13th Texas Symposium on Relativistic Astrophysics*,
    ed. M.P. Ulmer (Singapore: World Scientific), p. 467.

The Infrared Background From Stars at High Redshift

Jonathan C. McDowell

Nuffield Radio Astronomy Laboratories, Jodrell Bank, Macclesfield, Cheshire, England
SK11 9DL

As emphasized by Peebles and Partridge (1967), the light of the first stars will still
be present in the universe today as a faint background radiation. The expected energy
density of this background, if it is generated by a population of sources which convert
a fraction ϵ of their rest energy into radiation at a characteristic redshift z, may be
estimated from the simple formula

$$\Omega_R = \frac{\epsilon \Omega_*}{1+z}.$$

Here Ω_* and Ω_R are the comoving mass density of the sources and the present energy density of the radiation background, both measured in units of the cosmological critical
density.

For nuclear burning, $\epsilon \sim (1-3) \times 10^{-3}$ for stars above $10 M_\odot$. Thus for a mass of
stars comparable with the present visible mass density ($\Omega_* \sim 10^{-2}$) at a redshift of
a few, we expect $\Omega_R \sim 10^{-5}$. For comparison, the observed energy density of the 2.74K
blackbody microwave radiation is $\Omega_{CMB} = 1.0 \times 10^{-4}$ (Here and throughout I take $H_0 =$
50km/s/Mpc). If the dynamical dark matter is made up of the remnants of pre- or protogalactic massive stars (Truran and Cameron 1971, Carr, Bond and Arnett 1984), then we
expect $\Omega_* = 0.1$, and $z \sim 10 - 50$, again predicting $\Omega_R \sim 10^{-5}$.

The observed limits on the extragalactic background light were discussed in B.
Carr's paper in this volume. The most important recent development is the detection
of a cosmological far infrared background by the Berkeley–Nagoya collaboration (Matsumoto et al 1987). This rocket experiment measured the spectrum of the background in
six channels from 100 microns to 1100 microns. The three short wavelength channels measured the spectrum of the infrared cirrus, showing that it falls off faster than λ^{-1}
at long wavelengths. The three long wavelength channels measured background radiation
in a range where galactic contributions to the signal should be small, and the points
show a significant departure from the black body curve. Subtracting the 2.7K blackbody
(whose temperature is well determined in the Rayleigh–Jeans region) reveals an excess
background between 500 microns and 1 mm, with a total energy density about 20 per cent
of the microwave background, i.e. $\Omega_R \sim 2 \times 10^{-5}$.

Matsumoto et al note that simple Comptonization models cannot fit both the excess
and the Rayleigh–Jeans temperature of the microwave background, although more sophisticated models may be able to (Raphaeli, in preparation). However, the excess is well
fit by a simple dust emission law with a power law opacity. The peak of the excess is
significantly narrower than a black body and a λ^{-2} opacity law fits well. This point
is illustrated in Fig. 1 where the quantity plotted is $\Omega_R(\nu)$, the energy density per
unit logarithmic frequency interval. This fit leads naturally to the interpretation of
the excess as redshifted cosmological dust emission. In agreement with Matsumoto et al

159

N. Kaiser and A. N. Lasenby (eds.), The Post-Recombination Universe, 159–161.
© *1988 by Kluwer Academic Publishers.*

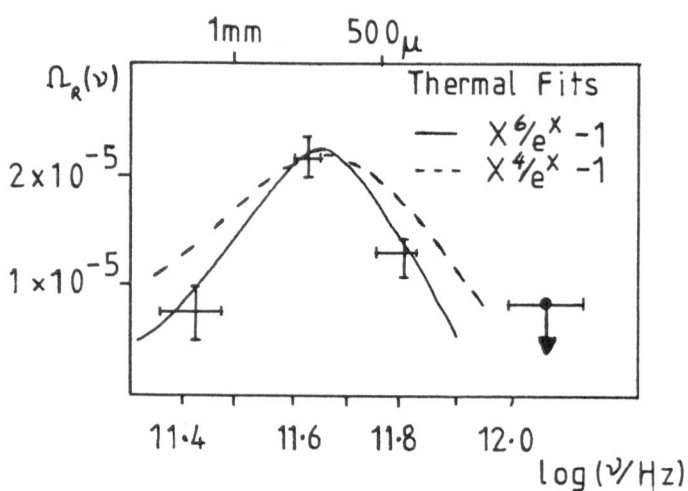

Far Infrared Background

(2·74K Black Body subtracted)

Fig. 1.

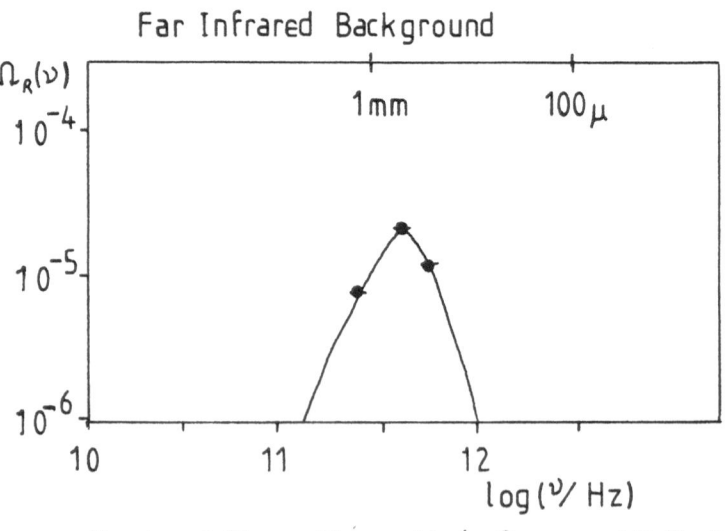

Far Infrared Background

Dust and Stars ($\Omega_{dust} = 10^{-6}$, $\Omega_{stars} = 0.5$, $Z = 60$)

Fig. 2.

(1987), I estimate the corresponding dust temperature to be (3.55 ± 0.1)K.

Perhaps the most likely scenario for producing redshifted dust emission is one in which a 'starburst era' occurs at a redshift of 5-7, with a large fraction of all protogalaxies undergoing merger and starburst events similar to objects like NGC 3256 (Graham et al 1984). However it will be difficult to find a model in which the process occurs at a sufficiently well defined epoch to give the observed narrow peak.

A more promising possibility is the reprocessing of an ultraviolet background created at high redshift by dust in the intergalactic medium or along the line of sight. Analytic (Bond, Carr and Hogan 1986) and numerical (Negroponte 1986, McDowell 1986b) models of cosmological dust emission indicate that the expected comoving dust emission temperature remains fairly constant over a wide redshift range and predicted that a far-infrared background would be produced with a rather well defined frequency peak. However the dust temperature from these models is rather higher than observed. Matsumoto et al (1987) have interpreted their measurements of the near-IR background at 2 microns as Lyman α emission at a redshift of 15, and suggested that the far IR background could be associated with it. However, simple models suggest that extreme dust properties would be required to fit the two measurements simultaneously.

I have developed a code for studying the cosmological background radiation produced by sources of arbitrary spectra in the presence of cosmologically distributed dust (McDowell 1986a,1986b). I use an empirical Galactic dust opacity law based on the observations of Savage and Mathis (1979) and the models of Draine and Lee (1984). The resulting model spectra are integrated over the observational bandpass to compare directly with the data. Simple analytic models suggest that the best fit to the new data should be obtained using Population III massive stars at a redshift of about 100. A good fit is indeed obtained with a model in which stars with $\Omega_* = 0.5$ burn from z=95 to z=60 (corresponding to a number of stellar generations) in a universe filled with dust with a density parameter $\Omega_d = 10^{-6}$. This tiny dust density is two orders of magnitude below the maximum amount allowed by quasar reddening observations. In this model (Fig. 2) the contribution from the silicate feature (M. Rowan-Robinson, private communication) is smeared out.

Further work is required to investigate the change in dust properties needed to lower the required emission redshift and so produce a less energetically extravagant scenario.

References

Bond, J.R., Carr, B.J., and Hogan, C., 1986. *Astrophys. J.* 306,428.

Carr, B.J., Bond, J.R., and Arnett, W.D., 1984. *Astrophys. J.* 277,445.

Draine, B.T., and Lee, H.M., 1984. *Astrophys. J.* 285,89.

Graham, J., Wright, G.S., Meikle, W.P.S., Joseph, R.D., and Bode, M.F., 1984. *Nature*, 310, 213.

Matsumoto,T. , Hayakawa, S., Matsuo, H., Murakami, H., Sato, S., Lange, A.E, and Richards, P.L., 1987, preprint.

McDowell, J.C., 1986a. Ph.D. thesis, University of Cambridge.

McDowell, J.C., 1986b. *Mon. Not. R. astr. Soc.* 223,763.

Negroponte, J.R., 1986. *Mon. Not. R. astr. Soc.* 222,19.

Peebles, P.J.E., and Partridge, R.B., 1967. *Mon. Not. R. astr. Soc.* 148,377.

Savage, B.D., and Mathis, J.S., 1979. *A. Rev. Astr. Astrophys.* 17,73.

Truran, J.W., and Cameron, A.G.W., 1971. *Astrophys. Spa. Sci.* .14,179.

LARGE-SCALE MICROWAVE BACKGROUND ANISOTROPIES
IN COSMOLOGICAL MODELS WITH EXOTIC COMPONENTS

Mirosław Panek
Copernicus Astronomical Center
Bartycka 18
00-716 Warszawa
Poland

ABSTRACT. The generalization of the Sachs-Wolfe formula for cosmological models containing exotic components of matter is presented.

In most of the cosmological scenarios the Universe during and after the hydrogen recombination is matter dominated. As it is not a constraint we may introduce dynamically important components other than nonrelativistic matter. These may be: a cosmological constant, relativistic particles, light cosmic strings etc.. They may affect the dynamics of the Universe and the growth of perturbations of barionic and dark matter.

It is rather complicated to find the large-scale microwave background radiation (MBR) anisotropies in such models. We present here a generalization of the standard Sachs-Wolfe formula for multicomponent models with nonstandard dynamics. We assume the Friedman-Robertson-Walker (FRW) geometry and treat components of matter as perfect fluids. The perturbations are described in the gauge-invariant formalism [1].

We deal with scalar perturbations. For a single mode described by scalar harmonic Q(k), where k is the wavenumber, the fluctuations of the MBR temperature at reception are given by [2]:

$$\left(\frac{\delta T}{T}\right)_R = \left(\frac{\delta T}{T}\right)_E + \int_E^R \left[\frac{1}{3}\left(\dot{\mathcal{E}}_b + k v_b\right) Q \right. \tag{1}$$

$$\left. + \frac{v_b}{k} Q_{|\alpha\beta} R^\alpha R^\beta \right] ds$$

where $\mathcal{E}_b$ and v_b are the gauge-invariant perturbation quantities for density and velocity of barionic fluid (Bardeen's $\mathcal{E}_m$ and V_s), the vertical bar denotes the covariant derivative with respect to the spatial part of the conformal FRW metric, R^α is a direction unit vector and the integral is calculated along the null geodesic from emission to

N. Kaiser and A. N. Lasenby (eds.), The Post-Recombination Universe, 163–165.
© 1988 by Kluwer Academic Publishers.

reception. The term $\left(\frac{\delta T}{T}\right)_E$ describes the MBR temperature fluctuations at the last scattering surface. If perturbations of radiation and barion densities are adiabatic then $\left(\frac{\delta T}{T}\right)_E = \frac{1}{3}\left(\varepsilon_b Q\right)_E$. We do not include this term in the following formulae as it is usually negligible in comparison with the integral term in (1).

The formula (1) is valid for any multicomponent FRW model. It should be accompanied by equations of evolution of barionic perturbations [2],[3]. These equations include the effect of all remaining components.

For any flat FRW model we can use the equations of evolution of perturbations to rewrite (1) in the form:

$$\left(\frac{\delta T}{T}\right)_R = -\frac{v_b}{k} R^\alpha Q_{1\alpha}\Big|_E^R - \frac{\dot{v}_b}{k} Q\Big|_E^R +$$
$$+ 2 \int_E^R \left[\ddot{v}_b + \left(\frac{\dot{S}}{S} v_b\right)^\cdot\right] Q \, ds \tag{2}$$

where the dot denotes the conformal time derivative and S is the scale factor. This is the generalization of the Sachs-Wolfe formula.

For the flat, dark matter models with adiabatic perturbations the evolution of perturbations in the matter dominated era is:

$$\varepsilon_b = \varepsilon_{bE}\left(\frac{\tau}{\tau_E}\right)^2 \quad , \quad v_b = -\frac{2\varepsilon_{bE}}{k\tau_E}\frac{\tau}{\tau_E}$$
$$\Phi_H = \frac{6\varepsilon_{bE}}{k^2\tau_E^2} \tag{3}$$

where τ is the conformal time. We have also written the value of one of the gauge-invariant gravitational potentials [1]. Putting the above solutions into (2) gives us the standard Sachs-Wolfe formula in the gauge-invariant notation:

$$\left(\frac{\delta T}{T}\right)_R = \frac{1}{3} \Phi_H \left[\left(\tau R^\alpha Q_{1\alpha}\right)\Big|_E^R + Q\Big|_E^R\right] \tag{4}$$

We can see that any departure from the matter-dominated law of evolution of the scale factor $(S \sim \tau^2)$ or any other than given by (3) evolution of perturbations would not give us the cancellation of the integral term.

We need values of observational quantities which describe the large-scale MBR anisotropies - multipole moments or the correlation function. To obtain them we add single mode effects according to adopted statistical model. In the standard case for Gaussian field of perturbations, random phases and scale-invariant n=1 spectrum the correlation function of the MBR fluctuations at angular scale θ is [4]:

$$W(\theta) = \frac{3\,a_2^2}{2\pi} \left(\ln \frac{2}{1-\cos\theta} - \frac{3}{2}\cos\theta - 1 \right) \qquad (5)$$

where a_2 is the quadrupole moment.

It is not possible to obtain the general analytical formula for $W(\theta)$ in more complex models as (2) is much more complicated than (4). In the standard case the amplitude of the MBR anisotropy is given by the growth factor of the density perturbations (the same for all scales). In the general case this amplitude is given by another function of perturbations, and in addition the growth factors of different scale perturbations are usually different. The relation of the comoving scale to the angular scale may also be changed. As a consequence of these changes we can expect that in the models with exotic components of matter the correlation function $W(\theta)$ may have a form different from (5), even if the initial spectrum of perturbations is scale-invariant.

In other words, two interpretations of the shape of $W(\theta)$ different from (5) are possible.

First, this may come from other than n=1 spectrum of perturbations in the Universe dominated by nonrelativistic matter. As we do not have any independent source of information on the spectrum at large scales, this interpretation can be supported or rejected only using theoretical arguments about the generation of perturbations.

Another interpretation is that our Universe contains exotic components influencing its dynamics. The formulae presented here may help to find $W(\theta)$ in such complicated models. As one can measure the present dynamics of the Universe then this interpretation is subject to an observational verification.

REFERENCES
[1] J.M. Bardeen, *Phys.Rev.D* 22, 1182 (1980)
[2] M. Panek, *Phys.Rev.D* 34, 416 (1986)
[3] L.F. Abbott, M.B. Wise, *Nucl.Phys.B* 237, 226 (1984)
[4] P.J.E. Peebles, *Astrophys.J.* 263, L1 (1982)

MICROWAVE BACKGROUND RADIATION OBSERVATIONS AT THE OWENS VALLEY RADIO OBSERVATORY

A. C. S. Readhead, C. R. Lawrence, S. T. Myers, W. L. W. Sargent
Owens Valley Radio Observatory 105-24
California Institute of Technology
Pasadena, CA 91125
USA

1. INTRODUCTION

For the last four years we have been engaged in an intensive program to look for intrinsic variations in the Microwave Background Radiation (MBR) at the Owens valley Radio Observatory (OVRO). We have not yet detected any variations but we have derived a very stringent upper limit. We believe that our limit is robust because different statistical tests give very similar results and because our data are consistent with gaussian statistics.

There are, of course, two major scientific motivations for this work. The first is that the observations provide important constraints on the density fluctuations at the epoch of recombination and hence on theories of galaxy formation, assuming that reionisation occurred at a late enough epoch. We hope eventually to measure the temperature fluctuation spectrum which would provide a firm observational foundation for theories of galaxy formation. The second motivation is to use these fluctuations as a probe of the Universe at epochs much earlier than the epoch of recombination.

We are told that the level of fluctuation we should expect on angular scales above a few arc minutes is in the range $\Delta T/T = 10^{-4}$ to 10^{-5} for isothermal and adiabatic fluctuations without dark matter (Wilson and Silk 1981). and that the level of fluctuation expected on models dominated by cold dark matter is around 10^{-5} or less (Bond and Efstathiou 1987). This means that, even with the best receivers at the best sites, weeks of observing time in the best weather are needed to make observations with the required sensitivity.

2. OBSERVATIONS AT THE OVRO

The OVRO is well suited to this kind of work since the atmosphere above the site is generally relatively dry, with less than 3 mm of precipitable water 20% of the time between November 1 and March 31, and we have a superb maser receiver which was designed specifically for MBR observations. This receiver is also used for Very Long Baseline Interferometry (VLBI), but the broad bandwidth, symmetric dual feed configuration and tunability away from the water vapour line at 22.3 GHz were all dictated by the requirements for MBR observations. The relevant characteristics of the 40m Telescope and this receiver are given in Table I.

The symmetric beam configuration enables us to eliminate most of the ground spillover and atmospheric effects by Dicke switching and alternating beams on the field of interest

N. Kaiser and A. N. Lasenby (eds.), The Post-Recombination Universe, 167–172.

TABLE I

CHARACTERISTICS OF THE 40 m ANTENNA AND RECEIVER

Beamwidth (FWHM)	110″
Beamthrow	7″.15
Beam efficiency	0.51
Observing frequency	20.0 GHz
Bandwidth	400 MHz
Dicke switch frequency	10 Hz
System temperature	40–50 K
Sensitivity on sky	$9 \text{ mK s}^{-1/2}$

every 20 seconds. Double switching schemes like this have been used by many workers in this field. But the atmosphere still remains the largest single source of error, and the ground spillover still causes significant effects when tracking over a large range of zenith angles. We minimise the effects of variation in differential ground spillover by observing fields 1° away from the north celestial pole and observing only near upper and lower culmination.

In order to minimise ground spillover effects the receiver is set up with two fields straddling the telescope axis in the horizontal plane so that the Dicke switching is done in azimuth. Thus, in a two-hour scan, the two reference fields describe 30° arcs about the main field, as shown in Figure 1.

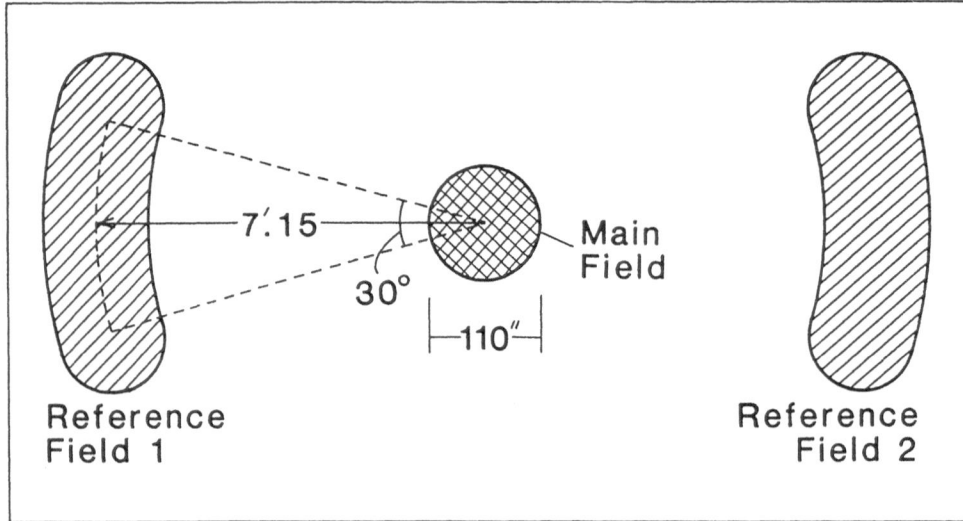

FIGURE 1. Beam pattern on the sky for two hours integration on a single field. Two beams of FWHM 110″ are aimed alternately every 20 s at the Main Field. The antenna moves only in azimuth. We measure $T = T_{\mathrm{MF}} - 1/2(T_{\mathrm{R1}} + T_{\mathrm{R2}})$.

Each day we observed eight fields centered at R.A. $= 1^{\mathrm{h}}, 3^{\mathrm{h}}, \ldots, 15^{\mathrm{h}}$ at upper culmination; the four fields centered at $5^{\mathrm{h}}, 7^{\mathrm{h}}, 9^{\mathrm{h}}$, and 11^{h} were also observed at lower culmination. Each scan lasted two hours.

We encountered a number of other problems in the course of these observations. These are summarised in Table II, together with the strategy we used for eliminating or minimising their effects.

TABLE II

Elimination of Systematic Errors

Noise Source	Size	Strategy
1. Atmosphere—different T_{H_2O} in two beams	0.1–5 K	*i.* Discard data for which there were clouds within 2 hr.
		ii. Discard data for which $\Delta T > 11$ K/air mass, and $\Delta T_{rms} > 0.4$ K/air mass.
		iii. Double switching.
		iv. Discard data for which $\Delta T > 2\times$ thermal limit.
2. Ground spillover	2 K	*i.* Minimise length of track by observing near NCP and at upper and lower culmination.
		ii. Double switching.
3. Receiver gain variations:		
A. Medium to long term:		
i. Changes in T_{R_x}	0.1–5 K	Dicke switch at 10 Hz.
B. Short term:		
i. $\Delta T \approx 3$ mK due to fluctuations in JT return pressure.	< 0.5 mK	Install large ballast tank in JT return line.
ii. Cycling of compressor fans.	< 20 mK	Run compressor at higher temperature.
4. Interference (RF pickup on IF cables between telescope and control building.	< 0.2 mK	A→D conversion at telescope.
5. Other.	< 50 μK	

The upshot is that we are now confident that there are no unknown sources of noise above a level of 50 μK. It is extremely difficult to detect, identify and eliminate sources of noise below 50 μK since a 3σ detection at this level requires a 28 hour observation in superb weather.

3. OBSERVATIONS AND DATA ANALYSIS

We have had five observing sessions since January 1984, and 170 days of 40 m Telescope time have been allocated to this program. We obtained data useful for our MBR observations for only about 15% of this time, and of the data lost 70% was due to weather.

The requirements for intrinsic anisotropy measurements are considerably more stringent even than those for Millimeter VLBI or optical observations, and much of the time lost to the MBR program was spent productively on less demanding programs. The remaining 30% of lost time was due to equipment problems.

After editing there remained 24 days of good MBR data which reduces to 17 days of good data on the sky after allowing for slewing, calibration, etc. We detected a weak source in one field (see below), leaving 15 days of good data on seven fields.

The average temperature difference for these seven fields is $\Delta T = -12 \pm 10$ μK. This noise level is 1.3 times the thermal noise limit for fifteen days integration.

We have made the following tests of the data to look for systematic effects:

1. *Effect of time of day:* We have sorted the data into 24 1-hour bins according to the time of day. No systematic effect is visible down to a level of about 30 μK.

2. *Increase of sensitivity with time:* We find that $\sigma \propto t^{-1/2}$ for all integration intervals up to 15 days. This gives us confidence that we will be able to improve substantially on the results reported here by continuing these observations.

3. *Long-term low level bias:* We have sorted the data into the five observing sessions and find a reduced chi-square of 1.7 for 5 degrees of freedom. The probability of a larger chi-square value, assuming gaussian statistics, is 12%. This is a bit low, but not so low as to cast serious doubt on our assumption of a gaussian distribution.

In both the hand editing and automatic editing of the data only the scatter in the data and deviations from the observed weighted means of the individual fields are used as a basis for discarding data. Neither of these procedures can introduce a bias into the data.

We conclude from these tests that there is no evidence of significant systematic errors biassing our data.

Our results on the eight fields are shown in Figure 2. It is clear that we have a significant detection in Field 7. We are convinced that this is due to a discrete radio source for two reasons. First, the antenna temperature of this field varied by more than a factor two over the course of the observations. Second, Pauliny-Toth *et al.* (1977) detected a source (#58) near this position in their 4.85 GHz survey of the NCP. We have also now detected a source in the direction of PT58 in 20 GHz observations on the 40m Telescope. We therefore do not consider this field further in the analysis of the data. It is clear from our observations of the remaining seven fields, for which the largest deviation from zero is -57 μK, that the rms temperature fluctuations are significantly smaller than 100 μK.

There is no universally accepted statistical procedure for deriving an upper limit on intrinsic anisotropy from the measurements in Figure 2. A detailed discussion of several procedures is given in the following paper. Using the likelihood-function and the likelihood-ratio-test methods described therein, we obtain upper limits (95% confidence) of 51 and 47 μK, respectively, for the beam pattern of Figure 1. If we then assume no correlation between the temperatures of the main fields and the reference fields, we find a limit on $\Delta T/T$ between *two* beams of 1.5×10^{-5} (95% confidence). Because of the crude assumptions involved, this limit is perhaps most useful for comparing with other published limits. Further details can be found in the following paper. It should already be clear that these observations rule out the standard adiabatic models without cold dark matter, and place interesting limits on isocurvature models.

4. FUTURE MBR OBSERVATIONS AT THE OVRO

We plan to continue MBR observations on the 40m Telescope at this frequency. We believe that it will be possible to achieve a factor two to three improvement over our present limit with this system, at which point discrete-source confusion will become a

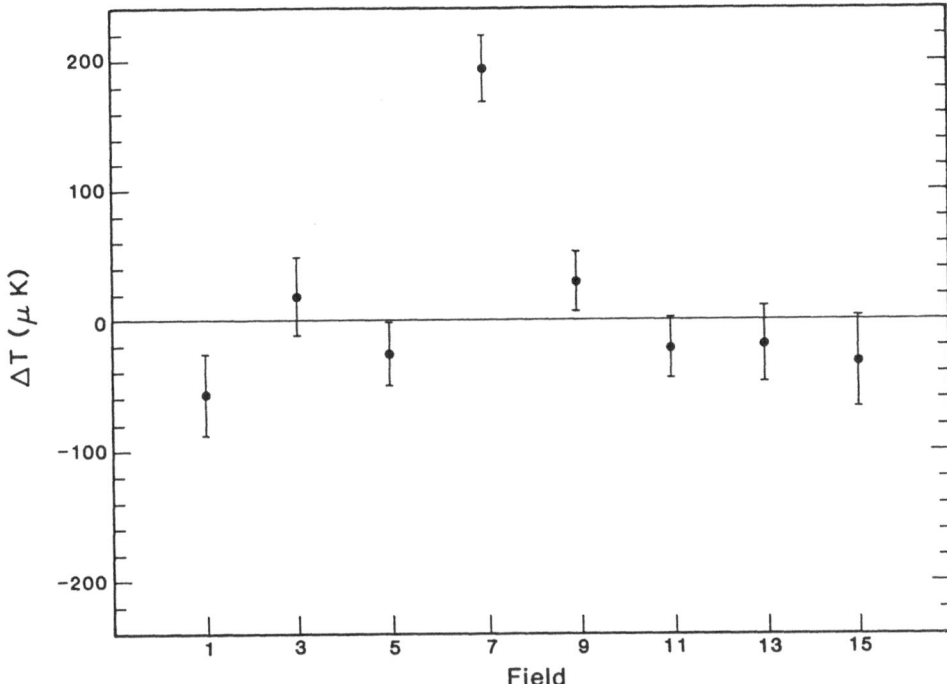

FIGURE 2. Results for eight fields at Decl.=89°, R.A.= "Field". Error bars are based on the scatter of the individual 80 s data points.

serious problem. We will observe the remaining four fields out of the dozen that we have selected, and continue observations on the present eight fields. Although it has taken four years to reach our present limit it is worth pointing out that weather conditions over the last four years have been atypical. We have not had the normal very cold, dry winter seasons in the Owens Valley, which is not surprising in view of the peculiarities in the global weather pattern over this period, and it is clear that even one really good observing season could yield a substantial improvement in our results.

In addition to this program on the 40 m Telescope, we are presently installing a 5.5 m Telescope dedicated to MBR observations, and building a 32 GHz maser receiver for use on this instrument. The expected characteristics of this system are given in Table III.

The operation of this telescope as a dedicated system for MBR observations should help in tracking down and eliminating low level sources of systematic noise, and we hope to achieve a sensitivity level of about $\Delta T/T = 3 \times 10^{-6}$.

Acknowledgements: We would like to thank Harry Hardebeck, Al Moffet, the conference organizers, and NATO for their various essential contributions to this presentation. Microwave background research at OVRO is supported by the IBM Research Fund, the Irvine Foundation, and NSF grants AST 85-09822 and AST 86-10693.

TABLE III

CHARACTERISTICS OF THE 5.5 m ANTENNA AND RECEIVER

Beamwidth (FWHM)	8'
Beamthrow	24'
Beam efficiency	0.7
Observing frequency	32.0 GHz
Bandwidth	750 MHz
Dicke switch frequency	1 kHz
System temperature	40–50 K
Sensitivity on sky	5 mK s$^{-1/2}$

REFERENCES

Bond, J. R., and Efstathiou, G. 1987, *M. N. R. A. S.*, **226**, 655.

Wilson, M. L., and Silk, J. 1981, *Ap. J.*, **243**,14.

Pauliny-Toth, I. I. K., Witzel, A., Preuss, E., Baldwin, J. E., and Hills, R. E. 1977, *Astron. Astrophys. Suppl.*, **34**, 253.

MICROWAVE BACKGROUND RADIATION OBSERVATIONS AT THE OWENS VALLEY RADIO OBSERVATORY: ANALYSIS AND RESULTS

C. R. Lawrence, A. C. S. Readhead, and S. T. Myers
Owens Valley Radio Observatory 105-24
California Institute of Technology
Pasadena, CA 91125
USA

1. INTRODUCTION

The admonition "If you don't know where you're going you're liable to wind up somewhere else" is appropriate at the beginning of a discussion on MBR data analysis, since the best procedure to use depends on whether we want to: 1) compare our results with previous observations; 2) compare them with generic models; or 3) test individual models.

For comparison with previous observations, we might determine an upper limit to $\theta_{\rm sky}^2$, the variance of the observed quantity

$$T = T_{\rm MB} - \frac{1}{2}(T_{\rm R1} + T_{\rm R2})$$

(see Figure 1 of Readhead *et al.*, this volume, hereafter Paper I), and then convert this to a true sky variance under two assumptions used often enough in the past to be considered standard. These are, first, that the telescope beams are Gaussian. Figure 1 of Paper I shows that this is not a particularly good assumption, but neither is it particularly bad. Second, the MBR consists of uncorrelated, uniform patches the size of the telescope beam, with a Gaussian fluctuation spectrum. That is, the correlation function C of the fluctuations is approximately zero for angles ϕ greater than the beam separation, and constant for angles less than the beam separation. With these assumptions, it follows from the expression for T above that $\theta_{\rm sky}^2 = 3/2\, C(0)$. The factor of 3/2 reflects the difference between a triple-beam experiment, such as ours, and a double-beam experiment.

For comparison with a generic model, most of the procedure would stay the same, but we would make a more realistic assumption about the sky. For example, we might assume that the correlation function of fluctuations is

$$C(\phi) = A^2 \exp\left\{-\frac{1}{2}\left(\frac{\phi}{\phi_c}\right)^2\right\}, \tag{1}$$

where ϕ_c is the *correlation length*. It can then be shown that

$$\theta_{\rm sky}^2 = \frac{3}{2}C(0) - 2C(\phi = \text{beamsep}) + \frac{1}{2}C(\phi = 2 \times \text{beamsep}) + \text{beamwidth correction}. \tag{2}$$

Finally, for testing individual models, we should use the full information obtained in a beam map plus whatever fluctuation spectrum is given by the model in question.

N. Kaiser and A. N. Lasenby (eds.), The Post-Recombination Universe, 173–181.

We will not discuss individual models here, but will concentrate on the first two procedures described above. In both, the first step is to determine an upper limit to θ_{sky}^2. This might seem to be a straightforward statistical problem, but there is no universally accepted method for doing it. Nevertheless, so long as we realize the dangers involved in deriving MBR limits, we can survive the statistical swamp into which we must now descend, and nudge science forward a bit.

In §2 and §3 we discuss two ways of deriving limits on θ_{sky}^2. Both are well-known in the statistical literature, and one has been used widely for MBR observations in the past.

2. LIKELIHOOD

As before, let θ_{sky}^2 be the variance of the (assumed Gaussian) distribution of fluctuations for the beam pattern of Figure 1 of Paper I. Since we have a beam switching experiment, we can take the mean to be zero.

The probability density for a single measurement $T \pm \sigma$, under the stated assumptions, is given by

$$p(T, \sigma) = [2\pi(\sigma^2 + \theta_{sky}^2)]^{-1/2} \exp[-T^2/2(\sigma^2 + \theta_{sky}^2)].$$

We can write the joint density for the seven observed fields (excluding NCP 7) as

$$L(\{T_i\}|\theta_{sky}) \equiv \prod_{i=1}^{7} p(T_i, \sigma_i) = \prod_{i=1}^{7} [2\pi(\sigma_i^2 + \theta_{sky}^2)]^{-1/2} \exp\left\{\frac{-T_i^2}{2(\sigma_i^2 + \theta_{sky}^2)}\right\}, \qquad (3)$$

where $L(\{T_i\}|\theta_{sky})$, called the *likelihood function*, can be interpreted as the relative probability of the set of seven measurements as a function of the assumed sky variance. This function, normalized to its peak value, is shown in Figure 1. Note that by plotting only $\theta_{sky}^2 \geq 0$, we specifically exclude mathematically allowable but physically nonsensical imaginary values of θ_{sky}.

Right away, we see that there is no evidence for a detection of MBR fluctuations. While the observations are "most likely" if $\theta_{sky} = 13$ μK, they are only slightly less likely for $\theta_{sky} = 0$. The problem now is to determine an upper limit. Even restricting consideration to just the likelihood function, this is still a task for which there is no obviously correct method. In fact, tossing Figure 1 into a roomful of statisticians and asking for an upper limit might produce the mathematical equivalent to Eris, Goddess of Discord, throwing an apple marked "for the most beautiful" into the wedding feast of Peleus and Thetis! One possibility is simply to choose θ_{sky} so that $L(\theta_{sky}) = e^{-2}$, 0.10, 0.05, or whatever. From Figure 1, these relative likelihood values occur at 54, 58, and 68 μK, respectively.

Another possibility is the Bayesian approach (see, e.g., Berger 1985). Bayes' formula gives the probability density of θ_{sky} given the observations T_i as

$$p(\theta_{sky}|T_i) \propto L(\{T_i\}|\theta_{sky}) \, p(\theta_{sky}),$$

where $L(\{T_i\}|\theta_{sky})$ is the likelihood function, and $p(\theta_{sky})$ and $p(\theta_{sky}|T_i)$ are known, respectively, as the *prior* and *posterior densities*. $p(\theta_{sky})$ represents our (possibly nonexistent) knowledge of θ_{sky} before the observations. Many arguments have been advanced why ignorance must be represented by one prior or another. For now, let us take the simplest prior we can imagine consistent with the one fact we *do* know (i.e., $\theta_{sky}^2 \geq 0$), namely $p(\theta_{sky}) = c$ for $\theta_{sky} \geq 0$, and $p(\theta_{sky}) = 0$ for $\theta_{sky} < 0$. Then the posterior density, i.e., the density of θ_{sky} determined from what we knew before the observations plus the observations themselves,

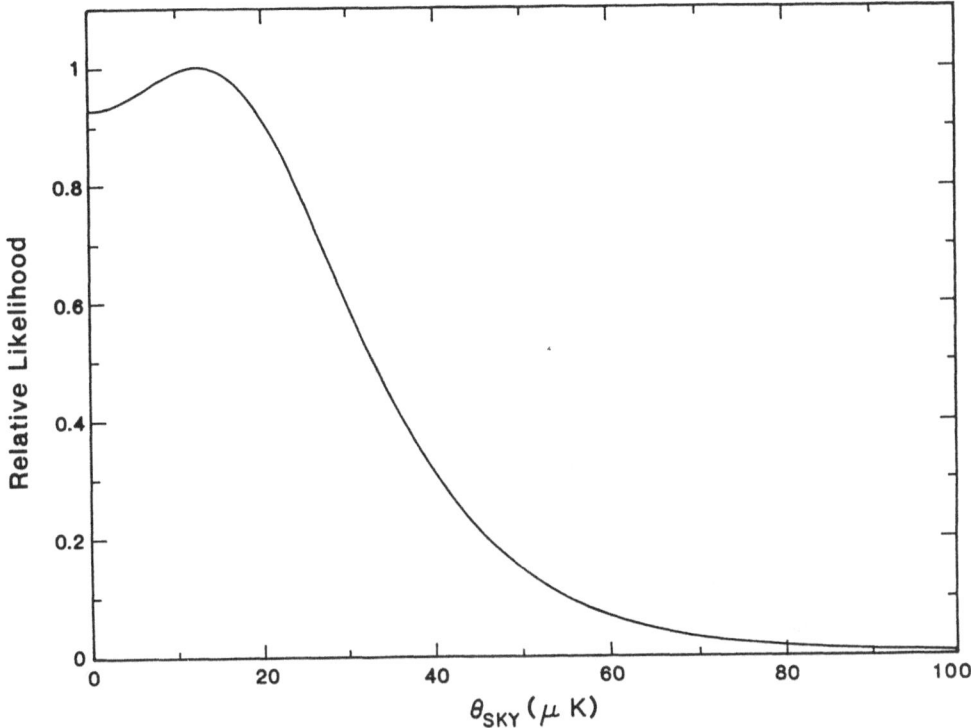

FIGURE 1. Likelihood function of Eq. 1 for the OVRO measurements of seven fields near the NCP.

will look just like Figure 1 except for normalization, and the area under the curve from 0 to θ^* gives the probability that $0 \leq \theta_{sky} \leq \theta^*$. Specifically, $P(\theta_{sky} < 51\ \mu K) = 0.95$. Thus 51 μK could be taken as a "95%" upper limit to θ_{sky}.

The born-again Bayesian will object that from consistency arguments the prior must be uniform in θ^{-1} rather than θ. We will not take up that argument here, except to say that the arbitrary cutoff that must be imposed to make the posterior distribution normalizable is unappealing, at best. It is clear from Figure 1, however, that the upper limit to θ_{sky} derived with a θ^{-1} prior will be *lower* than 51 μK, and dependent on the value of the cutoff chosen. We might say, then, that 51 μK is a conservative Bayesian limit from the OVRO observations.

3. LIKELIHOOD RATIO TESTS

3.1. Background

We turn now to a method from what is sometimes called the classical theory of hypothesis testing. Even though "likelihood" is part of the name of the method, and, indeed, the joint density from Eq. 3 will be required, this test seems quite different. One reason for this is that talking about LRTs requires many new words (see, e.g., Lehmann 1985, for a general discussion of likelihood ratio tests and definitions of the statistical terms used here). Likelihood ratio tests have been used for microwave background analysis by many authors, e.g., Boynton and Partridge 1973, Lasenby and Davies 1983, and Uson and Wilkinson 1984.

To begin, the problem is to choose between an hypothesis H and an alternative hypothesis K. Given a set of measurements whose distribution is known in terms of a parameter θ, we must find θ. (Assume that if we know θ, we know whether H is true.)

We need a "test," i.e., a rule that says accept H if such and such, and reject H if such and such. But first, some definitions. The *level of significance* α, is the probability of rejecting H if H is true, or the probability of a Type I error. The *power* β is the probability of rejecting H if K is true, so $1 - \beta$ is the probability of a Type II error. This is worth writing so it stands out:

$$\alpha \equiv P(\text{reject } H|H) \qquad \alpha = P(\text{Type I error})$$
$$\beta \equiv P(\text{reject } H|K) \qquad 1 - \beta = P(\text{Type II error})$$

Clearly, we want α small (ideally 0), and β large (ideally 1). But, as Mick Jagger said, you can't always get what you want, and since α and β are not independent, this is one of those times. The standard solution is to choose α small, and maximise β. The essential concept of how to do this follows.

Imagine that we have some multidimensional measurement space, with a subspace S called the *critical region*. Our test, or rule, will be: Reject H if our measurement $x \in S$, subject to the conditions $P(x \in S|H) = \alpha$ and $P(x \in S|K) = \text{maximum}$. Therefore S must consist of those points in the measurement space with the smallest values of

$$\lambda \equiv \frac{P(x|H)}{P(x|K)}.$$

In other words, we "grow" the critical region by starting with the point with the smallest value of λ, adding the next smallest, and the next, until the probability of measuring $x \in S$ when H is true is α. Then we stop. Remember that we are assuming that we know the distribution of x.

The above prescription sounds plausible. The details, for testing the simple hypothesis $H : \theta = \theta_0$ against the simple alternative $K : \theta = \theta_1$, are given by the Neyman-Pearson lemma. Specifically, it shows that the test: reject H if

$$\lambda^* \equiv \frac{P(x|H)}{P(x|K)} \leq k^*,$$

or accept H if $\lambda^* > k^*$, where k^* is given implicitly by the requirement that $P(\lambda^* \leq k^*|H) = \alpha$, is a *most powerful test at level* α.

In general, k depends on θ_1. If it doesn't, then the test is *uniformly most powerful*, and can decide between, e.g., $H : \theta \geq \theta_0$ and $K : \theta < \theta_0$. (Note that the test is the same as before, but the distributions satisfy additional requirements, and stronger theorems can be proved. We have titled this section likelihood ratio tests rather than Neyman-Pearson tests in order to emphasize this distinction.)

3.2. Results

We have seven measurements of $T_i \pm \sigma_i$. Then assuming Gaussian sky fluctuations with a triple-beam variance of θ_{sky}^2,

$$\frac{P(\text{meas}|H : \theta_{\text{sky}} = \theta_0)}{P(\text{meas}|K : \theta_{\text{sky}} = \theta_1)} = \frac{\prod_{i=1}^{7} p(T_i|\sigma_i, \theta_{\text{sky}} = \theta_0)}{\prod_{i=1}^{7} p(T_i|\sigma_i, \theta_{\text{sky}} = \theta_1)}$$

$$= \prod_{i=1}^{7} \left(\frac{\sigma_i^2 + \theta_1^2}{\sigma_i^2 + \theta_0^2} \right)^{1/2} \exp\left\{ \frac{T_i^2}{2} \left[\frac{\theta_0^2 - \theta_1^2}{(\sigma_i^2 + \theta_0^2)(\sigma_i^2 + \theta_1^2)} \right] \right\}.$$

This is rather messy. Fortunately, all essential information about θ is contained in

$$\lambda = \sum_i \frac{x_i^2}{(\sigma_i^2 + \theta_0^2)(\sigma_i^2 + \theta_1^2)},$$

which is *sufficient for* θ (for $\theta_0 > \theta_1$, a sign change is required). So we can use this simpler λ instead of the messy original, and all theorems apply.

Finally, we're ready to put an upper limit on θ_{sky}. Here's how it works.

1. Choose $H : \theta_{\text{sky}} = \theta_0$ and $K : \theta_{\text{sky}} = \theta_1$.
2. Choose a level of significance α.
3. Calculate the distribution of

$$\lambda = \sum_i \frac{x_i^2}{(\sigma_i^2 + \theta_0^2)(\sigma_i^2 + \theta_1^2)},$$

where x_i is Gaussian-distributed with variance $\sigma_i^2 + \theta_0^2$, and then determine the value of θ_0 for which the fraction α of the distribution is less than $\lambda_{\text{measured}} \equiv \lambda(x_i = T_i)$.

4. Show that for a given α, θ_0 is independent of θ_1, so that the test is UMP, and can be used for composite hypotheses.

5. Find the distribution of

$$\lambda = \sum_i \frac{x_i^2}{(\sigma_i^2 + \theta_0^2)(\sigma_i^2 + \theta_1^2)},$$

where, this time, x_i is Gaussian-distributed with variance $\sigma_i^2 + \theta_1^2$. Then the power $\beta(\theta_1)$ of the test against θ_1 is the fraction of the distribution below $\lambda_{\text{measured}}$.

Then we reject H in favor of K for all $\theta_1 < \theta_0$, and have the result that we need: $\theta_{\text{sky}} \leq \theta_0$ at level α, with power β against θ_1.

The distributions in steps 3 and 5 were found by Monte Carlo simulations, and are shown in Figure 2. The vertical line marks $\lambda_{\text{measured}}$. It is easy to see how the test works from the figure. If θ_{sky} were really 46.95 μK, then in measurements of different patches of sky we would expect to get λ smaller than we actually did only 5% of the time. But if θ_{sky} were really 0, then we would expect to get λ smaller than we actually did 72% of the time.

We say, then, that our upper limit on θ_{sky} from the likelihood ratio test is 47 μK (95% confidence, i.e., $\alpha = 0.05$), with power against $\theta_1 = 0$ of 0.72. The power decreases monotonically to α as $\theta_1 \to 47$ μK.

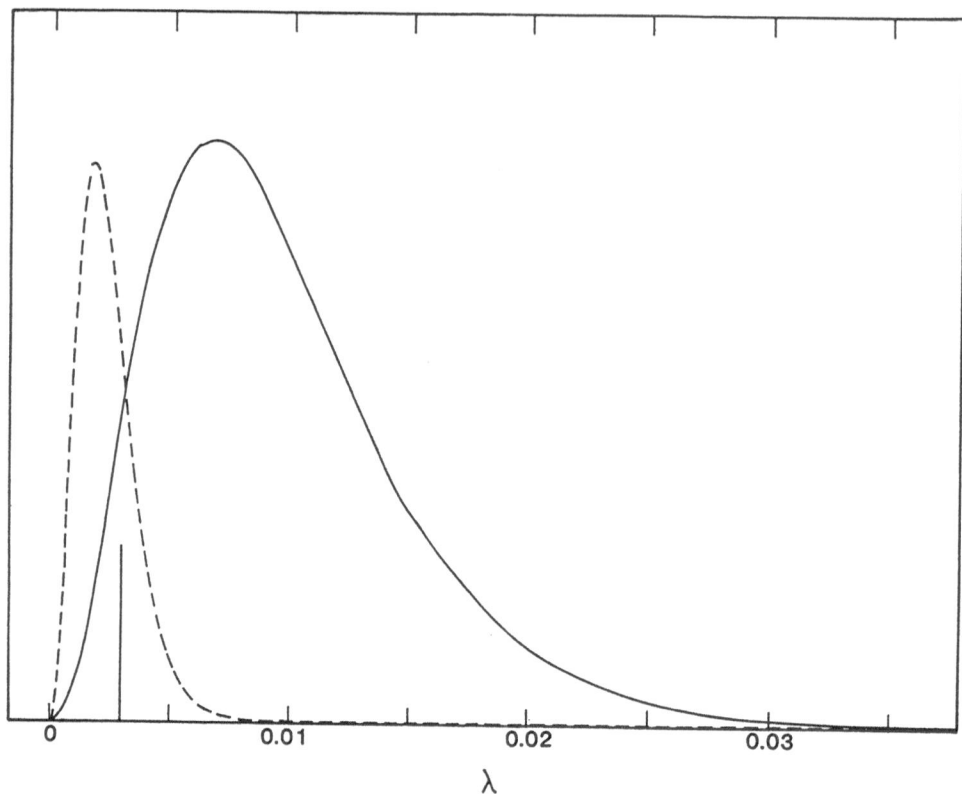

FIGURE 2. Distribution of $\lambda = \sum_i x_i^2/(\sigma_i^2 + \theta_0^2)(\sigma_i^2 + \theta_1^2)$ with $\theta_0 = 46.95, \theta_1 = 0$ from a Monte Carlo simulation. Each x_i was assumed to be Gaussian-distributed with variance $\sigma_i^2 + \theta_0^2$ (solid line) or $\sigma_i^2 + \theta_1^2$ (dashed line). In each case 300,000 sets of x_i were generated. The curves are independently normalized. $\lambda_{\text{measured}}$ is marked with a vertical line. The area under the solid curve to the left of $\lambda_{\text{measured}}$ is $\alpha = 0.05$, while the area to the left under the dashed curve is $\beta(\theta_1 = 0) = 0.72$.

There is a problem with LR tests that everyone who uses them should keep in mind. The theorems guarantee a "most powerful test of level α." But that's a little like saying that a certain pub has the "best food in town." It may be true, but doesn't guarantee a good meal! A test with a low significance level has a low probability of rejecting a true hypothesis; if the power is low, the test also has a low probability of rejecting a *false* hypothesis.

There is another problem with LR tests that is of particular importance in applications like ours where measurement errors contribute to the variance of the experimental quantity. In step 3, we calculate the distribution of λ under the assumption that $\text{var}(x_i) = \sigma_i^2 + \theta_0^2$, and change the distribution by changing the value of θ_0 until $\lambda_{\text{measured}}$ is at the α point of the distribution. Suppose now that the measurements are much closer to zero than expected for the size of the measurement errors (i.e., $\chi_\nu^2 \equiv$ reduced $\chi^2 \ll 1$), so close, in fact, that even for $\theta_{\text{sky}} = 0$ the value of $\lambda_{\text{measured}}$ is below the α point of the distribution. Then no test at level α will exist unless we allow $\theta_0^2 < 0$! This is not nice behavior, but there is no way of preventing it, because the data are unlikely *given the measurement errors alone*. We can hardly insist that the MBR make up for our bad luck. Fortunately, this situation is easy to recognize. The power of the LR test will be low, for when the measured values become small compared with the errors, the distributions found in steps 3 and 5 above move closer and closer to each other. This is illustrated in Figure 3, which is the same as Figure 2 but using a fake data set with low χ_ν^2.

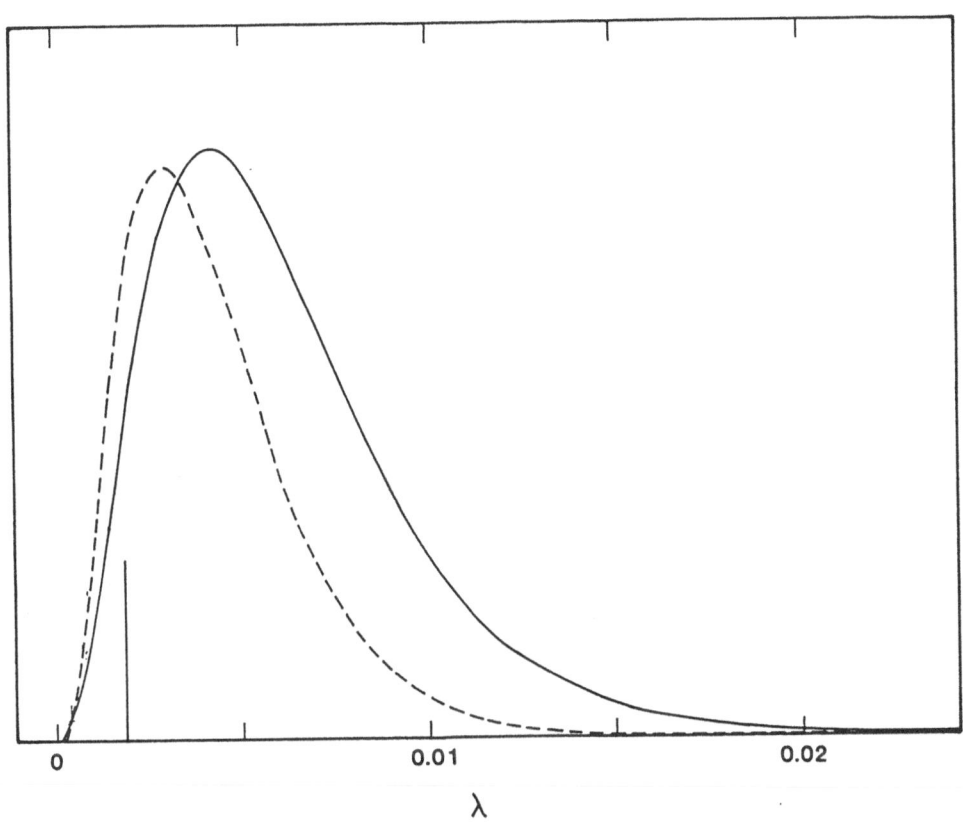

FIGURE 3. Illustration of the behavior of a LRT on data manufactured to have $\chi_\nu^2 \ll 1$. Each x_i was assumed to be Gaussian-distributed with variance $\sigma_i^2 + 21.59^2$ (solid line) or $\sigma_i^2 + 0^2$ (dashed line). In each case 300,000 sets of x_i were generated. The curves are independently normalized. The value of λ from the measured T_i is marked with a vertical line. The area under the solid curve to the left is $\alpha = 0.05$, while the area to the left under the dashed curve is only $\beta(\theta_1 = 0) = 0.12$. Such a test has so little power that it could barely make it up hills in the microwave background, and the limit of 22 μK that it produces isn't worth much. For comparison, the likelihood method with a uniform prior gives 44 μK on these data.

Since the troubles encountered with LR tests are revealed by the power, we simply say never trust a LR test without knowing the power.

4. THE MBR

The hard work is now done. From likelihood we have $\theta_{\text{sky}} < 51$ μK (95%), and from LRT we have $\theta_{\text{sky}} < 47$ μK (95% confidence, $\beta = 0.72$), all for a beam separation of $7\overset{.}{!}15$ and a beam width (FWHP) of $1\overset{.}{!}8$. The fact that both methods give about the same limit (when the power of the LRT is high) is reassuring, and lets us ignore questions about the "right" method in good conscience.

For comparison with previous observations, all that is left is to calculate

$$\frac{\Delta T}{T} = \frac{C(0)^{-1/2}}{T} = \frac{\sqrt{2/3}\,\theta_{\text{sky}}}{2.78} \le \begin{array}{ll} 1.5 \times 10^{-5} & \text{likelihood, 95\%} \\ 1.4 \times 10^{-4} & \text{LRT, 95\% conf, } \beta = 0.72. \end{array}$$

For comparison with the generic model of Eq. 1, we must calculate $C(0)^{-1/2}$ from Eq. 2 as a function of ϕ_c (see Davies *et al.* 1987 for a discussion of the beam correction). The result for $\theta_{\text{sky}} = 51$ μK is shown in Figure 4. Sensitivity is low at small ϕ_c because many fluctuations are averaged in a single beam, and at large ϕ_c because the three beams are too close together to reach from peaks to valleys. The minimum $\Delta T/T$ of 1.6×10^{-5} is achieved for $\phi_c \approx 2\overset{.}{!}5$, which gives the closest approximation to the uncorrelated-patches-the-size-of-the-beam assumption used above. Figure 4 is a more general, and therefore much better, summary of the OVRO results than a single number.

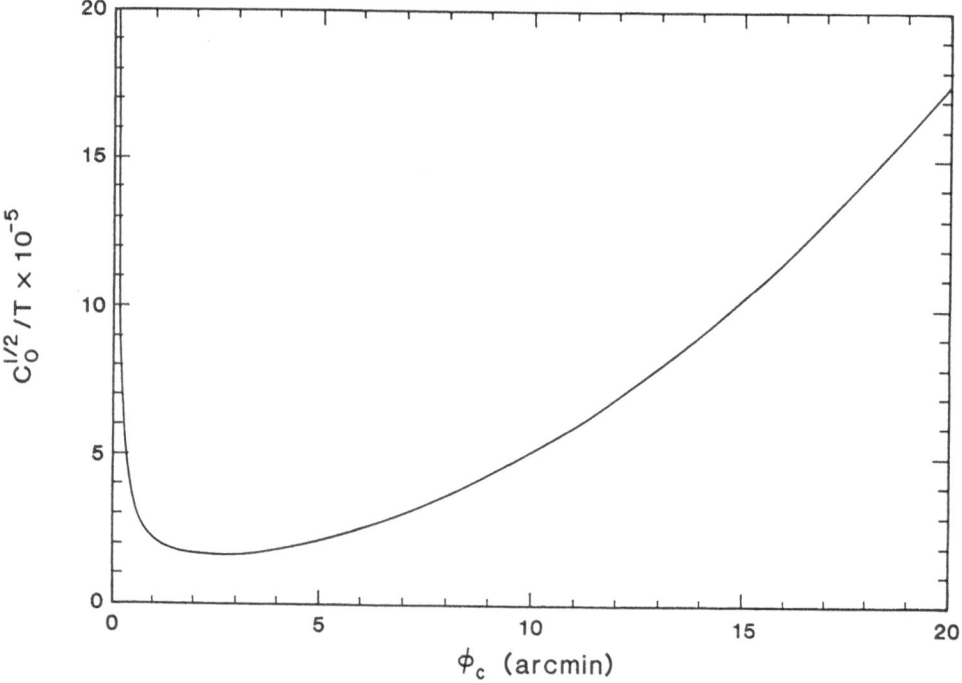

FIGURE 4.   Limits on $\Delta T/T$ from the OVRO observations, calculated from Eq. 2 for $\theta_{\text{sky}} = 51$ μK.

5. WARNING

Statistical results are not absolute truths. We have given our results at the 95% level, meaning that (under the assumption of Gaussian statistics) there is a 5% chance that the MBR fluctuations are larger than we say. This is customary for MBR work, but 5% corresponds to about 1.6σ, well short of the usual 3 or 5 or 10σ standard for believable data.

Acknowledgements: We would like to thank the organizers of this conference and many of the participants for discussions here and elsewhere who have contributed to our statistical education. MBR research at OVRO is supported by the IBM Research Fund, the Irvine Foundation, and NSF grants AST 85-09822 and AST 86-10693.

REFERENCES

Berger, James O. 1985, *Statistical Decision Theory and Bayesian Analysis* (New York: Springer-Verlag).
Boynton, P. E., and Partridge, R. B. 1973, *Ap. J.*, **181**, 243.
Davies, R. D., Lasenby, A. N., Watson, R. A., Daintree, E. J., Hopkins, J., Beckman, J., Sanchez-Almeida, J., and Rebolo, R. 1987, *Nature*, **326**, 462.
Lasenby, A. N., and Davies, R. D. 1983, *M. N. R. A. S.*, **203**, 1137.
Lehmann, E. L. 1985, *Testing Statistical Hypotheses* (New York: Wiley).
Readhead, A. C. S., Lawrence, C. R., and Myers, S. T., this volume (Paper I).
Uson, J. M., and Wilkinson, D. T. 1984, *Nature*, **312**, 427.

ON MEASUREMENTS OF THE ZELDOVICH-SUNYAEV EFFECT

Yoel Rephaeli
School of Physics and Astronomy
Tel Aviv University
Tel Aviv 69978
Israel

ABSTRACT. The full extent of beam dilution in measurements of the Zeldovich-Sunyaev (Z-S) effect can be appreciated when it is realized that the parameters characterizing intracluster gas density and temperature profiles assume values within a wide range. We discuss the possible role of beam dilution in previous measurements of the effect. We also estimate the contribution of hypothetical intra-supercluster hot gas to the Z-S signal along a line of sight to a cluster which is a memeber of a supercluster.

1. INTRODUCTION

Thomson scattering of the Cosmic Microwave Background (CMB) photons by hot electrons (Zeldovich and Sunyaev 1969) imprints a unique signature on the CMB spectrum. This signature is of considerable cosmological interest. The analogous scattering of the photons by hot electrons in the intracluster (IC) space of rich clusters of galaxies (Sunyaev and Zeldovich 1972) is extremely interesting from cosmological and astrophysical points of view (Gould and Rephaeli 1978, Fabbri, Melchiorri and Natale 1978, Danese and De Zotti 1980, Rephaeli 1980, White and Silk 1980, Sunyaev and Zeldovich 1980a, 1980b, Rephaeli 1981). Numerous attempts have been made to measure the effect, and there are by now few claimed detections (see references below and in the review by Sarazin 1986).

Very sensitive measurements are required for the detection of the Z-S effect. Here we discuss the significance of a suitable choice of the telescope beam size in order to avoid drastic dilution of the effect. Another essential requirement for detection of the effect is that the intensity along the reference lines of sight be (only) negligibly affected by, e.g., hot gas in the neighborhood of the cluster. We estimate the level at which hot intra-supercluster (ISC) gas can interfere with measurements of the effect towards a cluster which is a member of a supercluster (SC).

2. BEAM DILUTION OF THE Z-S EFFECT

The spatial dependence of the Z-S effect is contained in the Thomsonization parameter

$$y = \int (kT/mc^2) \, n \, \sigma_T \, dl \,, \tag{1}$$

N. Kaiser and A. N. Lasenby (eds.), The Post-Recombination Universe, 183–185.

where n and T are the electron density and temperature, σ_T is the Thomson cross section, and the integral is over a line of sight through the cluster. The gas density, temperature and their radial profiles are deduced from X-ray spectral and spatial measurements. The gas is usually modeled as a polytrope $T \propto n^{\gamma-1}$ (Gull and Northover 1975, Cavaliere and Fusco-Femiano 1976), and the density profile is commonly taken to be of the form $n = n_0(1+r^2/a^2)^{-\alpha}$, where γ, α and the core radius, a, are best fits to the X-ray data (e.g., Henriksen and Mushotzky 1986). In rich clusters the values of these parameters usually fall in the ranges a = 0.2 - 0.5 Mpc, $\gamma = 1 - 5/3$, and $\alpha = 1 - 1.5$ (e.g., Sarazin 1986). Since n(r) and T(r) are monotonically decreasing functions of (the radial coordinate) r, y is maximal for a line of sight through the center of the cluster. The effective value of y "seen" by a beam of a given size is obtained by integrating the product n(r) T(r), convolved with the beam profile, along the line of sight. Usually, measurements are made with telescopes whose beams have Gaussian shapes, for which the integration was carried out (Rephaeli 1987) for various values of θ_c, the angular core radius, $\alpha \gamma$ (dependence is only on this combination of α and γ) and the beam HWHM, θ_b. From the entries in Table 1 of Rephaeli (1987), we see that y can be up to 25 times lower than its maximum value for the above range of values of $\alpha \gamma$, if the Z-S effect is measured with a beam having $\theta_b/\theta_c = 4$.

The clusters along whose lines of sights significant intensity decrement was claimed to have been measured are A665, A2218 and 0016+16 (Birkinshaw and Gull 1984, Birkinshaw, Gull and Hardebeck 1984). These measurements were made with the OVRO 40 m telescope with $\theta_b \simeq 1.65$ and .9 arcmin at 10.7 and 20.3 GHz, respectively. A665 and A2218 are moderately distant clusters with redshifts of 0.18 and 0.174, respectively, while 0016+16 has z = 0.541 (Birkinshaw and Gull 1984). Taking a typical core radius a = 0.4 Mpc, it follows that for the former two clusters $\theta_b/\theta_c \leq 1.3$. However, in the measurements of 0016+16 this ratio assumes the high values 2.1 and 3.9, respectively. From Table 1 of Rephaeli (1987), we see that beam dilution may have been very significant in these measurements. Lasenby and Davies (1983) also attempted measuring the intensity decrement towards A576 and A2218. But the beam had $\theta_b = 4.5$ arcmin; for A2218, this implies the high value of about 3.5 for the above ratio. Again, beam dilution may have seriously affected detectability of the signal in this cluster.

While the limitation of finite beam size has been recognized all along, its potentially drastic implication becomes clear only when one takes the full observationally-deduced range of the relevant spatial cluster parameters.

3. THE Z-S EFFECT DUE TO INTRA-SUPERCLUSTER GAS

Clustering on the very largest scales is of basic importance in theories for the evolution of structure in the Universe. Such issues as the the behaviour of the mass-to-light ratio on SC scales, the baryonic content and the velocity field in SC are of very much cosmological interest. Of particular interest to us here is the amount of hot gas in the ISC space. The gas may be detectable through its X-ray emisssion. We (Persic, Rephaeli and Boldt 1987) have recently searched the HEAO-1 A-2 data base for emission from the Bahcall and Soneira (1984) complete (up to a redshift z = 0.08) sample of SC. No significant SC 2 - 10 keV emission was found, and our 3σ upper limit on the flux of a SC in the sample translates to a bound on the SC mass

$$M \leq 1 \ 10^{16} \ (R \ /10 \ \text{Mpc})^{3/2} \ (f \ /0.1)^{-1} \ (T_e \ /10^8 \ K)^{-1/4} \ h_{50}^{-1} \ M_\theta. \qquad (2)$$

In this expression, R, and f are the SC radius and baryonic mass fraction, respectively, T_e is the ISC gas temperature, and h_{50} is the Hubble constant in units of 50 km s⁻¹ Mpc⁻¹. Note that the temperature dependence is as shown only for $T_e = O(10^8$ K).

The SC Thomsonization parameter can be easily estimated for a uniform distribution of the ISC gas. It is

$$y \simeq 6 \ 10^{-6} \ (M/10^{16} \ M_\theta) \ (R/10 \ Mpc)^{-2} \ (f/0.1) \ (T_e \ /10^8 \ K) \ . \qquad (3)$$

so that in the Rayleigh-Jeans side of the spectrum the relative (CMB) temperature change, 2y, is $O(10^{-5})$ if the quantities in equation (3) assume the values to which they are scaled. For a comparison, in the Coma cluster $y \simeq 1 \ 10^{-4}$ (Rephaeli 1987). Thus, the Z-S effect is weaker in a SC than in a gas-rich cluster, unless M/R^2 is larger than its scaled value in equation (3). As this is a possibility, the additional distortion caused by ISC gas may be detectable in the richer and more compact SC. In such SC, the measurement of the Z-S effect towards a member cluster may be affected by ISC gas. In particular, this will be the case if the cluster position in the SC is such that the reference lines of sight are mainly in direction of the SC center, while the line of sight through the center of the cluster is away from that direction.

REFERENCES

Bahcall, N.A., and Soneira, R.M. 1984, Astrophys. J. (Lett.), 300, L35.
Birkinshaw, M., and Gull, S.F. 1984, Mon. Not. R. Astron. Soc., 206, 359.
Birkinshaw, M., Gull, S.F., and Hardebeck, H. 1984, Nature,309, 34.
Cavaliere, A., and Fusco-Femiano, R. 1976, Astron. Astrophys., 49, 137.
Danese, L., and De Zotti, G. 1980, Astron. Astrophys. 84, 364.
Fabbri, R., Melchiorri, F., and Natale, V. 1978, Astrophys. Sp. Sci., 59 223.
Gould, R.J., and Rephaeli, Y. 1978, Astrophys. J., 219, 12.
Gull, S.F., and Northover, K.J.E. 1975, Mon. Not. R. Astron. Soc., 173, 585.
Henriksen, M.J., and Mushotzky, R.F. 1986, Astrophys. J., 302, 287.
Lasenby, A.N., and Davies, R.D. 1983, Mon. Not. R. Astron. Soc., 203, 1137.
Persic, M., Rephaeli, Y., and Boldt, E. 1987, preprint.
Rephaeli, Y. 1980, Astrophys. J., 241, 858.
Rephaeli, Y. 1981, Astrophys. J., 245, 351.
Rephaeli, Y. 1987, Mon. Not. R. Astron. Soc., in press.
Sarazin, C. 1986, Rev. Mod. Phys., 58, 1.
Sunyaev, R.A., and Zeldovich, Y.B. 1972, Comm. Astrophys. Sp. Phys., 4, 173.
Sunyaev, R.A., and Zeldovich, Y.B. 1980a, Mon. Not. R. Astron. Soc., 190, 413.
Sunyaev, R.A., and Zeldovich, Y.B. 1980b, Ann. Rev. Astron. Astrophys., 18, 537.
White, S.D.M., and Silk, J. 1980, Astrophys. J., 241, 864.
Zeldovich, Y.B., and Sunyaev, R.A. 1969, Astrophys. Sp. Sci., 4, 301.

PROSPECTS FOR SUNYAEV-ZELDOVICH ASTRONOMY

Richard Saunders
Cavendish Laboratory
Madingley Road
Cambridge CB3 OHE
England

ABSTRACT. Observations of the Sunyaev-Zeldovich decrement towards galaxy clusters are reviewed. The redshift-independent nature of the decrement is emphasised and the possibility of the direct probing of cluster evolution via S-Z observations is considered. It is argued that there are good prospects for such observations with - and only with - the Cambridge enhanced 5-km telescope, whose salient features are described.

1. INTRODUCTION

There is a substantial body of theoretical work on the evolution of galaxy clusters. I'd like to stress that there is a good prospect of probing the evolution of galaxy clusters by *direct observation* via the Sunyaev-Zeldovich effect. First I'll briefly review the present status of S-Z astronomy, then point out why the effect is especially important for high-redshift studies, and finally sketch out the performance expected of the Cambridge enhanced 5-km telescope.

2. THE PRESENT STATUS OF S-Z ASTRONOMY

The major part of the successful S-Z observations to date is represented by Fig. 1, which has been kindly provided by Mark Birkinshaw. The figure shows the results of five winters' observing runs for the three clusters in which the S-Z effect has been positively detected. The observations were made with the Owen's Valley 40-m single-dish telescope at 20 GHz by a team including Birkinshaw, Gull, Hardebeck, Moffet & Myers. In the figure, ΔT is shown as a function of angular distance away from the supposed cluster centre along a north-south line; the error bars are $\pm 1\sigma$; the error boxes represent estimates of the maximum systematic errors present, the estimates being partly based on observations of reference fields. See Birkinshaw et al (1984) for more details of observing method and analysis.

187

N. Kaiser and A. N. Lasenby (eds.), The Post-Recombination Universe, 187–191.

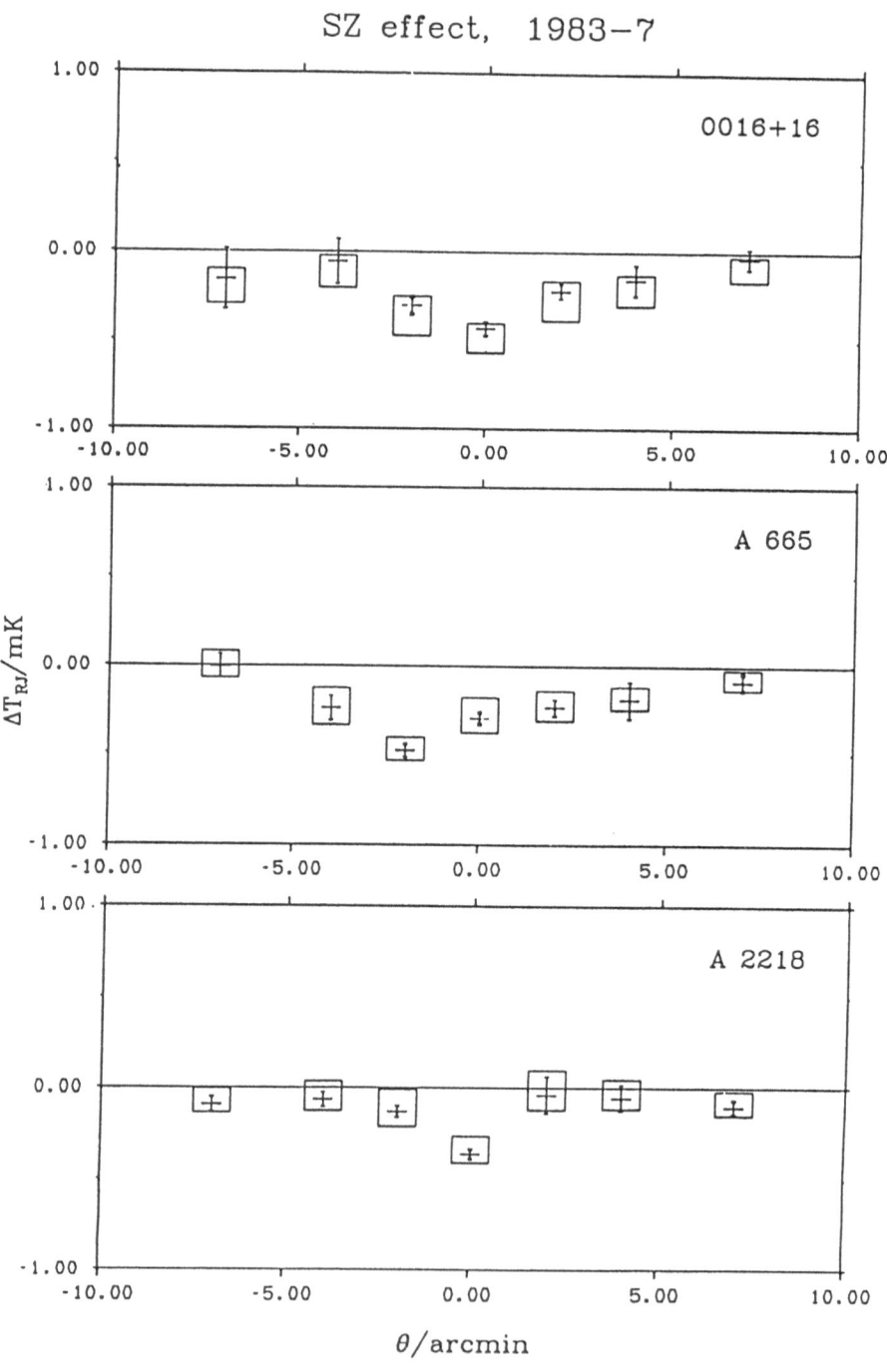

Figure 1

Two particular points to note about these results are that the offset centre in A665 has been confirmed by independent reanalysis of the X-ray data, and that the large error associated with the +2' point in A2218 is due to the uncertainties involved in subtracting a contaminating radiosource using VLA data extrapolated in frequency. But the key results here are:

hard evidence of the S-Z effect (complemented by very good agreement with Uson's more recent data (NRAO 140-foot single-dish - scanning technique different from OVRO) at the centres of A665 and 0016+16);

clear profiles of the spatially extended S-Z decrement, as expected from the observed extended X-ray emission.

The only problem with these excellent results is that, with a single dish, they took five years to get.

3. THE S-Z EFFECT AT HIGH REDSHIFT AND THE CONTINUOUS S-Z BACKGROUND

The magnitude ΔT of the S-Z decrement measured in the direction of a cluster depends on the cluster gas temperature, density, line-of-sight depth and physical constants - but it does not depend on the redshift of the cluster. This is made very clear by Sunyaev (1978). Yet the significance of this redshift independence for observational cosmology does not seem to be widely appreciated and I think it well worth emphasising. It means that a cluster at very high redshift (z) is as easy to spot as a similar cluster at low z if one uses the S-Z method, but certainly not if one uses X-ray, optical or radio synchrotron methods.

This leads to the possibility of directly probing the evolution of clusters. I illustrate this as follows. There are about 500 Abell clusters per steradian with 0<z<0.2. The average angular distance d separating a pair of neighbouring clusters is about 160 arcmin. Tabulated next are the values of d (for Ω=0 and 1) if the local population of Abell-type clusters continues at constant comoving number density to a redshift z, above which there are no clusters.

z	d (arcmin)	
	Ω=0	Ω=1
0.2	160	160
1	20	26
3	6	12
6	3	7

The point is that only a relatively small amount of sky has to be mapped in S-Z - and with only the same sensitivity as is needed to detect a nearby cluster - before we find out whether our ideas about cluster evolution are at all correct. (Shaun Cole has carried out sophisticated calculations on a similar theme, as reported elsewhere in these proceedings. Employing a Press-Schechter cluster evolution function, rather than my extremely crude assumption of no evolution

until a cutoff, he in fact predicts results similar to mine for z=3.)
    As S-Z decrements get closer together on the sky, the effect of
the angular size-redshift relation is to create an overlapping of
neighbouring S-Z profiles, whatever the redshifts of the clusters. A
continuous S-Z background would then be formed. Statistics on the
angular and temperature variations in the background would still
however place strong constraints on cluster evolution, particularly as
the variation would be very non-Gaussian given decrement profiles in
individual clusters that are deeper and narrower than those shown in
the low resolution data of Fig. 1.

4. THE CAMBRIDGE ENHANCED 5-KM TELESCOPE

The birth of microwave background astronomy has been slow and painful
- in spite of outstanding prospects - because there were no telescopes
designed to make the requisite sensitive observations sufficiently
free from systematic error. Single-dish telescopes are ultimately
limited by differential spillover of the two feeds used in
beam-switching. Interferometers on the other hand have several
advantages over single dishes for microwave background work (see
Saunders, 1986):

    they are not limited by differential spillover;

    all structures inside the primary beam are simultaneously
    accessible leading the the efficient production of a *map*;

    they reduce the tropospheric signal by resolving much of it out;

    they can provide a high resolution map at the same frequency to
    allow removal of contaminating radiosources.

    The difficulty of using the VLA for this work is that its large
dishes preclude the possibility of providing sensitivity to the arcmin
plus angular sizes important in microwave background astronomy at the
high frequencies ($>$15 GHz) at which source contamination is not
dominant. That is why we have been enhancing the Cambridge 5-km
telescope which potentially has the right performance. The
enhancement programme consists of three parts:
    (a) Fitting cryogenic systems to all 8 aerials. Constructing FET
amplifiers for 5 GHz and HEMT amplifiers for 15 GHz. Apart from the
15 GHz component, this has been completed and has improved sensitivity
by a factor of almost 3, to Tsys $\approx$45K.
    (b) Modifying cabling to enable the 4 moving aerials to be
operated in "dense-pack" mode. This has been completed.
    (c) Design and construction of a completely new broadband analogue
and digital receiving system. This will increase bandwidth from 10
MHz to 350 MHz and will correlate on all possible 28 baselines rather
than the previous 16. The large band must be cut into 15-MHz wide
channels to permit interference rejection and reduce chromatic

aberration; this is done partly with filtermixers in the analogue system and partly with a digital correlator similar in some respects to a Fourier Transform Spectrometer. This will give a further sensitivity improvement by a factor of about 7. Most of the design and much of the hardware has been completed. Commissioning at 15 GHz is expected to be done in 1989.

Particular attention is being paid to the problems of suppression of "correlator offsets" by, inter alia, suppressing interference generation and employing sophisticated phase-switching schemes.

The resulting sensitivity we expect is 100-200μK (depending on declination) in a day on scales up to 3 arcmin at 15 GHz. This performance is enough to provide a good map (not just a profile) of the S-Z decrement in one of the clusters in Fig. 1 in a fortnight – not a year. This will allow routine comparison of S-Z and X-ray maps, the study of gas content as a function of richness and z, and the direct probing of cluster evolution as discussed in section 3, as well as many other programmes. And with our planned Very Small Array we can do much more ....

5. REFERENCES

M. Birkinshaw, S.F. Gull, H. Hardebeck, 1984. *Nature* **309**, 34.

R.A. Sunyaev, 1978. In *IAU Symposium no. 79*, eds. M.S. Longair & J. Einasto, Reidel, Dordrecht. p.393.

R. Saunders, 1986. In *Highlights of Astronomy*, **7**, ed. J.-P. Swings, Reidel, Dordrecht. p. 325.

THE CONSTRAINTS ON SUPERCONDUCTING STRINGS

Bronisław Rudak*, Mirosław Panek#
Copernicus Astronomical Center
*Chopina 12/18, 87 100 Toruń, Poland
#Bartycka 18, 00 716 Warszawa, Poland

ABSTRACT. We analyse distortions of the spectrum of microwave background radiation caused by heating of cosmic plasma by the population of superconducting cosmic string loops. We find that the effect in the Rayleigh–Jeans part of the spectrum is at the observational level and may be used to constrain the parameters involved in the cosmological explosions scenario by Ostriker, Thomson and Witten (1986).

Loops of superconducting cosmic strings (SCS) proposed by Witten (1985) [1] can serve as seeds of cosmic explosions. This idea and the cosmological scenario based on it were presented recently by Ostriker et al. (1986) [2] (OTW). As the oscillating SCS loops radiate electromagnetic (em) energy they can heat up their surroundings and generate explosions if cooling processes are not effective. The explosions can blow spherical voids, with galaxies forming at overdensity shells around them. We deal with effect that originates mostly at epochs before hydrogen recombination and gives simple way for seeking constraints on main parameters in OTW scenario - the distortions of the MBR spectrum coming from heating of the plasma by the SCS loops. The SCS loops lose their energy via the gravitational and the em radiation as well as in the plasma dissipation. The em waves radiated by the SCS loops cannot freely propagate as their frequencies are much lower than the plasma frequency, and therefore their energy heats up the plasma. It is possible that some fraction of the energy dissipated due to direct interactions with the plasma is in form of e.g. bremsstrahlung radiation. We make an assumption that all the energy coming from the dissipation is also heating the plasma and call the sum of energies of the em radiation and of the dissipation the thermal energy. We also assume that the heating is homogeneous as for high redshifts the heat transport is effective. Any thermal energy transfer to the plasma after an epoch $z_T \sim 10^7$ should influence the MBR spectrum. Let $Q(t)$ denote the sum of the em and the dissipation losses of the population of SCS loops per unit volume per unit time at the moment t. The quantity (ρ_γ is the density of radiation):

$$q(z_1, z_2) = \int_{t(z_1)}^{t(z_2)} Q(t)/\rho_\gamma(t) dt, \qquad (1)$$

with appropriatly chosen z_1 and z_2 can be related to the observable distortions of different parts of the MBR spectrum (i.e. its deviations from the planckian form).

193

N. Kaiser and A. N. Lasenby (eds.), The Post-Recombination Universe, 193–195.
© 1988 by Kluwer Academic Publishers.

Our aim is to calculate the value of $q(z_1,z_2)$ and use the observational limits to constrain the values of parameters of the OTW model.

The basic parameters are μ (mass per unit length of the string) and ϵ (energy density of primordial magnetic field expressed as the fraction of the radiation density). The solutions of equations of motion for the SCS are not available and we have to make assumptions. First we assume that the gravitational radiation rate γ_g, the length – radius relation, the initial size of the loop and the initial number density of loops are the same as for ordinary cosmic strings. Second, we treat the rate of the em radiation γ_{em} as a free parameter. For the background we assume the $\Omega = 1$ Friedman universe with density in baryons $\Omega_b h^2_{50} = 0.1$ (h_{50} is the Hubble constant in units of 50 km/s/Mpc). Loops formed at moment t_i have radii $R_i = \partial t_i$, $\partial = 0.2$. The mass of the loop is $M = \mu L$ where the length $L = \beta R$, with $\beta = 9$ [3]. The loop oscillates with the period $T = L/2$. The power of gravitational radiation of the loop is $L_g = \gamma_g G \mu^2$ (we take $\gamma_g = 50$). The SCS loops in the model acquire currents due to cosmological expansion removing the stream of primordial magnetic field. We analyse the case of bosonic SCS only and the maximal possible current for them is $I_{max} = e\sqrt{\mu}/\hbar$. The loop with current j ($j = I/I_{max}$) emits em waves with the power $L_{em} = \gamma_{em}\alpha\mu j^2 = f_{em}L_g$. The initial value of f_{em}, $f_{emi} = 1.825 \cdot 10^5 \kappa \epsilon \mu_6^{-2}$ serves as a parameter in the OTW model ($\mu_6 \equiv (G\mu/c^2) \cdot 10^6, \kappa \equiv \gamma_{em}/\gamma_g$). As the loop shrinks the current grows as $j \sim R^{-1}$, until it reaches the maximum value. We assume that since this moment the power of the em radiation is given by the maximal power, found for $j = 1$. Some energy is dissipated by the interactions of cosmic plasma with the magnetosphere around the string. The power of the dissipation is $L_d = f_d L_g$, with f_d given in [4]. The OTW scenario requires the parameters to be within the ranges: $0.1 < \mu_6 < 10$, $10^{-4} < f_{emi} < 10^2$. OTW assumed that $\kappa = 1$. Further investigations [5] showed that this value can be much higher. We take $1 \leqslant \kappa \leqslant 10^3$. The evolution of the SCS loop is given by: $\dot{M} = \beta\mu\dot{R} = -(L_g + L_{em} + L_d)$, (2). The rate of the thermal energy production is:

$$Q(t) = \int_0^{R_{max}(t)} (L_{em} + L_d)\, n(R,t)dR, \qquad (3)$$

where the number density of loops is $n(R,t)dR = \nu\partial^{-5/2} t_i^{-4}(z/z_i)^3 dR_i$ with $\nu = 0.01$ [3], and $R_{max}(t) = \partial t$.

As the limits of integration in (1) we take $z_\mu = 3.6 \cdot 10^5 (\Omega_b h^2_{50})^{-1.2}$ and $z_a = 3.7 \cdot 10^4 (\Omega_b h^2_{50})^{-0.5}$ (for details see [6]). For z_1 and z_2 chosen as above, one can relate the total energy production per unit volume in this period expressed in term of radiation energy density $q(z_\mu, z_a)$ to the chemical potential attaining the maximum value μ_{ch0} in the R-J part of the MBR spectrum: $\mu_{ch0} = 1.4\, q(z_\mu, z_a)$. Latest observations of the R-J region, still rather uncertain, suggest that the chemical potential is smaller than $5 \cdot 10^{-3}$ for $\Omega_b h^2_{50} = 0.1$ [7], [8].

The final results for $q(z_\mu, z_a)$ are shown on Figure 1. Our results are functions of three parameters: μ_6, f_{emi} and κ. Their combination gives $\epsilon = 5.48 \cdot 10^{-6} f_{emi} \kappa^{-1} \mu_6^2$, the fundamental parameter. Formally for $0.1 < \mu_6 < 10$, $10^{-4} < f_{emi} < 10^2$ and $\kappa = 1$ the range of ϵ is ($5.5 \cdot 10^{-12} - 5.5 \cdot 10^{-2}$). We would like to notice that the values of $\epsilon \gg 5 \cdot 10^{-8}$ may not be allowed by the fact that the observations constrain the magnetic field in the intergalactic space to $B < 10^{-9}$Gs. If we assume that the density of magnetic energy evolves as the radiation density till today then these observations limit ϵ to the value given above.

We can see that for $\mu_6 = 10$, $10^{-3} < f_{emi} < 10 - 10^2$, and for $\mu_6 = 1$, $f_{emi} = 0.1 - 1$,

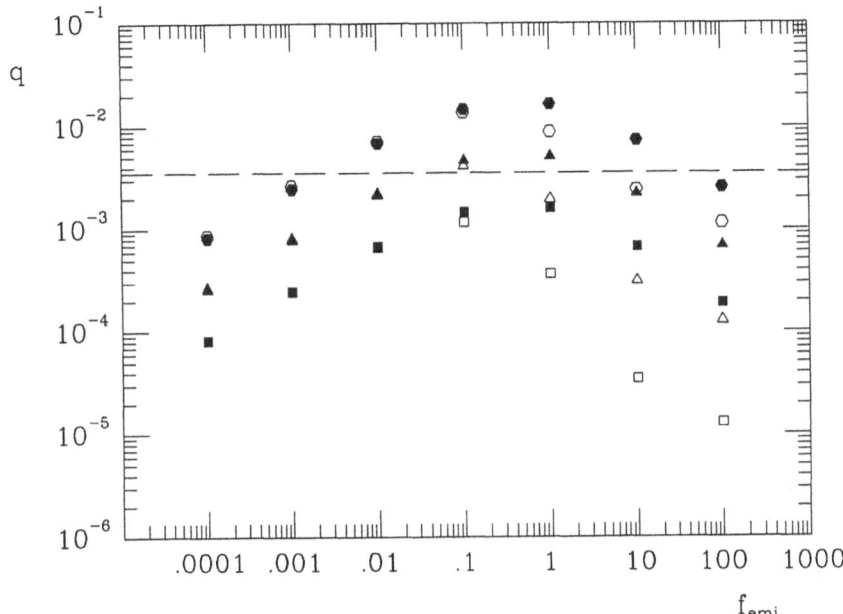

Figure 1. The total energy q generated by SCS loops via em + dissipation channels between redshifts z_μ and z_a expressed in terms of radiation energy density vs. the initial rate of the em radiation, f_{emi}. Squares, triangles, and hexagons are for μ_6 = 0.1, 1, and 10 respectively. Open symbols are for κ = 1, corresponding to γ_{em} = 50. Filled symbols are for κ = 10^5.

points representing values of $q(z_\mu, z_a)$ lie above the limit determined from μ_{cho}. Also according to suggestion in [9] that f_{em} = const is a better approximation that $f_{em} \sim j^2$ we have done calculations for this case. Here we are above the observational limit for μ_6 = 10, 0.1 ≲ f_{em} < 10.
Simple analytical expressions for $q(z_\mu, z_a)$ in both cases can be obtained if one neglects the dissipation in (2) and (3). This, as well as the discussion of the role of dissipation, is presented elsewhere [6].
To conclude, we can see that if we require the distortions of the MBR spectrum in the R-J region to be not in contradiction with observations then the following range of parameters is forbidden: μ_6= 10, 10^{-3}< f_{emi} < 10 - 10^2, μ_6 = 1, f_{emi} = 0.1 - 1 or μ_6 = 10, 0.1 ≲ f_{em} < 10 in case of constant f_{em}.

REFERENCES:
[1] E. Witten, *Nucl.Phys.B* **249**, (1985), 557.
[2] J.P. Ostriker, C. Thomson, E. Witten, *Phys.Lett.B* **180**, (1986),231.
[3] R.H. Brandenberger, A. Albrecht, N. Turok, *preprint NSF-ITP-86-15*, (1986).
[4] M. Panek, B. Rudak, *Phys.Lett.B* (1987) in press.
[5] A. Vilenkin, T. Vachaspati, *Tufts University preprint* (1986).
[6] B. Rudak, M. Panek, *Phys.Lett.B* (1987) submitted.
[7] L. Danese, G. De Zotti, *Astron.Astrophys* **84**, (1980), 364.
[8] G.F. Smoot et al., *Astrophys.J.* **291**, (1985), L23.
[9] J.P. Ostriker, C. Thomson, *preprint* (1987).

HORIZON-SIZE COSMIC STRINGS

Shoba Veeraraghavan
Astronomy Department
University of California
Berkeley, CA 94720, U.S.A.

Abstract. Cosmic strings of size comparable to the causal horizon could have cosmologically important effects. I describe below a method to calculate their gravitational effects and estimate the tensor Sachs-Wolfe temperature anisotropy due to an infinite string.

1. INTRODUCTION

Cosmic strings are topological defects predicted in some particle theories to arise during a phase transition in the early universe. The network of cosmic strings is characterized by a single parameter, the mass per unit length μ of any string segment. For strings in Grand Unified theories, $G\mu \sim$ few $\times 10^{-6}$ in units where $c = 1$. Strings are an atypical source term in cosmological perturbation theory in that the non-Newtonian terms can be quite important, because strings typically move at relativistic velocities. Fortunately, they are a weak perturbation upon the background metric (I will assume that $G\mu$ is of order 10^{-6}, because only these strings would be important in cosmology), and so linear perturbation theory about the background universe is adequate in calculating their gravitational effects.

2. COSMIC STRINGS IN AN EXPANDING UNIVERSE : 'BIG' STRINGS

The string network can be separated into 'small' and 'big' strings based on the relative sizes of the curvature radius of each string segment and the causal horizon. The expansion of the universe is negligible in determining the dynamics of a 'small' string whose curvature radius is much smaller than the horizon. This means that to a very good approximation the string is in an asymptotically Minkowski background. On the other hand, 'big' strings $i.e.$ infinite strings and the horizon-size loops, have a curvature radius comparable to the horizon size, so that their evolution must take into account the Hubble expansion. 'Smallness' and 'bigness' depend on size relative to the horizon size at each time and a 'big' string becomes 'small' at a later time.

The cosmological effects of 'big strings' could be quite significant. For instance, strings create fluctuations in the temperature of the cosmic microwave background radiation. However, estimates of their effects are usually made in a 'small' string approximation : Kaiser & Stebbins (1984) calculated that $\delta T/T$ induced by a moving infinite string in a cold dark matter universe is marginally below current detection levels, while Traschen, Brandenberger & Turok (1986) estimated that such strings should contribute no more than $10^{-7}\mu_6$ to the Sachs-Wolfe term in $\delta T/T$. A 'big' string theory, which treats strings as a perturbation upon a Freidmann universe is required to estimate $\delta T/T$ from 'big' strings properly.

N. Kaiser and A. N. Lasenby (eds.), The Post-Recombination Universe, 197–199.
© *1988 by Kluwer Academic Publishers.*

The energy-momentum tensor that is appropriate for a string network in a homogeneous, isotropic and spatially flat universe is obtained by extremising the string action and has the form,

$$T^{mn}(\mathbf{x}, \eta) = \frac{\mu}{a^4} \int d\sigma \epsilon \delta^{(3)}(\mathbf{x} - \mathbf{r}(\sigma, \eta)) \left[\dot{r}^m \dot{r}^n - \frac{r'^m r'^n}{\epsilon^2} \right].$$ (1)

In (1), η is conformal time, the x^i are spatial coordinates and a is the expansion factor. $\mathbf{r}(\sigma, \eta)$ describes the string trajectory in 4-spacetime, where η and σ are coordinates on the 2-dimensional string worldsheet. The string linear energy is measured by $\epsilon = \sqrt{\mathbf{r}'^2/(1 - \dot{\mathbf{r}}^2)}$ in the gauge $\dot{\mathbf{r}} \cdot \mathbf{r}' = 0$. The expression (1) is correct for any arbitrary string configuration, reduces to Turok's result (1985) in the Minkowski limit $\ddot{a} = \dot{a} = 0$, and describes a 1-dimensional $p = -\rho$ perfect fluid in the rest frame of each infinitesimal string element.

2.1. Compensation.

The creation of strings at the phase transition transfers energy from radiation (the 'Higgs radiation' of Traschen *et. al.*, 1986) to strings, and in time the strings chop up into loops which decay by emitting gravitational radiation. Statistically, one expects some anti-correlation between the existing string network (or its components) and radiation. On scales of interest to us, any 'Higgs radiation' compensation is irrelevant. It is difficult to quantify the compensation due to the gravitational radiation background from decayed loops in the absence of detailed understanding of the dynamics, evolution and correlations of the string network. I will instead assume that there is just a single infinite string in the universe, thus there is no gravitational radiation compensation.

2.2. Tensor Mode Perturbations from an Infinite Straight Static String.

I will further consider only the tensor mode perturbations to the metric tensor, h_{ij}, because these are by definition gauge-invariant. They couple only to τ_{ij}, the tensor part of the string energy-momentum tensor, T_{ij}. This restriction to tensor modes eliminates the danger of spurious gauge mode perturbations. For instance, in the case of $\delta T/T$ which is gauge-invariant because it is an observable, the gauge problem arises in determining the hypersurface of last scattering. The expression for T_{ij} is an integral one, and too complicated to be written analytically unless the string configuration is highly symmetric. So let the string be straight, lying parallel to the z-axis. Let it also be static (with zero *peculiar* velocity, $a\dot{\mathbf{r}} = 0$) because at late times $\dot{\mathbf{r}}$ decays as a^{-2}. The maximum $\delta T/T$ so obtained is an upper limit to that caused by a straight moving string.

The non-zero cartesian components of τ_{ij} are

$$\tau_{ij}(x, y, z) = \frac{\mu}{4} \delta(x)\delta(y) \begin{bmatrix} 1 & 0 & 0 \\ 0 & 1 & 0 \\ 0 & 0 & -2 \end{bmatrix} + \frac{\mu}{4\pi r^2} \begin{bmatrix} \cos 2\phi & \sin 2\phi & 0 \\ \sin 2\phi & -\cos 2\phi & 0 \\ 0 & 0 & 0 \end{bmatrix},$$ (2)

with $r = \sqrt{x^2 + y^2}$ and $\phi = \arctan y/x$. Notice that the stress-energy tensor is invariant along the z-axis. Now let us assume that the scale factor scales as a power-law with time, $a \sim \eta^\alpha$. Then in a radiation-dominated universe ($\alpha = 1$) or matter-dominated universe ($\alpha = 2$), the tensor perturbations to the metric are,

$$h_{ij}(r, \eta) = \frac{G\mu}{c^2} \begin{bmatrix} 1 & 0 & 0 \\ 0 & 1 & 0 \\ 0 & 0 & -2 \end{bmatrix} \times \begin{cases} 2\sqrt{1-u^2} - 2\ln\left(\frac{1+\sqrt{1-u^2}}{u}\right), & \alpha = 1 \\ 3\sqrt{1-u^2} - (2+u^2)\ln\left(\frac{1+\sqrt{1-u^2}}{u}\right), & \alpha = 2, \end{cases}$$ (3)

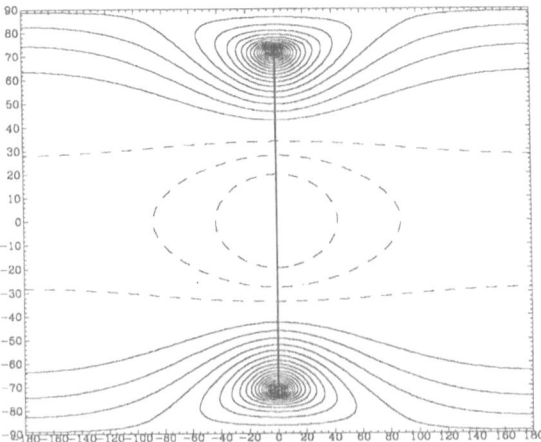

FIGURE 1. All-sky map of $\delta T/T$ around an infinite straight static string, at $z_s = 1$, calculated on a 1° grid in an instantaneous recombination approximation. The segment of string inside the observer's horizon lies along $\phi = 0$, $-\arccos u_R \leq \theta \leq \arccos u_R$. Solid lines are +ve-valued contours, dashed lines are −ve-valued. The range in $\delta T/T$ is −0.73 to 5.47 in units of $G\mu$.

where $u = r/\eta$ is the dimensionless ratio of the distance from the string to the horizon size. The metric perturbations vanish exactly at a horizon distance from the string ($u = 1$), and grow logarithmically near the string, $Z_{ij}(r, \eta) \sim (G\mu/c^2)\ln u/2$ as $u \to 0$. The perturbations are self-similar in that they can be described in terms of a single dimensionless variable. At a fixed physical distance from the string the perturbations grow larger with time.

It is straightforward to calculate the tensor Sachs-Wolfe term in $\delta T/T$ generated by the string. With no loss of generality, we can allow the observer to lie along the z-axis at a distance $R = u_R \eta_0$ from the string (η_0 is the present conformal time). Define an observer-centred coordinate system of latitudes θ and longitude ϕ such that the string is at $\phi = 0$, $-\arccos u_R \leq \theta \leq \arccos u_R$. The minimum redshift of the string is then $1 + z_s = (1 - u_R)^{-2}$. Figure 1 shows an all-sky map, on a 1° grid, of the predicted temperature fluctuation pattern $\delta T/T(\theta, \phi)$ from an infinite string at $z_s = 1$ in an instantaneous recombination assumption. The magnitude of the fluctuations is few$\times 10^{-6}\mu_6$, and it decreases as the string moves to higher redshift, or if the last scattering surface has a finite thickness. Although the metric perturbation diverges near the string, $\delta T/T$ itself is always finite. Full details of our work on the gravitational effects of horizon-sized string configurations and the temperature anisotropies will be found in Veeraraghavan, Stebbins & Silk (1987).

Acknowledgements : This work is being done in collaboration with Albert Stebbins and Joseph Silk, and I thank them for many useful discussions.

References

N. Kaiser & A. Stebbins, *Nature* **310** (1984), 391-393.
J. Traschen, N. Turok & R. Brandenberger, *Phys. Rev.* **D34** (1986), 919-930.
S. Veeraraghavan, A. Stebbins, & J. Silk, in preparation.

METAL–RICH ABSORPTION LINE SYSTEMS

Jacqueline Bergeron

Institut d'Astrophysique CNRS
98bis, bd Arago
75014 PARIS, FRANCE

ABSTRACT. We review the properties of metal–rich absorption line systems derived from statistical studies of homogeneous samples. There is some evidence of evolution in the MgII line density with redshift but larger samples are needed to confirm this trend. The absorption systems can differ by their ionization level and by the strength of their lines. There is a strong correlation between the opacity to UV radiation of the absorbing clouds and their ionization level which implies that the neutral and singly ionized elements are present only in transition regions of large optical depth to UV ionizing photons. The heavy element abundances can be roughly estimated for systems with large HI column densities. They are a few tenths of the solar values. Even in the clouds with largest HI column densities there is no detection of dust. We present results of our search for the galaxies giving rise to MgII absorption line systems at redshift $z<1$. We compare properties of these galaxies with those derived from statistical studies of MgII absorption systems. From our high identification rate of the absorbing galaxies we infer strong constraints on their luminosity function at $z\sim0.4$. These galaxies have very large gaseous envelopes and show signs of strong stellar formation activity. They are usually field galaxies.

1. INTRODUCTION

The absorption line systems in the spectra of quasars have been increasingly studied since it has been realized that they provide direct information on high redshift intervening objects not otherwise detectable. The large amount of observational material available reveals the great complexity of the absorbers and the existence of different classes of objects.

N. Kaiser and A. N. Lasenby (eds.), The Post-Recombination Universe, 201–223.

We discuss here only the metal–rich systems with absorption lines of small velocity dispersion, typically b<100 km s^{-1}. It was suggested already two decades ago by Wagoner (1967) and Bahcall and Spitzer (1979) that these absorbers could arise in intervening galaxies. This assumption has been confirmed by the statistical properties of the absorption line and by the identification of low redshift (z<1) absorbing galaxies.

The strongest metal absorption lines are the CIV and MgII doublets which, for a given observed spectral domain, sample different redshift ranges. We first present the CIV anf MgII samples currently available, give the average size of the absorbers and discuss possible cosmological evolution of the density per unit redshift for the CIV and MgII systems. We then outline the properties of the phases with different ionization level, and for each phase give the constraints on their ionization parameter and gas density as derived from photoionization models. Although not enough information is available to derive the heavy element abundances for most absorption line systems, it is shown that rough abundances can be estimated for absorbers of large HI column densities. Unsuccessful searches for dust associated with high column density absorbers are also presented.

To confirm the intervening origin of the metal–rich systems the most direct approach is to identify the galaxy responsible for typical CaII (z<0.15), MgII (z>0.15) or CIV (z>1.1) absorption systems. One can search for CaII absorption for a priori known galaxy–QSO pairs as first attempted by Boksenberg and Sargent (1978). We summarize results obtained for this type of searches. Another approach is to identify the galaxy associated with MgII absorption systems. To understand why the first attempts to identify MgII absorbing galaxies were negative, we have estimated the identification rate of these MgII absorbers predicted from statistical properties of the MgII absorption line systems. We then describe results of our search for these galaxies and outline their specific properties.

2. THE SAMPLES

We give bibliography of unbiased and homogenous absorption line surveys which can be used for statistical analyses. Homogeneous unbiased samples are obtained from observations of sight lines a priori unknown. They are rest equivalent width limited. Multiple systems spanning less than 300 km s^{-1} are counted only as one system. Systems with redshift larger than the quasar emission redshift or smaller by less than 3000 km s^{-1} are excluded since they could be associated with the quasar or its underlying galaxy. The wavelength range shortwards of Lyα in emission from the quasar is also excluded to avoid confusion with the Lyα forest systems.

The larger surveys for narrow metal–rich absorption systems were first obtained in the blue for the CIV doublet. The first sample observed by Weymann et al. (1979) cannot be fully used since it was not obtained with photon counting detectors and no reliable equivalent width could be measured. To the blue survey of Young et al. (1982), Bergeron and Boissé (1984) added observations from about ten sight lines, mostly from published studies of quasar pairs. For rest equivalent width limits equal to $w_{r,lim}$(CIVλ1548 and λ1550) of 0.3 and 0.2 Å this led to 38 systems in 39 sight lines. Most of the quasars from the early survey of Weymann et al. (1979) were reobserved by Foltz et al. (1986) and this added 17 systems. Combined with the previous samples this gives 55 systems in 70 sight lines at an average redshift $\bar{z}=1.7$. However the redshift range adequately sampled, zz1.40–2.05, is small. There is a new uniform sample which comprises about 130 CIV systems down to $w_{r,lim}=0.15$ Å(Sargent et al. 1988) for which only preliminary results are available (Sargent, 1988).

The MgII samples were also first build from blue surveys. Adding to the results of Young et al. (1982) and Foltz et al. (1986) new observations in the blue, Tytler et al. (1987) have obtained, for $w_{r,lim}=0.6$ Å, a sample of 8 MgII systems at $\bar{z}=0.5$ in 90 sight lines. For the subsample with $w_{r,lim}=0.25$ Å there are 14 MgII systems. Preliminary results are also available for a UV survey done in the wavelength range $\lambda\lambda$3150– 3950 by Boulade et al. (1988).

To study the ionization level of the metal systems Boissé and Bergeron (1985) observed 18 quasars in the red down to $w_{r,lim}$ (MgIIλ2796 or FeIIλ2382)=0.6 Å. Although the observations spanned a fairly large redshift range zz1.2–1.7, the survey was biased since were only observed quasars for which the MgII counterpart of a known CIV system was expected. More recently an unbiased red survey was completed by Lanzetta et al. (1987) who found 16 MgII systems at $\bar{z}=1.67$ in 31 sight lines down to $w_{r,lim}=0.6$ Å. For the subsample with $w_{r,lim}=0.3$ Å they get 10 MgII systems. This survey gives crucial new information to determine the cosmological evolution of MgII systems and to study relative properties of systems with different ionization levels. Further it has led to the discovery of low ionization MgII systems without CIV counterpart. Combining the blue and red surveys gives 24 systems, with $w_{r,lim}=0.6$ Å, spanning the large redshift range zz0.2–2.05. There is also a red survey by Caulet and York (1988) which is however biased since high spectral resolution observations have been obtained in only small wavelength ranges centred on expected MgII doublets from known CIV systems.

3. AVERAGE SIZE OF THE ABSORBERS AND COSMOLOGICAL EVOLUTION

For Friedmann models with zero cosmological constant and a decceleration parameter q_o, the number of absorption line systems per unit redshift is given by

$$\frac{dN}{dz} = \frac{c}{H_o} n_o \sigma_o (1 + z)^{1+\alpha} (1 + 2 q_o z)^{-0.5}, \qquad (1)$$

where n_o and σ_o are the density and projected cross–section on the sky of the absorbers at z=0, c is the speed of light, H_o the Hubble constant and α represents a possible cosmological evolution.

The $n_o\sigma_o$ product is estimated using a Schechter luminosity function and a Holmberg radius–luminosity scaling law

$$dn = \phi_o x^{-5/4} e^{-x} dx, \qquad (2)$$

$$\sigma = \sigma_* x^{5/6}, \qquad (3)$$

with $x = L/L_*, \phi_o = 3 \ 10^{-3}$ Mpc^{-3} and M$_*$(B)=-20.6, using H$_o$= 50 km s^{-1} Mpc^{-1}. Assuming that half of all galaxies, whatever their luminosity, are surrounded by spherical halos, one gets

$$n_o\sigma_o = \tfrac{\pi}{2}\Gamma(\tfrac{7}{12})\phi_o R_o^2. \qquad (4)$$

With the above assumptions the Holmberg radius of an L$_*$ galaxy (at z=0) is R_H=22 kpc.

The size of the absorbers is obtained assuming no cosmological evolution, $\alpha=0$, and an open universe, $q_o=0$. The values derived for R_o are given in Table 1. The dependence on H$_o$ is eliminated by normalizing to R_H. The product $n_o\sigma_o$ takes identical values if we assume either absorbing spherical halos around half of all galaxies or absorbing thick discs around every galaxies. The Lyman limit sample compiled by Bechtold et al. (1984) refers to absorbers optically thick at their Lyman limit (LLS).

Table 1. Average size of the absorption line systems

system	$w_{r,lim}(\text{Å})$	$\frac{dN}{dz}$	$\bar{z}$	R_o/R_H	references
MgII	0.6	0.26	0.51	2.8	1
	0.25	0.69	0.5	4.5	1
	0.6	0.85	1.67	3.8	2
	0.3	1.25	1.62	4.6	2
CIV	0.3	1.64	1.70	5.2	3
LLS		1.29	2	4.4	4

references. 1 : Tytler et al. (1987). 2 : Lanzetta et al. (1987). 3 : Bergeron and Boissé (1984). 4 : Bechtold et al. (1984).

The size of the absorbing gaseous envelopes is a function of $w_{r,lim}$, as clearly shown in Table 1 for the MgII systems. There is also an excess of small

CIV doublets, w_r(CIVλ1548) in the range 0.05 to 0.20 Å, in the data presented by Meyer and York (1987) and Bergeron and D'Odorico (1988). The radii of the ionized envelopes are at least equal to 4 to 5 R_H which is about 3 times larger than the average dimension, ~ 1.5 R_H, of HI discs around spiral galaxies. These HI discs have sharp edges at N(HI)$\sim$1 10^{19} cm^{-2} as recently discussed by Sancisi (1988). Similar information for the ionized absorbing envelopes will be obtained only when a break or turn–over will be detected at small w_r in the equivalent width distribution of the CIV or MgII systems.

Cosmological evolution of absorption line systems is usually expressed in terms of the index γ

$$\frac{dN}{dz} \propto (1+z)^\gamma, \tag{5}$$

with γ=1 for q_o=0 and γ=0.5 if q_o=0.5.

The redshift evolution of metal systems is difficult to determine since their density is not very high. Smaller uncertainties are obtained when a large redshift interval is sampled, as for the MgII systems. Results are presented in Table 2. The observed samples have been analyzed using a maximum likelihood method. For MgII, Kunth (1988) has combined a new UV survey (Boulade et al. 1988) to that of Tytler et al. (1987). For CIV, Boissé (1988) has reanalyzed the data compiled by Bergeron and Boissé (1984) and has also added observations by Foltz et al. (1986). A value of γ derived from the unpublished survey of Sargent, Boksenberg and Steidel is also quoted.

Table 2. Cosmological evolution of metal systems

ion	$w_{r,lim}$(Å)	z range	γ	references
MgII	0.6	0.2–2.05	2.4 $\pm$ 0.8	1
			2.0 $\pm$ 0.5	2
CIV	0.3	1.4–2.05	1.0 $\pm$ 2.0	3
			1.8 $\pm$ 1.5	3
	0.15	1.4–3.1	−1.2	4
LLS		0.5–3.3	1.3$\pm$1.5	5

references. 1 : Lanzetta et al. (1987). 2 : Kunth (1988). 3 : Boissé (1988). 4 : Sargent (1988). 5 : Bechtold et al. (1984).

A positive evolution is most probably present (2σ level) for the MgII systems. Since they should constitute a large fraction of the Lyman limit systems, the different values of γ found for the two samples may just reflect the large uncertainties involved. The possible opposite evolution derived from published CIV data and the new larger sample of Sargent et al. (1988) may not be only a consequence of uncertainties inherent to small samples but also to the different redshift ranges covered. A more thorough investigation will be made when this new sample will be published. Further an unambiguous answer will be obtained for the evolution of both the LLS and CIV systems when a UV absorption line survey will be done with the Hubble Space Telescope.

4. IONIZATION LEVEL

The ionization level of the absorbers is well defined only when low and high ions are observable in the optical thus for z>1.2. At these redshifts three phases are present, characterized essentially by the presence/absence of the CIV and MgII doublets. We will mainly outline properties related to the opacity of ionizing photons, multiplicity of the systems and velocity spread of the subcomponents, the velocity of the absorber relative to that of the quasar, and the redshift range thus a possible cosmological evolution. We first restrict the discussion to systems at z>1.2.

4.1. Systems of high ionization level (H)

In these systems apart from the H Lyman series, the lines usually detected are the CIV doublet and occasionnally the SiIV doublet. In a few cases NV is present even for $z_a < z_e$ systems. Lines from singly ionized and neutral elements are not detected even in high signal to noise spectra. The ionization level can be defined by the value of the line equivalent width ratio

$$\frac{CIV}{MgII} = \frac{w_r(CIV\lambda 1548)}{w_r(MgII\lambda 2796)}.$$

For high ionization level absorbers the lower limit of CIV/MgII can reach 10 to 40 (see e.g. Bergeron and D'Odorico 1986). These systems are also usually weak with $w_r(CIV\lambda 1548) < 1$ Å.

The opacity to ionizing radiation of the absorbers can be derived by comparing statistical properties of CIV and Lyman limit systems, as done by Bergeron et al. (1987a). From the rate of occurrence of optically thick Lyman limits in their LLS sample, Bechtold et al. (1984) infer that systems with $w_r(Ly\alpha) > 1.7$ have a discontinuity at the Lyman limit. Although the bulk of high ionization level systems have $w_r(Ly\alpha) <1.7$ Å, there are some cases ($\sim$ 10 to 20 %) for which $w_r(Ly\alpha)$ can reach 2 to 3 Å. It is yet to be investigated whether the latter show weak lines of CII and SiII, although MgII or FeII could remain undetected, as it is the case for the optically thick z=1.649 system in PKS0215+015 (Blades et al. 1985, Bergeron and D'Odorico 1986).

The HI density distribution is particularly difficult to determine for the high ionisation level systems. Most often the HI column density, N(HI), cannot be derived from studies of their optically thin Lyman limit, whereas the Lyα line is saturated and on the logarithmic part of the curve of growth. Thus it is usually necessary to estimate the velocity dispersion b from the CIV doublet, assume that the same value of b applies for HI and CIV, and be sure that individual components have been isolated which requires a spectral resolution of at least FWHM=1 Å (to resolve 2 components at z=2 spanning 80 km s^{-1}).

Some high ionization level systems are multiple, but the extreme cases of multiplicity do not belong to this class of absorbers. The velocity spread of the subcomponents is moderate, up to 300 km s^{-1} in some cases, whereas the velocity dispersion of individual components is within the range b=5 to 20 km s^{-1}. The multiple systems are usually those of larger w_r(CIVλ1548).

4.2. Systems of mixed ionization level (M)

For these absorbers low and high ions are present. The absorption lines are usually strong with w_r(CIVλ1548) and/or w_r(Lyα) > 2 Å.

The ionization parameter CIV/MgII has values within the range 0.3–2.5. Therefore there is a clear discontinuity of CIV/MgII between the high and mixed ionization level systems, and a definite tendency for the systems with increasing w_r(MgIIλ2796) to have a decreasing value of the ionization parameter as found by Boissé and Bergeron (1985) and confirmed by Lanzetta et al. (1987: see their Fig.11).

As mentioned above from statistical arguments, most if not all of these absorbers should have optically thick Lyman limits. This is verified when observations of the Lyman limit, in the optical or the UV, are available (Bechtold et al. 1984 ; see also the review of Bergeron et al. 1987a) .

These systems often have a complex velocity structure. Extreme cases show 10 to 15 components spanning up to 2000 km s^{-1}. Amongst the subsystems there is often a mixture of high and mixed ionization levels (Pettini et al. 1983, Foltz et al. 1986) and the M systems tend to span a smaller velocity range than the overall complex (Lanzetta et al. 1987). There is however some multiple M systems with all subcomponents of mixed ionization level, as the z=1.345 system in PKS0215+015 (Bergeron and D'Odorico 1986).

The $z_a \sim z_e$ systems, which are typically found in steep–spectrum radio–loud quasars, often show multiple structure (Foltz et al. 1986), including extreme cases as the two complexes in GC1556+335 (Morris et al. 1986). This tendency to get an excess of $z_a \sim z_e$ systems in steep–spectrum radio quasars is also present at z<1.2 : the 4 $z_a \sim z_e$ systems known at low z are in radio–loud quasars, 3 being steep–spectrum radio sources (Bergeron et al. 1987a). It is among the $z_a \sim z_e$ systems that the larger spreads of ionization level are found, usually from singly to four times ionized elements, even showing sometimes heavy neutral atoms together with NV as in the z=0.401 system in PKS1912–54 (Bergeron and Boissé 1986). However $z_a \sim z_e$ systems often belong to the H class.

4.3. Systems of low ionization level (L)

These absorbers have been discovered by Lanzetta et al. (1987) in their red survey. Absorption lines of CII, MgII and FeII are present but the CIV doublet is undetected. There are only 5 L systems in this red sample. The absorption lines are fairly strong with w_r(MgIIλ2796) in the range 0.4 to 1.7 Å.

The upper limit on the ionization parameter CIV/MgII is 0.5 in 4 cases and 0.2 for the z=1.636 system in PKS0237−233 using the CIV limit given by Boroson et al. (1978). Therefore there is no clear discontinuity between the M and L systems contrary to what was found for the M and H systems.

About half of the damped Lyα absorbers (N(HI)$\geq 10^{20}$ cm^{-2}) contain only metals in a low ionization level (Wolfe 1988), and belong to the L class. Do all L absorbers are damped Lyα systems ? One of the five systems found by Lanzetta et al. (1987) has a damped Lyα line. No information on Lyα is available for the remaining four cases, for which this line is shortward of 3300 Å. UV observations are needed to specify the link between L and damped Lyα systems.

Too few L absorbers are known and they have not been observed at high enough spectral resolution to conduct a proper study on their multiplicity. However for a few multiple systems of mixed ionization level there are components of low ionisation state.

In summary the stronger systems at z>1.2 are a) of mixed or low ionization level, b) optically thick to ionizing radiation, c) have large column densities and/or multiple components spanning a large velocity range. In our Galaxy only the M and L classes have been detected. The percentage of H, M and L systems at $\bar{z}\sim$1.7, in the survey of Lanzetta et al. (1987), equals 49, 37 and 14 % respectively. Thus contrary to our knowledge prior to this red survey, only about half of the metal−rich absorbers are of high ionization level.

4.4. Systems at low redshift

At z<1.2, systems of only mixed or low ionization level are known since no UV absorption line survey is yet possible. Four of these systems, observed with IUE, show a strong discontinuity at the Lyman limit and strong but no damped Lyα lines. However there are two fairly weak MgII systems which do not show any break at the Lyman limit. In one case the system is certain, with saturated MgII lines of moderate strength (z=0.852 in AO 0235+164), but the UV data are very noisy (Snidjers et al. 1982). Although UV observations of better S/N ratio are needed for confirmation, this suggests that at low redshift some of the weaker systems of low ionizing level are optically thick to ionizing radiation which should imply a cosmological evolution of the ionization state of the absorbers.

There is an excess of weak MgII systems in the samples of Tytler et al. (1987) and Boulade et al. (1988) at 0.25<w_r(MgIIλ2796) <0.6 Å, which is not present in the survey of Lanzetta et al. (1987). Although the latter does not go as deep as the lower z surveys, this could also point towards a cosmological evolution of the ionization state of the weaker systems.

Some of the low z MgII systems have an associated CIV absorption as shown by IUE observations of bright quasars (Bergeron et al. 1987a) but the low sensitivity of IUE prevents any conclusion when the CIV doublet is undetected (Bergeron and Kunth 1983). Further low redshift absorbers of high ionization level remain to be discovered.

Complex velocity structure, difficult to reconcile with motions within a single isolated galaxy, are also found in low z MgII absorbers as for the z=0.8526 system in PKS1327-206 with 6 components spanning 560 km s^{-1} (Bergeron et al. 1987b : see their Figs. 1 and 2). Multiple systems spanning at most 300 km s^{-1} are also often present in low z MgII absorbers (Bergeron and D'Odorico 1988).

5. PHOTOIONIZATION MODELS

The ionization state of the absorbing gas can be estimated assuming that photoionization is the dominant process. The metal absorption line widths imply a radiation ionizing source at least for ions of higher ionization level , as CIV, for which the observed velocity dispersions b=5 to 8 km s^{-1} are incompatible with collisional ionization by thermal electrons. However for MgII the smallest values found for b are 8 to 10 km s^{-1} which only gives a fairly high upper limit for the gas temperature, T(MgII)<1 10^5 K, and does not exclude thermal collisional ionization.

The basic assumptions introduced for the photoionization models are a) ionization by the diffuse UV background, b) transfer of the ionizing radiation in the regions of the absorber optically thick to the UV flux, c) given metal abundances. For these models, the ionization level of the elements is mainly a function of the ionization parameter alone

$$U=N(\nu > \nu_{LL})/(nc), \tag{6}$$

where N$(\nu > \nu_{LL})$ is the total flux of ionizing photons incident on the cloud and n the gas density. The values for n and for the size ℓ of the absorber are also a function of the absolute value of the ionizing radiation flux.

A grid of photoionization models were built by Bergeron and Stasinska (1986) for U in the range 10^{-5} to 10 and abundances relative to the solar ones from 10^{-4} to 1. The spectrum of the radiation flux was assumed to be a power law J$(\nu) \propto \nu^{-\alpha}$ with 1$\leq \alpha \leq$2. Most models involve clouds with constant gas density but constant gas pressure was also investigated.

The main results are :
1. the existence of a strong discontinuity of singly ionized element to CIV ionic ratios at N(HI)$\sim$ 2 10^7 cm^{-2} for U>1 10^{-3}. This result is independent of the

heavy element abundances and the power law index α. For a given value of U this discontinuity is more pronounced for MgII and FeII than for CII and SiII.

For a likely value of the diffuse UV background at $z\simeq2$–2.5 due to quasar emission (Sargent et al. 1980, Gondhalekar 1983).

$$J(\nu_{LL}) = 1 \ 10^{-21} \ \text{erg cm}^{-2} \ \text{s}^{-1} \ \text{Hz}^{-1} \ \text{st}^{-1} \tag{7}$$

the inferred value of n for an homogeneous cloud is small : $1 \ 10^{-2} \ \text{cm}^{-3}$ for $U=2 \ 10^{-3}$.

2. the possibility to derive directly either U or N(HI) from some ionic ratios. A few of these ratios are independent of N(HI), as N(SiIII)/N(SiIV) for $U\leq10^{-1}$ and N(NV)/N(CIV) for $U\leq3 \ 10^{-3}$ (see Figs. 4 and 5 in Bergeron and Stasinska 1986). Others remain roughly contant for $N(HI)>3 \ 10^{17} \ \text{cm}^{-2}$. This occurs when the ions considered originate essentially from the optically thin part of the absorbing cloud, which is the case for ions at least 3 times ionized whatever $U<10$. The N(SiIV)/N(CIV) can thus also be used to fix U.

For absorbers optically thin to ionizing radiation, U can be derived from any ionic ratio since they are all independent of N(HI). Knowing several ionic ratios would then help to determine α. Unfortunately these absorbers have typically small column densities and their metal absorption lines are weak and difficult to detect. We will come back to this point while discussing the heavy element abundances for the absorbers of high ionization level.

The N(OI)/N(CII) is a good indicator of large N(HI) and is of particular interest for smaller z absorbers if both Lyα and the Lyman limit cannot be observed. It is close to unity for $N(HI)>1 \ 10^{19} \ \text{cm}^{-2}$ for $U>10^{-4}$ (see Fig. 7 in Bergeron and Stasinska 1986).

3. the range of U found for both the high and mixte ionization level absorbers is two decades around $U\sim(1$–$3)10^{-3}$. This is true whatever $1\leq \alpha \leq 2$, but this range is slowly shifted towards higher values of U when α increases. The low ionization level systems, newly discovered, do not yet significantly increase the allowed domain for U since the lowest upper limit of 1/5 found for the N(CIV)/N(MgII) ratio is close to values observed for the M systems. For $\alpha=1.5$ this limit implies $U<1 \ 10^{-4}$ for $N(HI)=1 \ 10^{20} \ \text{cm}^{-2}$. Using the value of $J(\nu_{LL})$ given in eq. 5 still leads to low gas densities even in the optically thin case, $n>0.2 \ \text{cm}^{-3}$.

The H systems have in average an ionization parameter about one order of magnitude higher than the M systems. The domain of U common to both types of absorbers is a function of the abundances. It is small for clouds of constant density, roughly a factor of 2 for solar abundances $Z_\odot$, and no common range exists for $Z<Z_\odot/20$. For clouds with constant pressure the common range is increased by a factor of about 3.

4. the hydrogen ionization level, averaged over a whole cloud, is high even for large N(HI). Thus the size of the absorbing cloud is a function of the assumed ionization source and cannot be derived directly from N(HI) and n.

For high ionization parameters, the HII/HI ratio sharply drops at $N(HI)\sim2\ 10^{17}$ cm^{-2} and the size of the opaque core is small compared to the optically thin outer parts. This is no longer true for $U<10^{-3}$. Using $J(\nu_{LL})$ given by eq. 5, $\alpha=1.5$ and $U=2\ 10^{-3}$ leads to a size for absorbers of mixte ionization level of 10 kpc, whatever $10^{18}<N(HI)<10^{20}$ cm^{-2}, and the HII/HI ratio, averaged over the whole cloud, is then in the range 250 to 4.

Absorbers of low ionization level, without damped Lyα line, should have smaller sizes. For $U<1\ 10^{-4}$ and $N(HI)<1\ 10^{20}$ cm^{-2} clouds have dimensions $\ell<0.2$ kpc.

As mentioned above $N(HI)$ is not well determined for high ionization level absorbers, but some clouds should have $N(HI)\simeq3\ 10^{16}$ cm^{-2}. Using $U=1\ 10^{-2}$ leads to HII/HI=4 10^3 and $\ell=15$ kpc.

These size inferred for the M and L absorbers are somewhat smaller than derived from statistical analyses of MgII and CIV samples, from direct observations of the absorbing galaxies (see last section). This discrepancy would disappear if $J(\nu_{Ll})$ was roughly 2 to 3 times larger than given in eq. 7. A more complex spectral dependence of the ionizing flux should also be investigated.

6. HEAVY ELEMENT ABUNDANCES AND DUST CONTENT

There are a number of uncertainties involved in the determination of the heavy element abundances : a) the HI content of the absorbing cloud, b) the ionization state of the heavy elements, c) the velocity dispersion of individual components.

6.1 Absorption systems of high HI column density

For mixed and low ionization level systems the dominant ions are observed for some elements (in particular Si) thus no ionization correction factor is needed to determine the metal content. Further if $N(HI)$ is very large ($\geq 10^{20}$ cm^{-2}), the HII/HI ratio averaged over the whole cloud should be small : for the photoionization models discussed in the previous section $\overline{HII}/\overline{HI}<1$ for $N(HI)>1.5\ 10^{20}$ cm^{-2} and $U<0.3$ (constraint always satisfied for M systems at $z\sim2$).

An example of such cases is the $z=1.345$ system in PKS0215+015 for which data over a large wavelength range (Blades et al. 1985) and at high spectral resolution (Bergeron and D'Odorico 1986) are available. In this system the averaged HII/HI ionic ratio is roughly equal to 2 and the metal abundances are equal to 0.1-0.2 $Z_{\odot}$. Similar values have been derived for other large $N(HI)$ systems. These abundances are within the range found for the outer part of nearby galaxies from the analysis of HII regions at radial distances close to R_H.

Depletion of heavy elements onto grains in absorbing clouds of large $N(HI)$ is small, if present at all, contrary to usually found in Galactic interstellar clouds. The depletion factor due to dust can be best estimated from ionic ratios of metals with similar ionization structure since the elemental abundances are not accurately known. For the z=1.345 system in PKS 0215+015, with $N(HI)=$ 6 10^{19} cm^{-2}, the ionic ratios N(MgII)/N(FeII) and N(MgII)/N(MnII) are equal to 1.6 and 76 respectively (Bergeron and D'Odorico 1986). These values are close to those observed in the sun, 1.10 and 131, and much smaller than in Galactic interstellar clouds, 32 and 400, where Fe and Mn but not Mg are heavily depleted (De Boer et al. 1987). Meyer and York (1987) reach similar conclusions for the z=2.811 system in PKS 0528–250 with $N(HI)=$ 1.9 10^{21}cm^{-2}. They have identified NiII absorption from this damped Lyα system. Using Smith et al. (1979) estimate of the SII column density for this absorber gives a ionic ratio N(SII)/N(NiII) of at most 27 while it is equal to 8.3 for the sun and reachers 200 in Galactic interstellar gas (De Boer et al. 1987).

Searches for molecules in absorption line systems has led to very few identifications. The only confirmed case is the detection of molecular hydrogen in the z=2.811 system in PKS 0528–250 with a column density of 10^{18} cm^{-2} thus a low ratio $N(H_2)/N(HI) \simeq 10^{-3}$ (Foltz et al. 1988). If the dust to gas ratio in this absorber was the same as in our Galaxy the expected extinction would be $A_V = 0.8$ mag. The continuum of this quasar is not reddened compared to an average quasar spectrum implying $A_V < 0.3$ mag (Wolfe, private communication). Unsuccessful searches for H_2 at the sensitivity level of the above observations have been obtained for the z=2.309 system in PHL 957 with $N(HI)=$ 2.5 10^{21} cm^{-2} and a ratio $N(H_2)/N(HI) < 4\ 10^{-6}$ (Black et al. 1988) and the z=2.796 system in Q1337+113 with $N(HI)=$ 8 10^{20} cm^{-2} and $N(H_2)/N(HI) < 3\ 10^{-5}$ (Wolfe private communication). The presence of molecular hydrogen may be linked to higher gas density since among the above cases only the absorber with detected H_2 also shows CII* fine structure absorption (Sargent et al. 1980, Chen and Morton 1984).

Dust can also be identifed by the 2175 Å broad absorption feature if it has properties similar to those present in the Galaxy. Jura (1977) was the first to attempt observing this dust signature but his search was unsuccessful. A very high S/N ratio spectrum of the BL Lac object PKS 0215+015 was recently obtained by Boissé and Bergeron (1988). It does not show any deviation from a power law spectrum implying a dust to gas ratio in the z=1.345 absorbing clouds smaller than the Galactic value by a factor 7.5, but SMC type dust is compatible with the data. There is also no dust 2175 Å absorption feature associated with the z=2.811 system in PKS 0528–250 although some molecular hydrogen is present.

6.2 Absorption systems of low HI column density

Although the intervening assumption is strongly supported by the statistical analysis of CIV samples, it has not yet been unambiguously shown that, as for the M and L systems, they are associated with galaxies (this is a harder task since only H systems at z>1.1 are known). If they do, there should be some continuity in the metal abundances between those different classes of absorbers. This remains to be shown since not much is known about the heavy element abundances of the H systems. High spectral resolution data give the velocity dispersion, but the ionization degree of the metals and both the HI and HII content are difficult to ascertain. What are the ions most easily detectable aside from CIV for absorbers with $N(HI) < 2\ 10^{17}\ cm^{-2}$? Assuming a relative abundance between the heavy elements equal to the solar value, Bergeron and Stasinska (1986) found that whatever $U < 10^{-1}$: a) the $N(CII)/N(SiIV)$ ratio is always much larger than unity, b) the $N(CII)/N(NV)$ ratio is smaller than unity for $U > 7\ 10^{-3}$, with $N(CII)/N(CIV)$ $\simeq N(NV)/N(CIV) \simeq 0.1$ at $U = 7\ 10^{-3}$.

To get the ionization level of the H systems $(U > 10^{-3})$ optically thin to ionizing radiation, it thus appears necessary to detect either CII or NV. This requires high signal to noise ratio data with $w_{r,lim}$ roughly one order of magnitude smaller than $w_r(CIV\lambda1548)$. Such surveys are just beginning to become available (Meyer and York 1987, Bergeron and D'Odorico 1988).

For the weak $z_a < z_e$ systems of high ionization level, there is a possible confusion between the NV doublet and the Lyα forest which may account for the rareness of reported NV detections. If this bias was indeed preventing the discovery of NV doublets associated with CIV systems, this could explain why the range found for the ionization parameter, thus for the density of the absorbers, is only two decades for the H and M systems at $z \simeq 2$. The existence of some observational bias due in part to confusion with the Lyα forest for the CII absorption has already been demonstrated by the recent discovery of the L sytems.

Even if only CIV and HI are observed, it is however possible to place some rough limit on $C/C_\odot$ for the mere detection of CIV. In the optically thin case the $N(CIV)/N(HI)$ ratio goes through a maximum equal to 0.6 $(C/C_\odot)$ at $U = 2\ 10^{-2}$ (Bergeron and Stasinska 1986). If the velocity dispersion of both HI and CIV is given by the observations, the estimate of $N(HI)$ and $N(CIV)$ leads to a lower limit on $C/C_\odot$. Using the column densities derived by Atwood et al. (1985) for the $z = 2.24626$ and 2.26187 weak CIV systems in Q0420–388, one gets $C/C_\odot > 1.6\ 10^{-4}$ and $6.1\ 10^{-2}$. Although the uncertainties on $N(HI)$ are large, this shows that meaningful limits can be obtained.

6.3 The Ly α forest clouds

Although very metal deficient, the Lyα forest clouds may not be primordial gas. If so an important question would be the continuity of $Z/Z_\odot$ between the H metal-rich systems and the Lyα forest. The heavy element abundances can be more easily estimated for optically thick Lyα forest absorbers, since, whatever the ionization state in their optically thin outer regions, CIV and/or CII absorption should be present in the optically thick core and lines from these ions can be outside the Lyα forest range. Lyα forest absorbers of very large $N(HI)$ are rare but do exist and a good case is the $z = 2.0769$ system in Q2206–199N in which Robertson et al. (1983) have detected a weak CIV line. New high S/N ratio data of this quasar confirm the presence of a weak CIV doublet and could give some clue to the value of $Z/Z_\odot$ in Lyα forest clouds (Carswell et al. 1988).

7. PREDICTED PROPERTIES OF THE ABSORBING GALAXIES

The first attempt to detect the absorbing galaxy associated with low z MgII absorbers was made by Weymann et al. (1978), but the sensitivity of the detectors used then was too low to unambiguously find these galaxies. Later another unsuccessfull attempt was made by Carswell et al. (1984) for the z=0.356 and 0.359 MgII absorbers in Q1101–264, although more sensitive detectors were used. Recently a MgII system, z=0.430 in PKS2128–12, was successfully identified (Bergeron 1986).

To understand whether that detection was fortuitous or not, we have estimated the expected average magnitude of an absorbing galaxy and its average angular separation from the sight line to the quasar.

The fraction of galaxies giving rise to MgII absorption systems brighter than a given luminosity L can be derived from equations (1), (2) and (3)

$$\frac{dN(x)/dz}{dN(x=0)/dz} = \frac{\int_x^\infty \sigma dn}{\int_0^\infty \sigma dn} = \frac{\Gamma(x,7/12)}{\Gamma(7/12)}. \tag{8}$$

We find that 50 % of the absorbing galaxies are brighter than 0.3 L_* (M(B) = $-$19.3). For luminosities $L \geq L_*$ this percentage reduces to 19 % and it reaches 72 % for $L \geq 0.1$ L_*. Therefore most of the galaxies responsible for MgII absorption in quasar spectra are brighter than the LMC (x=0.174). The apparent magnitude of 0.3 L_* galaxies is given in Table 3 for different redshifts. Since K correction has not been applied in the determination of the apparent magnitude, the values of m in Table 3 correctly refer to an effective band width of $\Delta\lambda = (1+z)\Delta\lambda(B)$ centered on $\lambda_{obs}(B) = 4400(1+z)$Å.

Table 3. Apparent magnitude and size of 0.3 L_* absorbing galaxies

z	m	λ_{obs}(B) (Å)	θ (arcsec)
0.2	22.1	5280	8.6
0.4	22.8	6160	5.4
0.8	24.5	7920	3.8

The expected angular separation between the absorbing galaxy centre and the quasar can be estimated from equation (1), using the value of R_o given in Table 1 for $\bar{z}=0.51$ MgII sample with $w_{r,lim}=0.6$ Å. This angular separation, independent of H_o, can be written

$$\theta_o = 5.23\left(\frac{dN/dz'}{1+z'}\right)^{0.5}\frac{(1+z)^2}{z+z^2/2} \quad \text{arcsec,} \tag{9}$$

$$\theta(x) = \theta_o x^{5/12}. \tag{10}$$

Numerical estimates of $\theta(x=0.3)$ are given in Table 3.

The absorbing galaxies should thus be bright enough and well separated on the sky from the quasr image to be easily detectable with present day technics. Further the galaxy G2128–12, identified with the MgII absorber at z=0.430 towards PKS2128–12, is 8.6 arcsec away on the sky from the quasar and has a magnitude m(V)=21.5, properties close to the predicted ones. This led us to think that the negative attempts reported prior to 1985 were due in part to the low sensitivity of the instrumentation used.

8. THE MGII ABSORBING GALAXY SAMPLE

Our sample comprises two MgII absorption systems at the quasar emission redshift z_e and 9 at $z_a \ll z_e$. The redshift range covered by the absorbers is 0.16–0.79 and the rest equivalent width of MgIIλ2796 has values between 0.40 and 3.0 Å. When this survey started the homogeneous sample of MgII absorption systems of Tytler et al. (1987) was not yet available, and we selected published cases with data of good S/N ratio restricted to resolved MgII doublets. To our knowledge there is no obvious bias in our sample, at least for the z<0.5 absorbers. Searches for z>0.5 MgII absorbing galaxies were done only in optimum seeing conditions (FWHM $\leq$ 1.3 arcsec). All the observations were carried out with the 3.6m telescope at the European Southern Observatory (ESO), La Silla, Chile. Since at first we only got positive detections of the absorbing galaxy, we have started observing previous negative cases. Recent searches for the absorbing galaxies done by other researchers to a sensitivity level similar to that of our observations are also included in the sample.

We briefly summarize results for the $z_a \sim z_e$ systems. They provide information on the cluster usually associated with radio–loud quasars. We got observations only of two absorbers G2135–14 at z=0.20 and G1912–54 at z=0.401 (Bergeron and Boissé 1986). Their angular distance to the quasr is smaller by a factor of 8 and 2 respectively than expected from the distance of the nearest neighbour in the quasr sample of yee and Green (1984). Recent data obtained for the galaxy closest on the sky to the quasar PKS1912–54 (angular separation of 8.3 arcsec) show that its redshift is identical to that of the MgII absorption system. In the galaxy spectrum, there are strong stellar absorption lines of CaII,

HI and the G band and possibly a weak [OII] emission line (Bergeron and Boissé, 1988).

The $z_a \ll z_e$ sample comprises 13 systems out of which 10 with a clear identification of the absorbing galaxies. Some properties of the absorbers are given in Table 4. The angular separation between the galaxy centre and the quasar, θ, varies between 0.9 to 3.3 $R_H(L_*$ at $z=0)$ with a mean equal to 2.3 R_H for the 10 identified galaxies. For two MgII systems there is a possible candidate for which either inconclusive (G1101–26) or no (G1229–02) spectroscopic observations are available. These two possible absorbers have angular separation and magnitude similar to those derived for the identified cases.

All the MgII galaxies are intrinsically bright. This is not a consequence of an observational bias since these absorbers are about 2 magnitudes brighter (or more) than our detection limit m(r)$\simeq$ 24. Galaxies of the SMC and LMC type could indeed have been detected up to redshifts of 0.21 and 0.50 respectively. The

Table 4. Properties of MgII absorbing galaxies

quasar	z_a	type	θ (")	R/R_H	m(r)	M(r)	references
0151+04	0.160	em	6.4	1.1	19.1	−20.7	1
		em	10.9	1.9	20.2	−19.6	1
1127−14	0.313	em	9.5	2.7	19.3	−22.1	2
2128−12	0.430	em	8.6	2.9	21.0	−21.1	3
1511+10	0.437	em	6.7	2.4	22.0	−20.3	2
1038+06	0.441	em	9.6	3.3	21.0	−21.3	2
0109+20	0.535	abs	7.0	2.7	21.7	−21.0	2
2145+06	0.790	em	5.9	2.7	22.1	−21.7	2
0952+17	0.238	no galaxy found at z_a					2
1229−02	0.395[a]	—	8.6[b]	2.6	22.3	(−19.8)	2
1332+55	0.373	abs	5.0	1.6	20.7:	−21.3:	4
1209+07	0.393	HII	7.1	2.3	21.9	−20.2	5
0235+16	0.524[a]	HII	2.3	0.9			6
1101−26	0.356	—	11.0[b]	3.3	20.7	(−21.2)	7,2
	0.359						

[a] sytems with 21cm absorption detected
[b] possible candidates, not yet confirmed spectroscopically
references. 1 : Bergeron et al. (1988). 2 : this work, 3 : Bergeron (1986), 4 : Miller et al. (1987), 5 : Cristiani (1987), 6 : Smith et al. (1977), 7 : Carswell et al. (1984).

spread in M(r) is only of about 2 magnitudes with an average M(r)=−21.1. No K correction has been applied to derive M(r). Since the absorbers cover a wide redshift range, it is possible to derive an absolute B magnitude in the galaxy rest frame only for some of the galaxies (z≤0.5). For the overall survey we can derive an intrinsic M(λ3700) magnitude as done by Bergeron and Boissé (1988) for a sample somewhat larger than presented in this review.

The $z_a \ll z_e$ absorbing galaxies are not in dense clusters. For most fields we have at least the second galaxy closest on the sky to the quasar and our largest number of galaxies per field with identified redshfit is five. In only two cases there is a second object at $z \sim z_a$. Therefore most of the absorbing galaxies are field galaxies and sometimes they belong to loose groups.

A striking property of the galaxies with large gaseous envelopes is their spectral type. They all have blue continuum with $F(\lambda) \sim$ constant shortwards of λ3727, down to λ_r2140 for the z=0.79 absorber. Eight out of the ten galaxies show emission lines, essentially [OII]λ3727, among which two resemble giant HII regions with a weak continuum and very strong Balmer emission lines and forbidden [OII] and [OIII] lines (see Fig. 2 in Cristiani 1987). The whole gaseous envelope of these HII galaxies should have detectable extended [OII] emission as observed for G0235+16 (Cohen et al. 1987). The other six emission line galaxies have spectra similar to that of G2128−12 (see Fig. 3 in Bergeron 1986) with both emission lines and resolved stellar absorption lines. Their [OII]λ3727 rest equivalent width ranges from 15 to 30 Å whereas that of HII galaxies reaches about 100 Å. Thus all the galaxies with very extended gaseous envelopes identified so far show signs of recent or present strong stellar formation activity. The percentage of field galaxies at z∼0.5 with colors and spectra similar to those of the absorbing galaxies is not accurately known. Ellis (1988) gives a fraction of 20–30 % at z∼0.3 for galaxies with unusually strong [OII], w_r([OII]λ3727)>40 Å, which is comparable to the number given for cluster at similar redshift. From his multicolor photometric survey Koo (1986) suggests that the fraction of field galaxies at z∼0.4 intrinsically bluer than B−V=0.7 is ∼74 % while the local value is ∼ 40%. The galaxies of our sample are at least as blue as G2128−12 for which we got B−V=0.5.

If we consider the number given by Ellis we should multiply the values for R_* given in Table 1 by roughly $\sqrt{2}$ which would imply halos sizes larger than actually observed. Therefore our survey suggests that a large fraction (∼ 50 % or more if our adopted value of ϕ_o is an overestimate : see e.g. De Lapparent et al. 1986) of the field galaxies have properties similar to those observed for galaxies with very extended halos. To clarify this problem would require large spectroscopic surveys of field galaxies as those, unpublished, of Dressler, Ellis and Koo. Our test sample of field galaxies, which are not identified with the absorber, will also be of some value.

Extended gaseous envelopes associated with nearby galaxies have been detected by their CaII absorption in quasar spectra. The first discovery was made by Boksenberg and Sargent (1978) and there is now 5 identified cases presented in

Table 5. The angular separation between the galaxy centre and the quasar ranges between 0.7 and 1.3 R_H whereas for the MgII galaxies it could reach 3.3 R_H. The large CaII galaxy–quasar pair survey done recently by Morton et al. (1986) has

Table 5. Properties of the CaII absorbing galaxies

quasar	z_a	R/R_H	w_r(CaII K) (Å)	w_r(MgII 2796) (Å)	N(HI) (cm^{-2})	references
0955+32	0.0047	0.75	0.43	6.5 (D)	21cm abs	1,2
1327—20	0.0180	0.91	0.49		21cm abs	2,3
0154+04	0.0188	1.27	0.57			4
2020—37	0.0287	0.77	0.35		21cm abs	5,3
0446—21	0.0668	1.05	0.96			6
2128—12	0.4299	2.9	0.11	0.40	$2\ 10^{17}$—$8\ 10^{18}$	7,2
0454—22	0.4744	—	0.13	1.37		8
0002—42	0.8366	3.6^a	0.60	4.68		9,2

D refers to both lines of the doublet
^{a} possible candidates, not yet confirmed spectroscopically
references. 1 : Boksenberg and Sargent (1978). 2 : Bergeron et al. (1987b). 3 : Boissé et al. (1988). 4 : Bergeron et al. (1988). 5 : Boksenberg et al. (1980). 6 : Blades et al. (1981). 7 : Bergeron (1986). 8 : Robertson et al. (1988). 9 : Bergeron and Boissé (1988).

not revealed any new CaII envelopes. For most of their cases the angular separation is unfortunately larger than 3.5 R_H, distance at which we expect very few detection of gaseous envelopes from the results of our MgII galaxy survey.

For 3 out of the 4 CaII absorption systems at low z detected in spectra of radio–loud quasars there is an associated 21cm absorption (Table 5). The small sizes of these CaII envelopes are consistent with those usually observed for HI discs, which have been found recently to have sharp edges around 1.5 R_H (see e.g. Sancisi 1988). The identified MgII galaxy with detected 21cm absorption (G0235+16 in Table 4) is also of small extent.

Ca II absorption from MgII envelopes has been detected in a few cases reported in Table 5. They are weaker than observed for the CaII galaxies, except for a very strong MgII system. Further for one of the low redshift CaII galaxies a strong MgII absorption has been detected in the UV. The results presented in Table 5 may suggest a correlation between the radial distance to the galaxy centre and the absorption line equivalent width (or ionic column density). However there is only a weak trend of increasing projected separation with decreasing MgII rest equivalent width for the sample of identified MgII absorbers (Bergeron and Boissé 1988). Column densities should be used instead of w_r but they cannot always be determined from available data. Other parameters such as the galaxy luminosity (scaling R with L) should also be considered.

The magnitude M(r) of the galaxies with large MgII envelopes are within the range –20.3 to –22.1 with $<M(r)>$=–21.1. Including the two additional possible candidates increases only slightly this range towards the faint end to –19.8 and does not modify the value of $<M(r)>$. The average redshift of the galaxy sample equals 0.43 with an rms of 0.16. Thus the absolute magnitudes M(r) given in Table 4 are for most MgII galaxies fairly close to an intrinsic B magnitude, and the majority of absorbing galaxies are brighter than L_*.

Our success rate of the absorber identification is 77 % and increases to 92 % when including the possible candidates. As already mentioned, absorbers with luminosity as that of the LMC could have been detected up to z=0.50. Our results appear inconsistent with a MgII galaxy luminosity function without cut–off towards the faint end since a detection rate of about 80 % and no cut–off would imply that half of the identified absorbers be fainter than M(r)=–19.8 whereas none was so.

To determine more accurately the expected success rate one should not only consider a cut–off in the luminosity function at $x_c = L_c/L_*$ but also the effect of the seeing which prevents detecting galaxies too close on the sky to the quasar. The latter implies that galaxies of small size, thus intrinsically faint, cannot be identified from ground–based observations. However if we assume that objects closer on the sky than θ_s=2 arcsec from the quasar are not detectable, only very faint galaxies $x < 0.03$ are unobservable at z=0.4. The probability to identify the absorber with the above assumptions is

$$P = [\Gamma(7/12, x_c) - x_s^{5/6}\Gamma(-1/4, x_c)]/\Gamma(7/12, x_c), \tag{11}$$

where x_s is the normalized luminosity of a galaxy with a MgII envelope of radius θ_s. Through x_s P is a function of z. To reach a success rate of about 80 % at z=0.4 and 0.8 implies a cut–off at x_c=0.1 and 0.3 respectively. This is consistent with the faintest magnitude observed M(r)=–19.8 or x=0.48. Introducing a detectability limit, $x_V > x_c$, decreases further P which is still given by eq. (11) but with x_c replaced by x_V in the numerator.

A cut–off in the luminosity function should be taken into account when estimating the MgII galaxy radius. Values of R_o given in Table 1 should be multiplied by $[\Gamma(7/12)/\Gamma(x_c, 7/12)]^{0.5}$ which equals $\sqrt{2}$ for $x = 0.3$.

Another way to reconcile our observations with a luminosity function without cut–off would be to invoke a cosmological evolution of the galaxy luminosity. However this assumption is not supported by the galaxy surveys at z~0.4

of Koo (1986) and Ellis (1988).

It is unlikely that a faint galaxy right on the sight line to the quasar be responsible for the absorption system while the identified galaxy would only be its nearest neighbour. Both objects should have a redshift equal to z_a within $\sim 200\,\mathrm{km\,s^{-1}}$, i.e. be correlated in position and velocity space which would suggest that bright galaxies are always surrounded by faint satellite galaxies at a radial distance $R \sim 3\ R_H$. We can consider the sight line to quasars with absorption systems as a random point in the sky to estimate the expected angular separation between a faint undetected galaxy and a brighter one. Although in our sample the absorber and the quasar have unrelated redshifts, we can use results of the survey by Yee and Green (1984) for galaxies either possibly spatially correlated with quasars or in test fields. The former give, for radio–loud quasars, the density of galaxies in clusters and the latter the density of field galaxies. Their magnitude limit is roughly $m(\mathrm{Gunn\ r}) \sim 21$. Galaxies we identified as the absorbers at $z \sim 0.4$ would thus have been detectable in their survey but not LMC type galaxies. The average angular distance to the nearest bright neighbour is 18.4 arcsec in a cluster at $z = 0.30$ to 0.45 and ~ 31.5 arcsec for field galaxies (as determined for the radio-quiet quasar sample and the two test field surveys). For the same redshift range the average angular separation between the absorbing galaxies of our sample and an hypothetical faint galaxy on the sight line to the quasar is 7.0 arcsec, value incompatible with those derived by Yee and Green (1984).

In conclusion, galaxies giving rise to MgII absorption line systems in quasar spectra have (i) emission spectra with strong emission lines and/or a blue continuum which suggests on–going or recent stellar formation activity, (ii) large gaseous envelopes $R(\mathrm{MgII}) \sim 2.5\ R_H$ (in the assumption of spherical halos), (iii) they are bright $M(r) \sim -21$ (for $z \sim 0.4$ this is roughly an intrinsic B magnitude) (iv) the contribution of dwarf galaxies to absorption line systems is most probably negligible except if they always cluster around bright ones at a radial distance of $\sim 2.5\ R_H$.

There are two main questions which remain open. The first concerns the existence of bright field galaxies without large gaseous envelopes, problem possibly linked to the fraction of field galaxies without strong stellar formation activity. It will require a survey of high z galaxy–quasar pairs followed by a search for MgII absorption in the quasar spectra. The second is related to the range in ionization level of the absorbers, thus the existence of gaseous halos of very high ionization state, which calls for a low redshift CIV absorption line survey with the Hubble Space Telescope.

REFERENCES

Atwood, B., Baldwin, J.A. & Carswell, R.F. (1985). Ap. J. *292*, 58.

Bahcall, J.N. & Spitzer, L. (1969). Ap. J. Letters *156*, L63.

Bechtold, J., Green, R.F., Weymann, R.J., Schmidt, M., Estabrook, F.B., Sherman, R.D., Wahlquist, H.D. & Heckman, T.M. (1984). Ap. J. *281*, 76.

Bergeron, J. (1986). Astron. Ap. Letters *155*, L8.

Bergeron, J. & Kunth, D. (1983). Mon. Not. R. Astr. Soc. *205*, 1053.

Bergeron, J. & Boissé, P. (1984). Astron. Ap. *133*, 374.

Bergeron, J. & Boissé, P. (1986). Astron. Ap. *168*, 6.

Bergeron, J. & D'Odorico, S. (1986). Mon. Not. R. Astr. Soc. *220*, 833.

Bergeron, J. & Stasinska, G. (1986). Astron. Ap. *169*, 1.

Bergeron, J., Savage, B. & Green, R.F. (1987a). In *The Scientific Accomplishments of the IUE*. ed. Y. Kondon. Reidel, Dordrecht, p. 703.

Bergeron, J., D'Odorico, S. & Kunth, D. (1987b). Astron. Ap. *180*, 1.

Bergeron, J., Boulade, O., Kunth, D., Tytler, D., Boksenberg, A. & Vigroux, L. (1988). Astron. Ap., *191*, 1.

Bergeron, J. & D'Odorico, S. (1988). in preparation.

Bergeron, J. & Boissé, P. (1988), in preparation.

Black, J.H., Chaffee, Jr., F.H. & Foltz, C.B. (1988). preprint.

Blades, J.C., Hunstead, R.W. & Murdoch, H.S. (1981). Mon. Not. R. Astr. Soc. *194*, 669.

Blades, J.C., Hunstead, R.W., Murdoch, H.S. & Pettini, M. (1985). Ap. J. *288*, 580.

Boissé, P. & Bergeron, J. (1985). Astron. Ap. *145*, 59.

Boissé, P. (1988). In *High Redshift and Primeval Galaxies*. Eds J. Bergeron, D. Kunth & B. Rocca–Volmerange. Editions Frontières, in press.

Boissé, P., Dickey, J.M., Kazès, I. & Bergeron, J. (1988). Astron. Ap. J. *191*, 193.

Boissé, P. & Bergeron, J. (1988). Astron. Ap. *192*, 1.

Boksenberg, A. & Sargent, W.L.W. (1978). Ap. J. *220*, 42.

Boksenberg, A., Danziger, I.J., Fosbury, R.A.E. & Goss W.M. (1980). Ap. J. Letters *242*, L145.

Boroson, T., Sargent, W.L.W., Boksenberg, A. & Carswell, R.F. (1978). Ap. J. *220*, 772.

Boulade, O., Kunth, D. & Tytler, D. (1988). in preparation.

Carswell, R.F., Morton, D.C., Smith, M.G., Stockton, A.N., Turnshek, D.A. & Weymann, R.J. (1984). Ap. J. *278*, 486.

Carswell, R.F., Robertson, J.G. & Shaver, P.A. (1988). in preparation.

Caulet, A. & York, D.G. (1987). In *QSO Absorption Lines: Probing the Universe, A Collection of Poster Papers*, eds. J.C. Blades, C.A. Norman & D. Turnshek (STScI Publications), p.76.

Chen, J.S. & Morton, D.C. (1984). Mon. Not. R. Astr. Soc. *208*, 167.

Cohen, R.D., Smith, H.E., Junkkarinen, V.T. & Burbidge, E.M. (1987). Ap. J. *318*, 577

Cristiani, S. (1987). Astron. Ap. Letters *175*, L1.

De Boer, K.S., Jura, M.A. & Shull, J.M. (1987). In *Scientific Accomplishments of the IUE*, ed. Y. Kondon. Reidel, Dordrecht, p.485.

De Lapparent, V., Geller, M.J., Huchra, J.P. (1986). Ap. J. Letters *302*, L1.

Ellis, R. (1988). In *High Redshift and Primeval Galaxies*. Eds J. Bergeron, D. Kunth & B. Rocca–Volmerange, Editions Frontières, in press.

Foltz, C.B., Weymann, R.J., Peterson, B.M., Sun, L., Malkan, M.A. & Chaffee, Jr., F.H. (1986). Ap. J. *307*, 504.

Foltz, C.B., Chaffee, Jr. F.H. & Black, J.H. (1988). Ap. J. *324*, 267.

Gondhalekar, P.M. (1983). Mon. Not. R. Asrr. Soc. *204*, 997.

Jura, M. (1977). Nature *266*, 702.

Koo, D.C. (1986). Ap. J. *311*, 651.

Kunth, D. (1988). In *High Redshift and Primeval Galaxies*. Eds J. Bergeron, D. Kunth & B. Rocca–Volmerange. Editions Frontières. in press.

Lanzetta, K.M., Turnshek, D.A. & Wolfe A.M. (1987). Ap. J., *322*, 739.

Meyer, D.M. & York, D.G. (1987). Ap. J. Letters *315*, L5.

Miller, J.S., Goodrich, R.W. & Stephens, S.A. (1987). Astron. J., *94*, 633.

Morris, S.L., Foltz, C.B., Weymann, R.J., Schectman, S., Price, C., Boroson, T.A. & Turnshek, D.A. (1986). Ap. J. *310*, 40.

Morton, D.C., York, D.G. & Jenkins, D.B. (1986). Ap. J. *302*, 272.

Pettini, M., Hunstead, R.W., Murdoch, H.S. & Blades, J.C. (1983). Ap. J. *273*, 436.

Robertson, J.G., Shaver, P.A. & Carswell, R.F. (1983). XXIV Colloque International de Liège, ed. J.P. Swings, p.602.

Robertson, J.G., Morton, D.C., Blades, J.C., York, D.G. & Meyer D.M. (1988). Ap. J., in press.

Sancisi, R. (1988). In *QSO Absorption Lines : Probing the Universe*. Eds C. Blades, C. Norman & D. Turnshek, Cambridge University Press, in press.

Sargent, W.L.W., Young, P.J., Boksenberg, A. & Tytler, D. (1980). Ap. J. Suppl. *42*, 41.

Sargent, W.L.W. (1988). In *QSO Absorption Lines: Probing the Universe*. Eds C. Blades, C. Norman & D. Turnshek, Cambridge University Press, in press.

Sargent, W.L.W., Boksenberg. A. & Steidel, C.C. (1988). in preparation.

Smith, H.E., Burbidge, E.M. & Junkkarinen, V.T. (1977). Ap. J. *218*, 611.

Smith, H.E., Jura, M. & Margon, B. (1979). Ap. J. *228*, 369.

Snijders, M.A.J., Boksenberg, A., Penston, M.V. & Sargent, W.L.W. (1982). Mon. Notr. R. Astr. Soc. *201*, 801.

Tytler, D., Boksenberg, A., Sargent, W.L.W., Young, P. & Kunth, D. (1987). Ap. J., Suppl. *64*, 667.

Wagoner, R. (1967). Ap. J. *149*, 465

Weymann, R.J., Boroson, T.A., Peterson, B.M. & Butcher, H.R. (1978). Ap. J. *226*, 603.

Weymann, R.J., Williams, R.E., Peterson, B.M. & Turnshek, D.A. (1979). Ap. J. *234*, 33.

Wolfe, A.M. (1988). In *QSO Absorption Line : Probing the Universe*. Eds
    C. Blades, C. Norman & D. Turnshek, Cambridge University Press, in press.
Yee, H.K.C. & Green, R.F. (1984). Ap. J. *280*, 79.
Young, P., Sargent, W.L.W. & Boksenberg, A. (1982). Ap. J. Suppl. *48*, 455.

THE PERIODICITY IN THE REDSHIFT DISTRIBUTION OF PRIMORDIAL HYDROGEN CLOUDS

Yaoquan Chu and Xingfen Zhu
Institute for Astrophysics, University of Bonn, F.R.G.
The Center for Astrophysics, University of Science and Technology of China, China.

A great number of absorption lines observed on the short-wavelength side of the Ly-alpha emission line (i.e. the Ly-alpha forest) in high redshift quasars is now generally believed as being due to Primordial Hydrogen Clouds (PHC). It is widely accepted that these PHC are located at cosmological distances given by their redshifts. The PHC become a new class of objects which have high redshifts comparable with quasars.
The aim of this research is to compare the distribution of PHC with that of quasars. It has been discussed for a long time that there maybe exist a large scale inhomogeneity in the space distribution of quasars (e.g., a periodicity in the distribution of emission line redshifts, see, Fang, Chu, et. al. 1982). If the inhomogeneous distribution of quasars is a real cosmological effect and not due to the selection effect, it is reasonable to predict that a similar feature should also exist in the distribution of other high redshift objects. Indeed, the results we present here show some evidence in favor of the existence of periodicity in the distribution of redshifts of PHC.
1) The Redshift Distribution of PHC
We use a quite reasonable homogeneous sample of Ly-alpha absorption lines available in the literature with high resolution (0.8 -1.5 Å FWHM) compiled by Murdoch et. al. (1986), which includes 11 quasars and 277 Ly-alpha absorption lines. The observed absorption redshifts range from 1.510 to 3.715.
At first we plot a histogram of all redshifts of PHC . Before we try to interpret such distribution we must point out that there is a strong selection effect due to the different wavelength coverages of different quasars. A spectrum of one quasar can only contribute the Ly-alpha absorption line count between emission line redshift Z_m and the low limit of Ly-alpha absrption redshift Z_o which we accept in our sample. Therefore we could correct the initial histogram by multiplying a factor that is propotional to the number of quasar spectra which cover the redshift bin we study. The corrected histogram is shown in Fig.1.
Several obvious features can be easily seen in Fig. 1.
First, there is a very significant peak at $Z= 2.9$. A similar feature also exists in the distribution of emission line redshifts at $Z= 2$,

N. Kaiser and A. N. Lasenby (eds.), The Post-Recombination Universe, 225–227.
© *1988 by Kluwer Academic Publishers.*

which have been discussed for about 20 years. (e.g. Burbidge, 1968)
The reality of the peak at Z= 2 in the distributin of emission line redshifts of quasars is doubted by many authors due to the fact that there
is a strong selection effect in the identification of quasars (i.e.
preference for identification of Ly-alpha lines). But in the case of
absorption lines, the peak appearing in the corrected histogram is a new
result and we don't think it also could be interpreted using the same
selection effect. The similar feature in both distributions of quasars
and the PHC may imply that the existence of a peak is a real feature,
which reflects an interesting character of the large scale structure in
the universe.
Secondly, we find that the number of PHC increases with increasing of
redshift.
Thirdly, the number of PHC in the range of Z › 2.9 is significantly
larger than that in the range of Z ‹ 2.7. It means that the PHC mostly
exists in the early universe. Moreover similar things also happen for
quasars ,that is the number of quasars also have an obviously change
around Z= 2.5, but in opposite direction (Veron, 1987).
2) Power-Spectrum Analysis
    Besides the peak at Z= 2.9, there are some other peaks in Fig. 1. So
a natural question is that does the periodicity also exist in the
distribution of redshifts of PHC?
We have found a periodicity component in the distribution of emission
line redshifts for quasars, using the power-spectrum analysis. (Fang,
Chu, et. al. 1982). Therefore it is interesting to analyse the
power-spectrum of the redshift distribution for PHC.

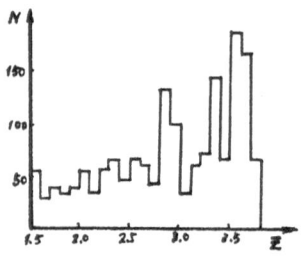
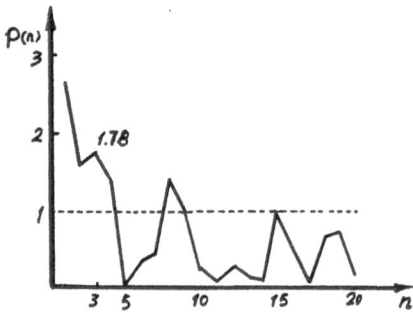

Fig.1   Fig.2

Fig. 2 shows the power-spectrum for all PHC redshifts in which we take
the argument x= Ln (1+Z). The redshift range is 1.5 ‹ Z ‹ 3.7. We find
a periodic component at n= 3, which corresponds to the length of period
of 0.2. Although the significant confidence is not very high, about 83%,
    but the length of period is just the same as we found in the
distribution of emission line redshifts of quasars (0.205)!
3) The Peaks in the Distribution of R
The feature in the distribution of PHC, which we disscussed above,
should also appear in statistics of other observational quantities of
PHC. One of such quantities for absorption lines is R=(1+Zem)/(1+Zabs).

R is also related to the ejective velocity which has been widely discussed in the study of absorption lines.

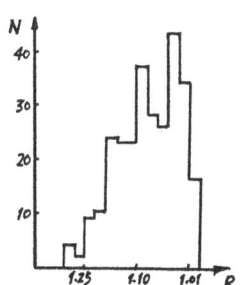

In Fig.4 we show the histogram of R for all quasars in our sample. We can find that several peaks obviously appear in the distribution of R. Such peaks can interpreted as follows:
If the periodicity is a real cosmological feature, it should exist in both quasars and PHC redshifts. As discussed by Chu, Fang and Liu(1984), for a set of objects located in the same direction, the distribution of redshifts should be periodic and have a series of peaks. The redshift values corresponding to the peaks are given by
$Ln(1+z) = \Lambda n + C$
Here Λ is a constant, C is a function of direction of the object and n equals some positive integer.

For emission line redshifts of quasars we have:
$Ln(1+Z_{em}) = \Lambda n + C1$
for PHC we also have :
$Ln(1+Z_{ab}) = \Lambda n' + C2$
Since the quasars and the PHC are located on the same line of sight, the initial phases are the same, C1= C2, so we have:

$$R = \frac{1+Z_{em}}{1+Z_{ab}} = e^{\Lambda \Delta n}$$

where $\Delta n = 0,1,2,3,\ldots\ldots$
This indicates that the histogram of R should have a series of peaks. If we take the value of Λ as 0.05 we find:
$\Delta n = 1$, R= 1.05
$\Delta n = 2$, R= 1.10
$\Delta n = 3$, R= 1.16
These results are consistent with the futures in Fig.3. It is interesting to point out that a similar result have been found by Chu, Fang and Liu (1984) in the distribution of relative velocity , using a large sample of metal absorption lines in quasars from the Hewitt and Burbidge Catalog(1980).

REFERENCES
Burbidge,G., 1968, Ap. J. letter, vol.154, L41.
Chu, Y., Fang, L.Z., and Liu, Y., 1984, Astrophys. letters, 24, 95.
Fang, L.-Z., Chu, Y., Liu,Y., Cao, Ch., 1982, Astron. Astrophys. vol.106, 287.
Murdoch,H.S., Hunstead,R.W., Pettini,M., Blades,J.C. 1986, Ap.J. vol.309, 19.
Veron,P. Astron. Astrophys. vol.170, 37.

QUASAR CLUSTERING AND GRAVITATIONAL LENSES

P.A. Shaver
European Southern Observatory
Karl-Schwarzschild-Str. 2
D-8046 Garching bei München
Federal Republic of Germany

ABSTRACT. The evidence for quasar clustering and its evolution is reviewed. The problem of distinguishing physical pairs from gravitationally lensed quasars is discussed, and it is shown that lensing probably contributes significantly to the incidence of very close pairs at high redshifts.

1. INTRODUCTION

Until recently, quantitative information on the large-scale structure of the universe was only available at $z \sim 0$ (galaxies) and $z \sim 1000$ (microwave background). With increasing interest in the evolution of structure and the formation of galaxies, more emphasis has been placed on observations pertaining to the intervening 99% of the history of the universe. Fortunately it is now possible to make deep, large-scale quasar surveys, and our knowledge of the clustering of quasars is rapidly improving. At the same time, deep galaxy surveys extend our knowledge of galaxy clustering out to $z \sim 0.5$, and studies of quasar absorption lines provide relevant information up to $z \sim 4$. With such observational material available, we may begin to piece together the evolution of structure over the history of the universe.

A parallel development over the last decade has been the discovery of several possible gravitational lenses. Lensing can distort our view of the distant universe, and confuse our understanding of high-redshift structure, at least on small scales. Gravitationally lensed quasars can be very difficult to distinguish from physical pairs of distinct objects, but it is important for our understanding of both gravitational lensing and physical clustering that this distinction be made.

A review is given here of our present knowledge of the clustering of quasars, of other evidence related to structure at high redshifts, and of the likely "contamination" by gravitational lensing.

229

N. Kaiser and A. N. Lasenby (eds.), The Post-Recombination Universe, 229–243.

2. THE CLUSTERING OF QUASARS

2.1. Small-Scale Clustering

Systematic searches for physical clustering of quasars began earlier this decade, when suitable quasar samples first became available (Osmer, 1981; Webster, 1982). The results were negative, in part because of the small sizes of these samples. The first tentative detections were made using a large but inhomogeneous quasar catalogue (Shaver, 1984), and a small but deep UVX survey (Boyle, 1986; Shanks et al., 1986, 1988). Larger homogeneous samples are now being assembled, and quasar clustering has recently been confirmed at the 5σ level.

The clustering properties of a sufficiently homogeneous sample can be examined by comparison with a randomly generated sample which is subject to exactly the same selection effects in both sky coordinates and redshift. The correlation function results from a comparison of the observed and random samples as a function of linear separation of quasar pairs. This has been done for two samples of 376 and 354 quasars respectively by Iovino & Shaver (1988) and Shanks et al. (1988), and in both cases clustering on comoving scales < 10 h^{-1} Mpc is detected at the 4-5σ level. When these results are combined, and allowance is made for overlap in the two samples, the significance level for the clustering found is 5.5σ. Furthermore, the clustering at $z < 1.5$ appears to be twice that at $z > 1.5$, with a significance of $\sim 2\sigma$. The clustering of quasars is therefore established beyond any doubt, and there is some indication of possible evolution.

Other recent estimates of quasar clustering have been made using different techniques, by Kruszewski (1987) and the author. Kruszewski has analysed nine homogeneous samples totalling 629 quasars, using random reassignment of redshifts within each sample to generate the comparison sample, and finds clustering at a high significance level, predominantly at lower redshifts. A new method, "normalization to large scales" (Shaver, 1984), allows clustering analysis of heterogeneous catalogues, unaffected by observational selection effects. As any clustering is expected to occur predominantly on small scales, it can be detected by comparing the incidence of quasar pairs of small projected and/or radial separation with those of large separation; both groups are subject to the same selection effects, which therefore cancel in the comparison. Recent applications of this technique by both Kruszewski and the author to the Véron catalogue (Véron-Cetty & Véron, 1987), containing about 3500 quasars, reveal significant (6σ) redshift-dependent clustering.

Figure 1 shows the correlation functions at low and high redshifts from Iovino & Shaver (1988) and Kruszewski (1987). A positive signal is prominent only at comoving separations < 10 h^{-1} Mpc, and at low redshifts. The redshift dependence of the correlation amplitude at 10 h^{-1} Mpc, from several different studies, is shown in fig. 2. It appears to be faster than expected for stable clustering, although this remains to be confirmed.

An extrapolation to the lowest redshifts would be well in excess

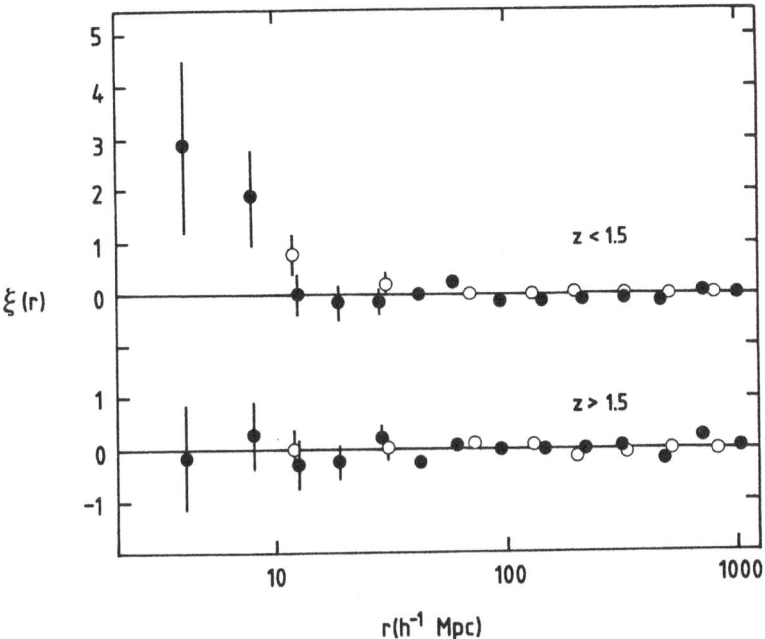

Figure 1. - Quasar two-point correlation function at low (z < 1.5) and high (z > 1.5) redshifts, with q_0 = 0.5. The filled circles are from Iovino & Shaver (1988), and the open circles are from Kruszewski's (1987) analysis of 9 homogeneous samples.

of the correlation amplitude for galaxies - closer to that for radio galaxies. This would be consistent with the similarity beween the radio galaxy and quasar cross-correlation amplitudes with galaxies (Table 1 below), if there is a common relationship between the autocorrelation amongst specific types of objects and the cross-correlation between those objects and galaxies generally. The ratio of the cluster-cluster and cluster-galaxy correlation amplitudes (cluster richness ~ 1) is about 2.6 from Table 1, and for radio galaxies (Fanaroff-Riley class I) it is about 4.5. If a similar ratio holds for quasars, then the expected correlation amplitude at low redshift in fig. 2 would be in the range 1-3, consistent with the direct measurements of quasar clustering.

2.2. Large-Scale Clustering

Clustering analyses such as those reported above are particularly well suited for the study of clustering which is concentrated to small scales, but non-random three-dimensional distributions covering a range of larger scales can be imagined which would not be prominent in the distribution of pair separations, and structures on a scale comparable to that of the sample studied could

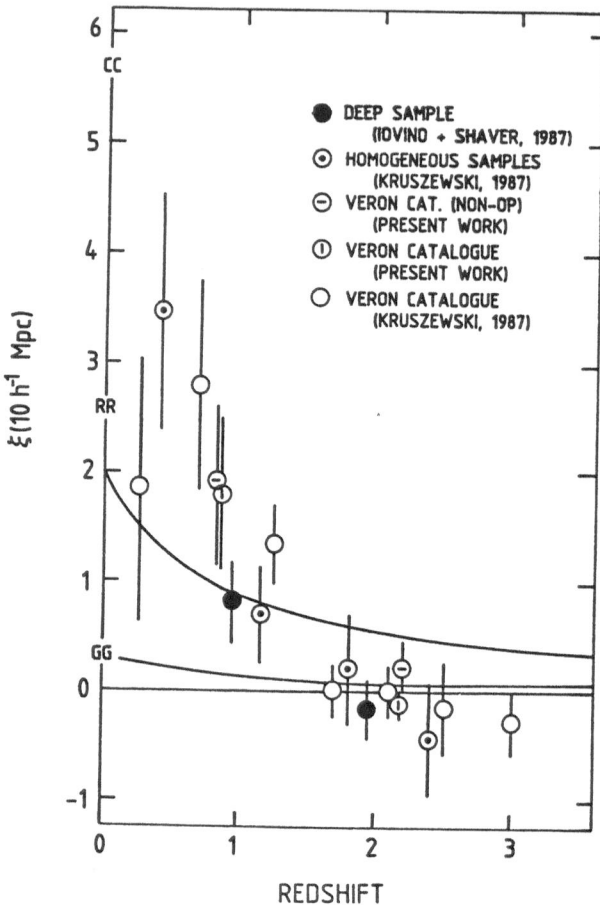

Figure 2. - Amplitude of the quasar correlation function at 10 h^{-1} Mpc
(comoving, q$_0$ = 0.5), as a function of redshift, from
several different studies as indicated. On the left axis are
marked the amplitudes of the correlation function at z $\lesssim$ 0.1
for galaxies (GG), clusters (CC - Bahcall & Soneira, 1983),
and radio galaxies (RR - Peacock et al., 1988). The two
curves show the expected evolution for stable clustering.

go undetected. Furthermore, the deep homogeneous samples that have
been used in clustering analyses are thin needles, extended
significantly only in redshift. They are also certainly well suited
for studies of small-scale clustering, but the only large-scale
structure to which they are sensitive is in the radial direction; if
this structure evolves it will show up predominantly at one epoch, and
clustering analyses in radial coordinates may wash it out. Such
structure is best studied using homogeneous samples covering large

areas at the relevant redshifts. Unfortunately the sky coverage of quasar surveys is extremely patchy, and such large-scale homogeneous samples do not exist.

There is one set of data, however, which can be used somewhat indirectly to search for structure on very large scales: the radio quasars. At relatively high radio power the radio luminosity function has the critical slope of -1.5. If it were much flatter the strong sources would dominate, resulting in a Hubble law; if it were much steeper the weak local sources would dominate, resulting in an inverse Hubble law. At some intermediate slope these two effects must cancel each other, and that is the critical slope of -1.5. The redshift distribution of strong radio quasars is therefore relatively independent of flux density. This means that, while the surface density of radio quasars will vary over the sky as a function of survey sensitivity and completeness, the redshift distribution should remain approximately constant. Thus it should be possible to search for structure in the distribution of quasars over the sky in redshift intervals, after normalizing by the total number in each direction.

Figure 3 shows the redshift distribution of 760 radio quasars from the Véron catalogue, with 6 cm flux density > 0.1 Jy and confirmed redshifts. These come from 19 bins of typical size 0.5 steradian containing 40 radio quasars each. The solid curves are from Monte Carlo simulations of a random distribution; they show the 25th, 50th, 75th, and 95th percentiles of the expected distribution of the number of quasars per bin in each redshift interval (that is, 25, 50, 75, and 95 percent respectively of the bins would contain fewer quasars than specified by the relevant curves at each redshift). Significant structure on these scales would broaden this distribution, as shown by the dashed lines, for which the number density deviates from random by a factor of two in 20 percent of the bins. The dots

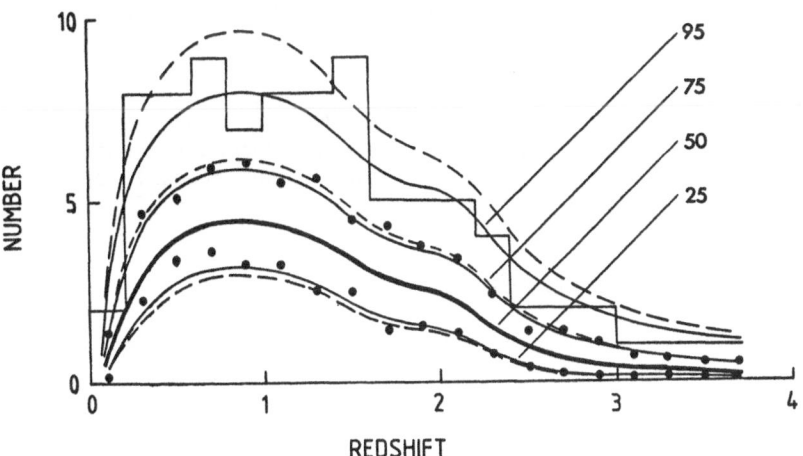

Figure 3. - Simulated and observed redshift distribution for 760 radio quasars from the Véron catalogue. See text for details.

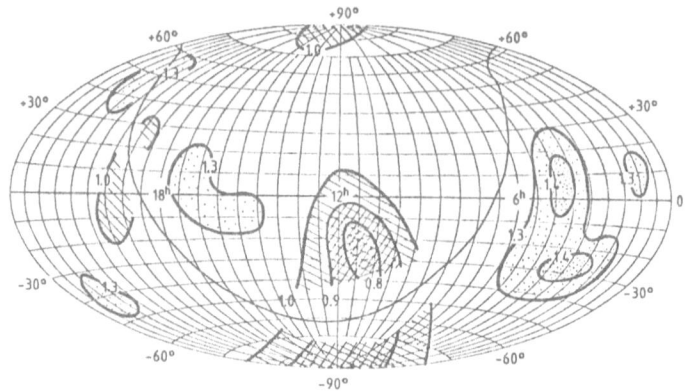

Figure 4. - Sky distribution in celestial coordinates of median
         redshift for all quasars in the Véron catalogue with 6-cm
         flux density > 0.1 Jy and confirmed redshifts from slit
         spectroscopy. The bin size is 0.5 steradian, containing
         typically 30-40 quasars. Hatched and dotted regions have
         median redshifts < 1.0 and > 1.3 respectively.

show the interpolated 25th and 75th percentile points from the
observed distribution, and the histogram shows the maximum number per
bin in each redshift interval.

    The observed distribution is close to random expectation, and
only a small fraction (< 10-20%) of the bins could have deviations as
large as a factor of two. The largest single deviation, which occurs
at z ~ 0.4-0.6, is also prominent as a minimum in an all-sky map of
median redshift for radio quasars (fig. 4), and has been discussed in
Shaver (1987). Some of the other major features have been examined,
but none of them gives the compelling impression of being a physically
real concentration. A range of bin size at different redshifts should
be studied, and this would best be done by making maps of normalized
surface density in discrete redshift intervals; a problem at present
is the limited and variable statistics. Maps such as these (and
fig. 4) can be used to identify large-scale concentrations if they
exist, or to set limits on such concentrations, redshift anomalies,
etc.

3. THE CLUSTERING AROUND QUASARS

It is well known that low-redshift quasars are very often found to
have companion galaxies, and are frequently located in groups or
clusters of galaxies (e.g., Stockton, 1978, 1982; Hutchings &
Campbell, 1983; Yee & Green, 1984; Heckman et al., 1984; Gehren et
al., 1984; Hutchings et al., 1984; Gilmore, 1984; Hintzen, 1984;
Stockton, 1986; Tyson, 1986; Chen & Zou, 1986; Surdej et al., 1986;
Yee & Green, 1986, 1987; Boyle et al., 1988; Stockton & MacKenty,
1987; Guzzo et al., 1988; Vader et al., 1987). Heckman et al. (1984)

find that 95% of galaxies located within a projected separation of 50 h^{-1} kpc of low-redshift quasars have the same redshifts as the quasars, within 1000 km s^{-1}. The companions are often distorted or compact, and there are sometimes extended emission-line regions reminiscent of debris from tidal interactions. The associated groups or clusters are compact, with core radii $\sim$ 60 h^{-1} kpc and velocity dispersions of $\lesssim$ 400 km s^{-1}. The amplitude of the quasar-galaxy correlation function lies between those for galaxies and clusters, with a similar power law. The strongest quasar-galaxy correlation is for radio quasars, and it appears to increase rapidly with redshift - by a factor of three from z = 0.4 to 0.6, where the associations are comparable to Abell clusters (Yee & Green, 1986, 1987; Yee, 1987).

At high redshifts there is also evidence, though generally less direct, that quasars are often located in groups or clusters of galaxies. There is a very high incidence of heavy-element absorption systems near the emission redshift of steep-spectrum radio quasars (Foltz et al., 1986; Anderson et al., 1987), but not for radio-quiet quasars (Sargent, 1988); this difference is strikingly similar to that at lower redshifts with regard to associated galaxies and clusters. The density of Lyα absorption systems decreases markedly near the emission redshift (Hunstead et al., 1986), again particularly for radio quasars (Bechtold, 1987). These facts are consistent with radio quasars being located in particularly dense clusters, inhospitable to the Lyα absorbers in the same way that they are to spiral galaxies today (Haynes & Giovanelli, 1986). One may also deduce that high-redshift quasars are located in dense environments from the high incidence of associated absorption in quasar pairs (Shaver & Robertson, 1984; Phillipps, 1986), or from the distortions and small sizes of the radio emission associated with high-redshift quasars (Barthel, 1986; Swarup et al., 1986). The associated absorption could also be due to debris from interactions and mergers, as suggested for the large emission-line regions around 3C-type galaxies at high redshifts (Djorgovsky et al., 1987). Finally, there is the direct evidence of a possible quasar-galaxy pair at z = 3.2 (Djorgovsky et al., 1985; Hu & Cowie, 1987; Djorgovsky et al., 1987).

It thus appears that quasars are located in groups or clusters of galaxies at all redshifts, and a plausible inference is that they are activated and/or fuelled by interactions (e.g. de Robertis, 1985; Gaskell, 1985; Roos, 1985; Byrd et al., 1987; Pringle et al., 1987).

Table 1 shows a hierarchy of correlation amplitudes at low and high redshifts. It suggests a number of inferences which might be drawn, connections between the observed features at low and high redshifts. The Lyα forest absorbers, for example, may be in some way associated with dwarf galaxies - both are numerous and weakly clustered. The heavy-element absorbers may be associated with more massive galaxies, particularly well-evolved ellipticals in clusters. The difference between the clustering properties of radio-loud and radio-quiet quasars appears to be the same at low and high redshifts, consistent with the radio-quiet quasars being located in spiral galaxies with some companions, and the radio-loud quasars in elliptical galaxies in clusters. The quasars themselves are more

Table 1. Normalized Correlation Amplitudes

$z \lesssim 0.1$			$z \sim 2$			
			$\alpha = -1.2$		$\alpha = -3$	
			$q_0 = 0$	$q_0 = 0.5$	$q_0 = 0$	$q_0 = 0.5$
CC[1]	18					
QQ[2]	3–15	QQ[2]		<5		<36
RR[3]	9					
CG[4]	3–10					
RG[4]	1–3					
$Q_R G$[5]	2–3	$Q_R M$[10]	6	4	43	29
$Q_0 G$[6]	1–2	$Q_0 M$[11]	<2	<1.3	<14	<9
		MM[12]	7	3	51	22
GG[7]	1	GG[7]	1	1	1	1
II[8]	0.2					
DD[9]	0.2	LL[13]	0.2	0.06	1.4	0.4

The symbols are as follows: G - galaxy, D - dwarf galaxy, I - IRAS galaxy, R - radio galaxy, C - cluster of galaxies, Q - quasar, Q_R - radio-loud quasar, Q_0 - radio-quiet quasar, M - metal-line (CIV) absorption system, L - Lyα-line absorption system. The correlation amplitudes are normalized to the galaxy-galaxy correlation amplitude, which is assumed to decrease with redshift as $(1+z)^\alpha$, with $\alpha = -1.2$ (stable clustering) and $\alpha = -3$.

References are as follows: 1 - Bahcall & Soneira (1983), 2 - this paper (fig. 2), 3 - Peacock et al. (1988), 4 - Prestage & Peacock (1987), 5 & 6 - Yee & Green (1986), 7 - Davis & Peebles (1983), 8 - Rowan-Robinson (1987), 9 - Davis & Djorgovski (1985), 10 - Weymann et al. (1979), 11 - Young et al. (1982), 12 - Sargent et al. (1980), 13 - Webb & Carswell (1987).

strongly clustered than galaxies in general, as may be expected from their affinity for companions. Such conclusions are only tentative at present, but it is clear that the clustering properties of quasars provide important clues to their nature and relationship to other objects, as well as to the evolution of structure in general.

4. QUASAR CLUSTERING VERSUS GRAVITATIONAL LENSING

Gravitational lenses are potential contaminants in studies of physical clustering, and physical pairs are potential contaminants in studies of gravitational lensing; it is clearly important to distinguish between them. This is difficult, however. Both are expected to be most prominent at the smallest angular or projected separations. The spectra of different quasars can be quite similar (Shaver, Wampler, & Cristiani, 1987), and the spectra of the images of a gravitationally lensed quasar can differ. Other criteria can be brought to bear, such as the properties of any radio or X-ray emission, broad absorption lines, polarization, the presence of a candidate lens, and especially temporal variations and forbidden-line redshifts, but it will probably remain difficult to distinguish with certainty between gravitational lensing and physical clustering in individual cases.

It has been suggested that gravitational lensing may even occur on arcminute scales due to intervening massive objects such as cosmic strings or supermassive black holes (e.g. Vilenkin, 1984; Hogan & Narayan, 1984; Gott, 1985; Paczyński, 1986a,b; Gott, 1986). In two specific cases of suggested arcminute-separation gravitational lenses (Surdej et al., 1983; Turner et al., 1986), subsequent evidence indicated that physical clustering was more likely (Surdej et al., 1986; Shaver & Cristiani,1986; Huchra, 1986). It is unlikely in any case that gravitational lensing dominates on these scales; the quasar clustering correlation amplitude appears to increase towards smaller redshifts, whereas gravitational lensing should be more prevalent at high redshifts.

It is on the smallest (arcsecond) scales that confusion between lenses and physical pairs is expected to be greatest. Lenses with masses of galaxies and clusters would produce separations up to 5-10 arcsec, corresponding to 50-100 h^{-1} kpc (comoving) projected separations at $z \sim 2$. The core radii of groups and clusters of galaxies associated with quasars at low redshifts are also ~ 60 h^{-1} kpc (Yee & Green, 1984, 1986, 1987); thus, physical pairs of quasars, which may mutually trigger activity in each other, may also occur preferentially on these scales. In both cases, therefore, a large excess of close (< 10 arcsec) quasar pairs with small redshift difference may be expected.

Such an excess is indeed found. Table 2 lists the known quasar pairs with projected separation $r_p \lesssim 100$ h^{-1} kpc; 11 or 12 of the 14 have small redshift difference ($\Delta V < 1000$ km s^{-1}). Table 3 shows the distribution of quasar pairs as a function of projected separation and redshift difference. On the largest scales, where clustering is weak or absent, less than one percent of quasar pairs have $\Delta V < 1000$ km s^{-1}. If there were no clustering or lensing, the same proportion would apply at small separations. The observed excess of quasar pairs with $r_p < 100$ h^{-1} kpc and $\Delta V < 1000$ km s^{-1} is significant at the 9σ level.

This excess is most unlikely to be due to a selection effect. Quasar pairs of small angular separation are undoubtedly preferentially searched for and rushed into press when found, but there is no observational selection effect so fine-tuned that it can pick out only those pairs with $\Delta V < 1000$ km s^{-1} and reject all others, and for the excess of close pairs with $\Delta V < 1000$ km s^{-1} to be due to a publication selection effect would imply that there are ~ 1000 quasar pairs with angular separation < 10 arcsec and $\Delta V > 1000$ km s^{-1} which

Table 2. QSO Pairs with $r_p \lesssim 100$ h^{-1} kpc

Pair	z	θ (")	r_p§ (h^{-1} kpc)	ΔV<1000 km s^{-1}?	Cand. Lens?	No. of Images	Ref.
1234+016AB	0.6,0.7	14	89	–	–	2	1
0023+171AB	0.9	5	40	✓	–	2	2
1548+114AB	0.4,1.9	5	45	–	–	2	3
1145−071AB	1.3	4	42	✓	–	2	4
0957+561AB	1.4	6	62	✓	✓	2	5
1115+080AB	1.7	2	17	✓			
1115+080AC	1.7	2	21	✓ }	✓	4	6
1115+080BC	1.7	1	11	✓			
2237+050AB	1.7	2	25	?	✓	2	7
1634+267AB	2.0	4	49	✓	–	2	8
1343+264AB	2.0	10	123	✓	–	2	9
2345+007AB	2.1	7	92	✓	–	2	10
0142−100AB	2.7	2	31	✓	✓	2	11
2016+112AB	3.3	3	48	✓	✓	3	12

§ Comoving, q_0 = 0.5.

References are as follows: 1 - Boyle et al. (1985),
2 - Burke (1986), 3 - Wampler et al. (1973), 4 - Djorgovski
et al. (1987c), 5 - Walsh et al. (1979), 6 - Weymann et al.
(1980), 7 - Huchra et al. (1985), 8 - Djorgovski & Spinrad
(1984), 9 - Crampton & Cowley (1987), 10 - Weedman et al.
(1982), 11 - Surdej et al. (1987), 12 - Lawrence et al.
(1984).

are known but unpublished! It is implausible not only that so many
very close pairs have actually been discovered, but also that many of
them have not been published - close quasar pairs with significant
redshift differences are of great interest in other contexts, e.g. the
study of common and associated absorption (Shaver & Robertson, 1984).
It can be concluded, therefore, that there is a real and very strong
excess of close quasar pairs with small redshift differences -
physical pairs, gravitational lenses, or both.

The distribution of the function $w^*(r_p)$ in Table 3 for low-
redshift ($\bar{z}$ < 1.5) quasars is similar to that for galaxies, although
with a higher amplitude. It may be somewhat steeper, possibly
reflecting the greater affinity of quasars for compact clusters (as is
the case for elliptical galaxies - Davis & Geller, 1976), or
conceivably some contribution from gravitational lensing. The
distribution for high-redshift ($\bar{z}$ > 1.5) quasars is quite different,
however; the clustering amplitude on Mpc scales is lower, consistent
with results reported in previous sections, but there is a sharp

Table 3. Distribution of Quasar Pairs

		r_p (h^{-1} Mpc)[§]			
		0.01-0.1	0.1-1	1-10	10-100
$\bar{z} < 1.5$:	N($\Delta V < 1000$ km s^{-1})	1(3)	2	9	53
	N($\Delta V > 1000$ km s^{-1})	2	24	499	7176
	w*(r_p)	67(202)	10	1.4	0.0
$\bar{z} > 1.5$:	N($\Delta V < 1000$ km s^{-1})	6(8)	0(1)	5	45
	N($\Delta V > 1000$ km s^{-1})	0	16	366	5489
	w*(r_p)	∞	-1(7)	0.7	0.0
Galaxy Pairs:	w*(r_p)	32	4	1.0	0.0

[§] Comoving, $q_0 = 0.5$

Quasars are from the Véron catalogue (Véron-Cetty & Véron, 1987), excluding those for which only objective prism redshifts are available. Numbers in parentheses include pairs listed in Table 2 but not yet included in the catalogue.

The galaxy data are from the Durham/AAT redshift survey (Peterson et al., 1986), and have been corrected for the effects of low-redshift crowding using Monte Carlo simulations.

w*(r_p) is defined by $x(r_p)/x(10\text{-}100$ h^{-1} Mpc$) - 1$, where $x = $ N($\Delta V < 1000$ km s^{-1})/N($\Delta V > 1000$ km s^{-1}).

increase at $r_p < 100$ h^{-1} kpc where the excess of pairs with $\Delta V < 1000$ km s^{-1} amounts to 8σ. The fact that the relative incidence of close pairs of small redshift difference is so much greater at high redshifts, where quasar clustering appears to be weaker, strongly suggests that gravitational lensing is a significant contributor. An alternative explanation, in which very close quasar pairs are more prevalent at high redshifts as a result of the frantic activity in dense clusters during the "quasar epoch", is also possible, but it runs counter not only to the probable evolution of clustering on larger scales, but also to the evidence favouring luminosity evolution over density evolution (c.f. Boyle et al., 1987).

This evidence for gravitational lensing of high redshift quasars is supported by an examination of the list of close pairs in Table 2. The best cases for lensing are 0957+561AB, 1115+080ABC, 2237+050AB, 0142-100AB, and 2016+112AB, based on criteria such as the presence of a candidate lens and multiplicity of images; all but the first are at z > 1.5, and they contribute 6 of the 8 close pairs in Table 3. Without them the incidence of close pairs at z > 1.5 is reasonably consistent with some modest clustering, or even a random distribution.

240

5. CONCLUSIONS

The physical clustering of quasars has been confirmed at a high level
of significance, and there is evidence of rapid evolution. This, and
information about the clustering of other objects at intermediate and
high redshifts, is beginning to tell us about the interrelationships
between these various objects, the environments of quasars, and the
evolution of structure over the history of the universe.

A high incidence of very close pairs of quasars is found, and
could be due to physical clustering, gravitational lensing, or both.
An examination of the known close pairs, particularly their redshift
distribution, suggests that many of those at high redshifts may indeed
be gravitational lenses. Though difficult, such a distinction between
physical pairs and gravitational lenses is important for our
understanding of both.

REFERENCES

Anderson, S.F., Weymann, R.J., Foltz, C.B., Chaffee, F.H. Jr. 1987,
Astron. J. **94**, 278.
Bahcall, N.A., Soneira, R.M. 1983, Astrophys. J. **270**, 20.
Barthel, P.D. 1986, in Quasars (ed. G. Swarup & V.K. Kapahi; Reidel),
p. 181.
Bechtold, J. 1987, in High Redshift and Primeval Galaxies, in press.
Boyle, B.J. 1986, thesis, University of Durham.
Boyle, B.J., Fong, R., Shanks, T., Clowes, R.G. 1985, Mon. Not. R.
astr. Soc. **216**, 623.
Boyle, B.J., Shanks, T., Fong, R., Peterson, B.A. 1987, in
Observational Cosmology (IAU Symp. No. 124; ed. A. Hewitt,
G. Burbidge, L.-Z. Fang; Reidel, Dordrecht), p. 643.
Boyle, B.J., et al. 1988, in Evolution of Large Scale Structures in
the Universe (IAU Symp. 130; eds. J. Audouze & A. Szalay; Reidel,
Dordrecht).
Burke, B.F. 1986, in Quasars (IAU Symp. No. 119; eds. G. Swarup & V.K.
Kapahi; Reidel, Dordrecht), p. 517.
Byrd, G.G., Sundelius, B., Valtonen, M. 1987, Astron. Astrophys. **171**,
16.
Chen, J.-S., Zou, Z.-L. 1986, Chinese Astron. Astrophys. 10, 190.
Crampton, D., Cowley, A.P. 1987, Bull. Amer. astr. Soc. **19**, 700.
Davis, M., Geller, M.J. 1976, Astrophys. J. **208**, 13.
Davis, M., Peebles, P.J.E. 1983, Astrophys. J. **254**, 437.
Davis, M., Djorgovski, S. 1985, Astrophys. J. **299**, 15.
Djorgovski, S., Spinrad, H. 1984, Astrophys. J. **282**, L1.
Djorgovski, S., Spinrad, H., McCarthy, P., Strauss, M.A. 1985,
Astrophys. J. **299**, L1.
Djorgovski, S., Spinrad, H., Pedelty, L., Rudnick, L., Stockton, A.
1987a, Astron. J. **93**, 1307.
Djorgovski, S., Strauss, M.A., Perley, R.A., Spinrad, H., McCarthy, P.
1987b, Astron. J. **93**, 1318.

Djorgovski, S., Perley, R., Meylan, G., McCarthy, P. 1987c,
    Astrophys. J. 321, L17.
Foltz, C.B., Weymann, R.J., Peterson, B.M., Sun, L., Malkan, M.A.,
    Chaffee, F.H. 1986, Astrophys. J. 307, 504.
Gaskell, M. 1985, Nature 315, 386.
Gehren, T., Fried, J., Wehinger, P.A., Wyckoff, S. 1984, Astrophys. J.
    278, 11.
Gilmore, G. 1984, Mon. Not. R. astr. Soc. 211, 25p.
Gott, J.R., III 1985, Astrophys. J. 288, 422.
Gott, J.R., III 1986, Nature 321, 420.
Guzzo, L., Danziger, I.J., Cristiani, S., Shaver, P.A. 1988, in
    Evolution of Large Scale Structures in the Universe (IAU Symp.
    130; eds. J. Audouze & A. Szalay; Reidel, Dordrecht).
Haynes, M., Giovanelli, R. 1986, Astrophys. J. 306, 466.
Heckman, T.M., Bothun, G.D., Balick, B., Smith, E.P. 1984, Astron. J.
    89, 958.
Hintzen, P. 1984, Astrophys. J. Suppl. 55, 533.
Hogan, C., Narayan, R. 1984, Mon. Not. R. astr. Soc. 211, 575.
Hu, E.M., Cowie, L.L. 1987, Astrophys. J. 317, L7.
Huchra, J.P. 1986, Nature 323, 784.
Huchra, J.P., Gorenstein, M., Kent, S., Shapiro, I., Smith, G.,
    Horine, E., Perley, R. 1985, Astron. J. 90, 691.
Hunstead, R.W., Murdoch, H.S., Pettini, M., Blades, J.C. 1986,
    Astrophys. Sp. Sci. 118, 505.
Hutchings, J.B., Campbell, B. 1983, Nature 303, 584.
Hutchings, J.B., Crampton, D., Campbell, B. 1984, Astrophys. J. 280,
    41.
Iovino, A., Shaver, P.A. 1988, in Evolution of Large Scale Structures
    in the Universe (IAU Symp. 130; eds. J. Audouze & A. Szalay;
    Reidel, Dordrecht).
Kruszewski, A. 1987, preprint.
Lawrence, C.R., Schneider, D.P., Schmidt, M., Bennett, C.L., Hewitt,
    J.N., Burke, B.F., Turner, E.L., Gunn, J.E. 1984, Science 223,
    46.
Longair, M.S., Seldner, M. 1979, Mon. Not. R. astr. Soc. 189, 433.
Osmer, P.S. 1981, Astrophys. J. 247, 762.
Paczyński, B. 1986a, Nature 319, 567.
Paczyński, B. 1986b, Nature 321, 419.
Peacock, J.A., Miller, L., Collins, C.A., Nicholson, D., Lilly, S.J.
    1988, in Evolution of Large Scale Structures in the Universe (IAU
    Symp. 130; eds. J. Audouze & A. Szalay; Reidel, Dordrecht).
Peterson, B.A., Ellis, R.S., Efstathiou, G., Shanks, T., Bean, A.J.,
    Fong, R., Zen-Long, Z. 1986, Mon. Not. R. astr. Soc. 221, 233.
Phillipps, S. 1986, Mon. Not. R. astr. Soc. 223, 173.
Prestage, R.M., Peacock, J.A. 1987, Mon. Not. R. astr. Soc., in press.
Pringle, J.E., Lin, D.N.C., Rees, M.J. 1987, BAAS 19, 695.
Robertis, M. de 1985, Astron. J. 90, 998.
Roos, N. 1985, Astrophys. J. 294, 486.
Rowan-Robinson, M. 1987, in Observational Cosmology (ed. A. Hewitt,
    G. Burbidge & L.-Z. Fang; Reidel), p. 229.

242

Sargent, W.L.W., Young, P.J., Boksenberg, A., Tytler, D. 1980, Astrophys. J. Suppl. **42**, 41.

Sargent, W.L.W. 1988, in Evolution of Large Scale Structures in the Universe (IAU Symp. 130; eds. J. Audouze & A. Szalay; Reidel, Dordrecht).

Shanks, T., Fong, R., Boyle, B.J., Peterson, B.A. 1986, in Quasars (ed. G. Swarup & V.K. Kapahi; Reidel), p. 37.

Shanks, T. et al. 1988, in Evolution of Large Scale Structures in the Universe (IAU Symp. 130; eds. J. Audouze & A. Szalay; Reidel, Dordrecht).

Shaver, P.A. 1984, Astron. Astrophys. **136**, L9.

Shaver, P.A. 1987, Nature **326**, 773.

Shaver, P.A., Robertson, J.G. 1984, in Frontiers of Astronomy and Astrophysics (ed. R. Pallavicini), p. 201.

Shaver, P.A., Cristiani, S. 1986, Nature **321**, 585.

Shaver, P.A., Wampler, E.J., Cristiani, S. 1987, Nature **327**, 40.

Stockton, A. 1978, Astrophys. J. **223**, 747.

Stockton, A. 1982, Astrophys. J. **257**, 33.

Stockton, A. 1986, Astrophys. Sp. Sci. **118**, 487.

Stockton, A., MacKenty, J.W. 1987, Astrophys. J. **316**, 584.

Surdej, J., Swings, J.-P., Henry, A., Arp, H., Kruszewski, A., Pedersen, H. 1983, in Quasars and Gravitational Lenses (24th Liège Int. Astrophys. Colloq.), p. 355.

Surdej, J., Arp, H., Gosset, E., Kruszewski, A., Robertson, J.G., Shaver, P.A., Swings, J.P. 1986, Astron. Astrophys. **161**, 209.

Surdej, J., Magain, P., Swings, J.-P., Borgeest, U., Courvoisier, T.J.-L., Kayser, R., Kellermann, K.I., Kühr, H., Refsdal, S. 1987, Nature, in press.

Swarup, G., Saikia, D.J., Beltrametti, M., Sinha, R.P., Salter, C.J. 1986, in Quasars (ed. G. Swarup & V.K. Kapahi; Reidel), p. 195.

Turner, E.L., Schneider, D.P., Burke, B.F., Hewitt, J.N., Langston, G.I., Gunn, J.E., Lawrence, C.R., Schmidt, M. 1986, Nature **321**, 142.

Tyson, J.A. 1986, Astron. J. **92**, 691.

Vader, J.P., da Costa, G.S., Frogel, J.A., Heisler, C.A., Simon, M. 1987, Astron. J., in press.

Véron-Cetty, M.-P., Véron, P. 1987, ESO Scientific Report No. 5.

Vilenkin, A. 1984, Astrophys. J. **282**, L51.

Walsh, D., Carswell, R.F., Weymann, R.J. 1979, Nature **279**, 381.

Wampler, E.J., Baldwin, J.A., Burke, W.L., Robinson, L.B., Hazard, C. 1973, Nature **246**, 203.

Webb, J.K., Carswell, R.F. 1987, in preparation.

Webster, A. 1982, Mon. Not. R. astr. Soc. **199**, 683.

Weedman, D.W., Weymann, R.J., Green, R.F., Heckman, T.M. 1982, Astrophys. J. **255**, L5.

Weymann, R.J., Williams, R.E., Peterson, B.M., Turnshek, D.A. 1979, Astrophys. J. **234**, 33.

Weymann, R.J., Latham, D., Angel, J.P.R., Green, R.F., Liebert, J.W., Turnshek, D.A., Turnshek, D.E., Tyson, J.A. 1980, Nature **285**, 641.

Yee, H.K.C. 1987, in Observational Cosmology (ed. A. Hewitt,
    G. Burbidge, L.-Z. Fang; Reidel), p. 685.
Yee, H.K.C., Green, R.F. 1984, Astrophys. J. 280, 79.
Yee, H.K.C., Green, R.F. 1986, in Quasars (ed. G. Swarup & V.K.
    Kapahi; Reidel), p. 481.
Yee, H.K.C., Green, R.F. 1987a, Astrophys. J. 319, in press.
Yee, H.K.C., Green, R.F. 1987b, Astron. J. 94, 618.
Young, P., Sargent, W.L.W., Boksenberg, A. 1982, Astrophys. J. Suppl.
    48, 455.

X-RAYS AND THE LARGE SCALE STRUCTURE OF THE UNIVERSE

Xavier Barcons
Institute of Astronomy
Cambridge University, U.K.
and
Departamento de Física Moderna
Universidad de Cantabria, Spain

ABSTRACT. We show how the study of the granularity of the X-ray background can yield important information on the large scale structure of the Universe at moderate redshift. Presently available limits on the fluctuations of the X-ray background constrain either the contribution to the background or the clustering properties of the cosmic X-ray sources.

1. INTRODUCTION

Unlike the Cosmic Microwave Background (CMB), the origin of the X-Ray Background (XRB) is still unknown. The contribution to the XRB by known classes of sources is very uncertain, ranging from 20 to 60 per cent, depending on evolution (Boldt & Leiter 1984, Boldt 1987, Giacconi & Zamorani 1987).

A crucial piece of information concerning the XRB is its isotropy. Aside the contribution from the galaxy and a dipole roughly aligned with the CMB one, the fluctuations of the XRB on scales of 5° are < 2.3 % (Shafer & Fabian 1983). This limit (and future ones) is very important for our understanding of the low redshift Universe (z < 10) because we see luminous matter to be very inhomogeneously distributed.

Using the link between the fluctuations of the XRB and the clustering properties of X-ray sources, as presented in Barcons & Fabian (1987), we examine the constraints on the large scale structure of the X-ray Universe coming from the isotropy of the XRB.

2. POINT SOURCES

QSOs have often been claimed to be the main contributors to the XRB. In addition to spectral problems for this mo-

N. Kaiser and A. N. Lasenby (eds.), The Post-Recombination Universe, 245–247.

del (the QSO spectra do not match the XRB spectrum), the fluctuations that QSOs imprint in the XRB (specially if they cluster) constrain their maximum contribution to this background. In a number of recent works (see the review by Shaver 1988) it has been shown that QSOs cluster on scales of about 10 h^{-1} Mpc. For the XRB, this means a decrease in the effective number of sources and so an increase in the fluctuations.

Assuming a gaussian QSO-QSO correlation function and a redshift distribution as the one given in the ASIAGO cata log for X-ray QSOs (this could be a suitable description of the known population of QSOs), these sources cannot make up more than 25 % of the XRB, to keep fluctuations at the observed level. This maximum contribution goes down if a power-law correlation function is assumed.

These conclusions would be relaxed for a new population of faint objects at redshift z > 2. In this case, clustering on scales $\sim$ 10 h^{-1} Mpc will not prevent these objects to make up the whole XRB. However, a correlation scale greater than $\sim$ 50 h^{-1} Mpc is excluded for the sources that produce the XRB, even at these redshifts.

Notice that the result presented by Hamilton & Helfand (1987) that QSOs produce a fraction between 40 and 60 per cent of the XRB, must be relaxed and these numbers taken as upper limits if there is clustering.

3. HOT GAS MODELS

As the spectrum of the XRB, in the energy range 3 -300 keV looks like thermal bremsstrahlung at a temperature of $\sim$ 30 keV (Marshall et al. 1980), there is the possibility that most of it originates in a hot gas. If this gas is uniform, about $\Omega_B \sim 0.25$ baryons are needed to reproduce the observed intensity (Guilbert & Fabian 1986, Barcons 1987). This number conflicts with standard theories of primordial nucleosynthesis which require $\Omega_B < 0.19$ (Boesgaard & Steigman 1985).

Aside the possibility of a non-standard model for nucleosynthesis (e.g. the one presented by Applegate, Hogan Scherrer 1987, which can reproduce primordial abundances with $\Omega_B \simeq 1$), the hot gas could be clumpy, and so it could be possible, in principle, to produce the XRB with $\Omega_B < 0.19$.

Independently of the model for the clumping, the high density regions, where the XRB is produced, cannot have sizes greater than $\sim$ 7 h^{-1} Mpc (taking $\Omega_B \simeq 0.1$) to keep the fluctuations at the observed level. Any model with bigger blobs, such that introduced by Daly (1988), needs anticlustering to have less fluctuations than those predicted by Poisson statistics.

Moreover, a nonuniform hot gas produces fluctuations in the CMB through Compton effect. For the Guilbert &

Fabian (1986) model in which the high density clumps are pressure confined by a yet hotter diffuse gas, the CMB fluctuations are greater than the observed limits unless these blobs have sizes < 20 h^{-1} kpc.

The conclusion is that if most of the XRB has been produced by hot gas, this gas must be either very uniform ($\Omega_B \sim$ 0.25) or high density regions must be very small (for Ω_B < 0.1).

4. OUTLOOK

The isotropy of the XRB appears to be a very interesting piece of information for the study of the structure of the Universe. When data on angular scales $\sim$ 1 arcmin become available in the future with the use of modern X-ray telescopes (BBXRT, AXAF, XMM,...) our understanding of the z < 10 Universe will be much better.

I am grateful to Andy Fabian with whom much of this work was done.

REFERENCES

Applegate, J.H., Hogan, C.J. & Scherrer, R.J., 1987, Phys. Rev. **D35**, 1151

Barcons, X., 1987, Ap. J. **313**, 547

Barcons, X. & Fabian, A.C., 1987, MNRAS (in the press)

Boesgaard, A.M. & Steigman, G., 1985, Ann. Rev. Ast. Astrop. **23**, 319

Boldt, E. & Leiter, D., 1984, Ap. J. **276**, 427

Boldt, E., 1987, Phys. Rep. **146**, 215

Daly, R.A., 1988, these proceedings

Giacconi, R. & Zamorani, G., 1987, Ap. J. **313**, 20

Guilbert, P.W. & Fabian, A.C., 1986, MNRAS **220**, 439

Hamilton, T.T. & Helfand, D.J., 1987, Ap. J. **318**, 93

Marshall, F.E., Boldt, E.A., Holt, S.S., Miller, R., Mushotzky, R.F., Rose, L.A., Rothschild, R. & Serlemitsos, P.J., 1980, Ap. J. **235**, 4

Shafer, R.A. & Fabian, A.C., 1983, in IAU Symposium 104: Early Evolution of the Universe and its Present Structure. Ed. G. Abell & G. Chincarini. Dordrecht:Reidel

Shaver, P., 1988, these proceedings

MONTE CARLO SIMULATIONS OF LARGE-SCALE STREAMING VELOCITIES

Edmund Bertschinger
Department of Physics
MIT, Room 6-207
. Cambridge, MA 02139
U.S.A.

ABSTRACT. Monte Carlo simulations are used to calculate the large-scale streaming velocities expected in the biased cold dark matter model. Galaxies are selected and "observed" in large simulations and dipole streaming motions are determined using a maximum likelihood method designed to simulate the procedures followed by Dressler et al. (1987). The streaming velocities found are about 2.5 times larger than the velocity averaged over a uniform sphere of radius $60\,h^{-1}\,\mathrm{Mpc}$. The results confirm earlier work by Kaiser and imply that unbiased cold dark matter is consistent with large-scale flows but that biased models have troubles.

1. INTRODUCTION

Large-scale deviations from uniform Hubble flow present a strong challenge to present models for the formation of structure. Deviations of $\gtrsim 500\,\mathrm{km\,s}^{-1}$ on scales $\gtrsim 50\,h^{-1}\,\mathrm{Mpc}$ have been reported by several groups using different distance indicators (Collins, Joseph, and Robertson 1986; Dressler et al. 1987; Mould 1988; see Lynden-Bell et al. 1988 for additional references). Theoretical studies of bulk flows expected in the standard cold dark matter model (Bond 1986; Vittorio, Juszkiewicz, and Davis 1986; Vittorio and Turner 1987) predict much smaller velocities. These differences between observations and theoretical predictions are not necessarily irreconcilable, however. The theoretical predictions have generally been made using a very different analysis method than have the observations. Typically the theorists (e.g., Vittorio et al. 1986) compute the rms velocity averaged uniformly or with Gaussian radial weighting over a sphere, computed using linear perturbation theory. However, the observers (e.g., Dressler et al. 1987) select a magnitude-limited sample of galaxies, their distance indicators have random errors, and they employ a maximum likelihood procedure to find the best fit model to the large-scale flow. It should be expected that such different methods of analysis yield different results.

Recently, Kaiser (1987) has shown that the observational selection and likelihood analysis employed by Dressler et al. (1987) biases the streaming velocity determination enough to be perhaps consistent with theoretical predictions of cold dark matter. Kaiser used directly the galaxy distribution of Dressler et al. (grouped into clusters for convenience) and used a maximum likelihood analysis with linear perturbation theory. His result is important but requires confirmation since it is based on a posteriori analysis of the observations rather than a priori predictions from the theory.

Here I present a preliminary report of a study of large-scale streaming velocities predicted from theory. Using Monte Carlo simulations, I have attempted to model the selection and analysis procedure used by Dressler et al. (1987). My guiding principle is to "work in the observer's plane."

N. Kaiser and A. N. Lasenby (eds.), The Post-Recombination Universe, 249–251.

2. PROCEDURE

The procedure used here is as follows. First, a random realization of Gaussian initial conditions with a cold dark matter power spectrum (from Davis *et al.* 1985, hereafter DEFW) is generated on a 64^3 grid of 200 h^{-1} Mpc on a side. For $\Omega = 1$, $h = 0.5$, the particle mass is $1.7 \times 10^{13} M_\odot$, larger than a galaxy. The smallest scale resolved by the simulations (3.1 h^{-1} Mpc) is not very nonlinear by the present. The Zel'dovich (1970) approximation is therefore used to evolve the particles rather than an accurate N-body code. In the present case this approximation should be excellent since the evolved power at the Nyquist frequency is only 1.25 times the white noise level. A bias factor $b = A^{-1} = 2.5$ is assumed for the normalization of the power spectrum, where $A^2 = \langle (\Delta M/M)^2 \rangle$ is computed using linear perturbation theory with uniform weighting in a sphere of radius 8 h^{-1} Mpc. This normalization is the same as that used by DEFW in their biased cold dark matter simulations. Note, however, that streaming velocities scale with A in the linear regime and that many other workers set $A = 1$.

Despite the large mass, each particle is treated as a galaxy. A magnitude-limited sample with $B_T < 13.0$ is selected by choosing galaxies with probability determined from the Schechter (1976) luminosity function with $\phi^* = 0.015\, h^3$ Mpc^{-3}. Galactic dust extinction is included following Fisher and Tully (1981). Because of the sparseness of points, galaxies are slightly undersampled (by a total of 5%) in nearby groups. This is compensated for by increasing the weight of the undersampled galaxies in the analysis. About 2200 galaxies are included in each sample. To select "ellipticals" and to include the effects of "biased" galaxy formation, a threshhold of $\delta\rho/\rho = 1.5$ on the grid scale is applied to the initial conditions to select 15% of galaxies in the densest regions. The galaxies are not grouped into clusters in the analysis.

Distances and peculiar radial velocities (in the frame of the cosmic microwave background) are calculated for each "galaxy" including a 25% random distance error. No Malmquist correction (cf. Lynden-Bell *et al.* 1988) has been applied. The maximum likelihood method was used to determine a dipole streaming velocity $\vec{V}$ and field velocity dispersion σ_f and standard errors.

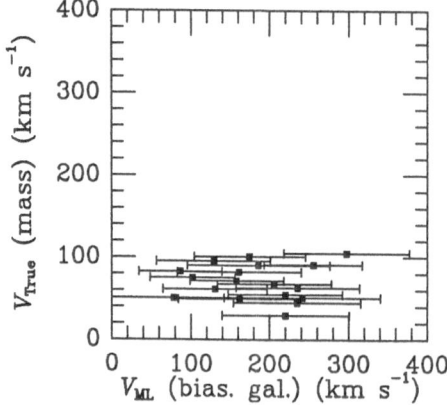

Figure 1. Streaming velocity determined from the biased "galaxy" sample by maximum-likelihood analysis (abcissa, with standard errors) vs. the true net streaming velocity (ordinate) for all points within a sphere of radius 60 h^{-1} Mpc in the same realization. Eighteen realizations are shown.

3. PRELIMINARY RESULTS

Figure 1 shows the main result of this work. The "empirical" velocities are, on average, about 2.5 times as large as the naive theoretical estimates. The effect of biasing (the selection of "galaxies" in the densest regions) alone accounts for a factor of about 1.5. Most of the enhancement in observationally determined streaming velocities comes from two sources: (1) a $B_T < 13.0$ sample is effectively much shallower than $60\,h^{-1}$ Mpc (this was accounted for in most previous theoretical studies by choosing a smaller sphere); and (2) the maximum likelihood procedure gives less weight to distant galaxies because of the linearly increasing distance (hence peculiar velocity) errors.

4. CONCLUSIONS

Despite the large increase in streaming velocities obtained by following the empirical selection and analysis procedures, the biased cold dark matter model with $b = 2.5$ appears to have difficulty explaining the $599 \pm 104\,\mathrm{km\,s}^{-1}$ streaming velocity determined by Dressler et al. (1987). However, the streaming velocities are expected to scale with b^{-1} (from the normalization of the power spectrum), so that a bias factor of $b = 1$ is probably consistent. This result confirms earlier work of Kaiser (1987). Unfortunately, without biasing the cold dark matter model has trouble on smaller scales (DEFW). Since the physical mechanisms for bias are not thoroughly understood, it is conceivable that large and small scale simulations can still be made consistent with observations.

The present work is preliminary: the analysis can and will be improved in several ways. The Zel'dovich method will be tested against N-body codes; the power spectrum normalization and the bias level will be varied; galaxies will be cataloged into groups; and the Malmquist bias will be corrected. Simulations will also be made including Virgocentric infall and a "Great Attractor," using the constrained random noise algorithm of Bertschinger (1987). Monte Carlo simulation methods are an excellent tool for including all of these effects in a comparison of theory and observations.

REFERENCES

Bertschinger, E. 1987, Ap.J. (Letters), 323, in press.
Bond, J. R. 1986, in Galaxy Distances and Deviations from Universal Hubble Expansion, ed. B. F. Madore and R. B. Tully (Dordrecht: Reidel), p. 255.
Burstein, D., and Heiles, C. 1978, Ap.J., 225, 40.
Collins, C. A., Joseph, R. D., and Robertson, N. A. 1986, Nature, 320, 506.
Davis, M., Efstathiou, G., Frenk, C. S., and White, S. D. M. 1985, Ap.J., 292, 371 (DEFW).
Dressler, A., Faber, S. M., Burstein, D., Davies, R. L., Lynden-Bell, D., Terlevich, R. J., and Wegner, G. 1987, Ap. J. (Letters), 313, L37.
Fisher, J. R., and Tully, R. B. 1981, Ap.J. Suppl., 47, 139.
Grinstein, B, and Wise, M. B. 1987, Ap.J., 320, 448.
Kaiser, N. 1987, preprint.
Lynden-Bell, D., Faber, S. M., Burstein, D., Davies, R. L., Dressler, A., Terlevich, R. J., and Wegner, G. 1988, Ap.J., 326, in press.
Mould, J. 1988, in Proc. IAU Symp. 130, ed. J. Audouze and A. Szalay (Dordrecht: Reidel), in press.
Schechter, P. 1976, Ap.J., 203, 297.
Vittorio, N., Juszkiewicz, R., and Davis, M. 1986, Nature, 323, 132.
Vittorio, N., and Turner, M. 1987, Ap.J., 316, 475.
Zel'dovich, Ya. B. 1970, Astron. Ap., 5, 84.

A test on the Gaussian nature of primeval fluctuations

S. A. Bonometto

Dept. of Physics of the University of Perugia - Via Elce di Sotto - Perugia (Italy).
I.N.F.N.- Sezione di Padova - Via Marzolo 8 - 35100 Padova (Italy).

According to their origin and to the extent of their non linear evolution, density fluctuations can have a Gaussian or a non-Gaussian character. Methods to determine such character would therefore furnish a substantial insight into the type of primeval fluctuations which produced the present state of the Universe. E.g., the discovery of a Gaussian or nearly-Gaussian behaviour would substantially decrease the possibility that primeval topological faults (strings) played a role in the formation of the actual fluctuation spectrum. This would certainly not inhibit their possible presence, but they would probably be not the only origin of the present large scale structure in the Universe.

In the frame of biased theories of galaxy formation (Kaiser, 1984; Politzer and Wise, 1984; Bardeen *et al.*, 1986), a direct probe into the Gaussianety is provided by the study of the n-point (n>2) functions. Both for Gaussian and non-Gaussian fluctuations the joint probability that 3 *objects* are in the volumes δV_1, δV_2, δV_3 reads

$$\delta^3 P = \delta V_1 \delta V_2 \delta V_3 \, n^3 \, g_{123} \, (1+\xi_{12})(1+\xi_{23})(1+\xi_{31}) \qquad (1)$$

where n is the average density of such *objects*, ξ_{ij} is the 2-point function and g_{123} is a known function (Matarrese *et al.*, 1986; Bonometto *et al.*, 1987). In the Gaussian case only $g_{123}=1$. In such case, therefore, the 3-point function reads

$$\xi^{(3)}_{123} = q_1(\xi_{12}\xi_{23} + \xi_{13}\xi_{32} + \xi_{21}\xi_{13}) + q_2 \, \xi_{12}\xi_{23}\xi_{31} \qquad (2)$$

with $q_1 = q_2 = 1$. Former analyses of the 3-point function seem to yield $q_1 \approx 1$ but $q_2 \approx 0$. Most of such data are obtained from angular correlations translating then data from 2 to 3 dimensions on the basis of suitable extensions of the Limber equation (see, e.g., Peebles, 1980). A basic requirement for the validity of such equations is that the distributions on space and luminosity are independent, i.e. that $\xi(r)$ has no luminosity dependence. In biased theories of galaxy origin, instead, one expects such dependence to be present. The impact of such dependence on the applicability of the Limber equations seems difficult to be determined *a*

N. Kaiser and A. N. Lasenby (eds.), The Post-Recombination Universe, 253–255.
© *1988 by Kluwer Academic Publishers.*

priori. We shall however formulate here a conjecture and support it by a number of observa
tional outputs. If such conjecture is true, some results obtained from the Limber equation
might be, at least partially, misleading. In turn, this would enable to work out simple test
on the Gaussian nature of the fluctuation fields.

Let $\varepsilon(x)$ be the density fluctuation field (x is a point in the ordinary 3-space) and le

$$\eta(\Omega) = \int_0^{r_\infty} dr \, \varepsilon(r, \Omega) \tag{3}$$

be the *projection* of the field $\varepsilon(x)$ onto the celestial sphere, built in order that - if light trac

mass - $\eta(\Omega)$ would represent the apparent luminosity fluctuations due to the intrinsical lumi

nosity fluctuations $\varepsilon(x)$. For the sake of simplicity we shall assume here that the spatial metric is flat, while the cut-off at r_∞ is a rough way to indicatethe presence of an evolution. Ou

conjecture is then that *if $\varepsilon(x)$ is distributed in space in a Gaussian way, $\eta(\Omega)$ is distributed in
a Gaussian way on the celestial sphere.*

According to Bonometto *et al.* (1987), an apparent-magnitude limited catalog can
be introduced just applying the *bias* criterion to the field $\eta(\Omega)$. In this way one selects am
ongst the luminosity peaks those exceeding a certain thershold, expressed in terms of the
quadratic average of luminosity fluctuations. This threshold directly furnishes the limiting
magnitude which defines the sample.

If $\eta(\Omega)$ is Gaussian the angular 3-point correlation should have a form of the
kind introduced in (1)-(2). As usual we shall replace ξ by w to indicate angular functions
this implies that the joint probability of finding three *objects* on the celestial sphere is

$$\delta^{(3)}P = \delta\Omega_1\delta\Omega_2\delta\Omega_3 \, n_\Omega \, g_{123} \, (1+w_{12})(1+w_{23})(1+w_{31}) \tag{4}$$

If angular analyses indicate then that $g_{123}=1$, the Gaussian character of the matter
distribution is established. Current conclusions on the values of q_1 and q_2 in (2) mostly
come from angular function analyses. In fact it is clear that (4) implies an angular 3-point
function

$$w^{(3)}_{123} = q_1 \, (w_{12}w_{23}+w_{23}w_{31}+w_{31}w_{12}) + q_2 \, w_{12}w_{23}w_{31} \tag{5}$$

again with $q_1 = q_2 = 1$. The observational outputs indicating that $q_1 \sim 1$ and $q_2 \sim 0$ lead to
similar conclusions on q_1 and q_2, via the Limber equation. However, if $w < 1$, the level of
Poisson noise present in data can make the determination of q_2 fairly unprecise.Only from
sets of data for which $w \sim 1$ (i.e. not too deep) reliable determinations of q_2 can be perform
ed. The Zwicky set of data (Zwicky *et al.* 1961 - 1968) is perfectly suitable for these aims.

In this context we whish to report here the results of specific fits operated on the
Zwicky sample. This was done both assuming (4) to hold and fitting $<g_{123}>$ and on the
basis of the expression

$$\delta^{(3)}P = \delta\Omega_1 \delta\Omega_2 \delta\Omega_3\, n_\Omega^{\;3}\, (1 + w_{12} + w_{23} + w_{31} + w^{(3)}_{123}) \tag{6}$$

with $w^{(3)}_{123}$ given by (5), taking there $q_2 = 0$. Errors were estimated dividing the sample in two subsets A and B. We shall not give here details on the fitting procedure, which will be reported elsewhere (for the method, see Sharp et al., 1984). The results are given in table 1 and seem to indicate: i) A 3-point function expression containing a term $\propto w^3$ provides a much better fit of data than an expression without it (in the former case errors are $\sim 2\%$, in the latter one $\sim 16\%$). ii) The distribution is Gaussian.

Table 1

	$<g_{123}>$	γ	q_1	γ
whole sample	1.017	1.67	0.99	1.67
subset A	0.979	1.68	1.15	1.44
subset B	1.027	1.71	0.84	1.86
A-B average	1.003	1.70	1.00	1.65

The fits provide also a value for the slope γ of the 2-point function. The first 2 columns refer to a fit of $<g_{123}>$ (with cubic term) and yield 1.00 0.02 (γ=1.70±0.02). The other 2 column refer to a fit of q_1 and yield 1.00±0.16 (γ=1.65±0.21). The errors are obtained from the split into two subsets and in the identical way for the two cases.

Attempts were also performed to extend the above analysis to the 4-point function. The results, which will be reported elsewhere, seem again to support the above conclusions.

References

Bardeen J. M., Bond J. R., Kaiser N., Szalay A. S. (1986) **Ap.J. 304**, 15
Bonometto S. A., Lucchin F., Matarrese S. (1987) **Ap.J.** (in press)
Kaiser N. (1984) **Ap.J. 284**, L9
Matarrese S., Lucchin F., Bonometto S. A. (1986) **Ap.J. 310**, L21
Peebles P.J.E. (1980) *'The large scale structure of the Universe'* (Princeton University Press - *Princeton*)
Politzer H.D., Wise M.B. (1984)**Ap.J 285**, L1
Sharp N., Bonometto S.A., Lucchin F. (1984) **A.& A. 130**, 79
Zwicky F., Herzog E., Wild P., Karoowicz M., Kowal G.T. (1961 - 1968) *'Catalogue of galaxies and clusters of galaxies '* (Caltec - *Pasadena*)

CAN EXPLOSIONS GENERATE FILAMENTS AND VOIDS?

B. J. Carr
School of Mathematical Sciences
Queen Mary College
Mile End Road
London E1 4NS

ABSTRACT. The explosion scenario can naturally explain the existence of filamentary structure and large-scale voids providing the shells generated by the explosions fill the Universe when they have a radius of about 10 Mpc. The scenario is particularly attractive if the shells first fragment when they collide since, in this case, most of the gas is swept up into sheets and one may make a lot of baryonic dark matter through high pressure cooling flows. A shell radius of 10 Mpc could arise naturally if one assumes that the shells grow as detonation waves until the end of the Compton-cooled era.

1. INTRODUCTION

In recent years it has become clear that the Universe exhibits filamentary or cellular structure, with thin elongated superclusters surrounding low density voids. For example, the Local Supercluster is a pancake with a diameter of about $20h^{-1}$Mpc (Tulley 1982), where h is the Hubble constant in units of 100 km/s/Mpc. The Perseus-Pisces Supercluster is a filament with a length of about $100h^{-1}$Mpc (Giovanelli et al. 1986). A $6°x120°x15h^{-1}$Mpc slice near Coma exhibits bubbles with an average radius of $12h^{-1}$Mpc and one especially round bubble with a radius of $25h^{-1}$Mpc (de Lapparent et al. 1986). Bootes has a void of radius $30h^{-1}$Mpc (Kirshner et al. 1981). Finally, there is the possible evidence for large-scale streaming motions (Dressler et al. 1987).

It is rather difficult to explain these sorts of features with the usual sort of cosmological scenario, in which all large-scale structure arises from the action of gravity on the density fluctuations left over after decoupling. This suggests that a more exotic mechanism for generating large-scale structure may be required. In particular, the evidence for bubbles may point to the importance of explosions. In fact, the explosion scenario, originally proposed by Ostriker & Cowie (1981) and Ikeuchi (1981), predates the discovery of bubbles and was not originally conceived as producing the scale of structure discussed above. Nevertheless, we will argue here that explosions could naturally explain large-scale filaments and voids.

N. Kaiser and A. N. Lasenby (eds.), The Post-Recombination Universe, 257–262.
© 1988 by Kluwer Academic Publishers.

2. THE SCALE OF STRUCTURE GENERATED BY COSMIC EXPLOSIONS

The explosion scenario suggests that at some redshift z a small fraction of the Universe goes into exploding seeds (eg. stars or clusters of stars). Each explosion generates a shock-wave which sweeps up a shell of gas. Since the shell may have much more mass than the seed , this provides a natural amplification mechanism for producing large-scale structure. The shell may eventually fragment, after which the fragments continue to expand but cease to sweep up   gas. Thus the amount of amplification is determined in part by when fragmentation occurs and this, in turn, depends upon the dominant cooling mechanism.

At early times the dominant cooling mechanism is the Compton effect of the microwave background. The associated cooling timescale is $\tau_{cool} \approx 2 \times 10^{12}(1+z)^{-4}$y and this exceeds the cosmic expansion timescale at redshifts exceeding $z_c \approx 9\Omega^{0.2}h^{0.4}-1$, where Ω is the total density parameter in units of the critical density. The maximum amplification factor in this period can be shown to be

$$ \eta \equiv \frac{M_{shell}}{M_{seed}} \simeq 4 \times 10^5 \ (1+z)^{-1.7} \ \left[\frac{\epsilon}{10^{-4}} \right]^{0.6} \left[\frac{M_{seed}}{10^6 M_\odot} \right]^{-0.4} \tag{1} $$

(Carr & Ikeuchi 1985) where ϵ is the efficiency with which the seed's rest mass is turned into explosive energy (normalized to the sort of value associated with supernovae) and the seed mass M_{seed} is normalized to the mass of the first objects which are expected to arise in many cosmological scenarios. We note that η is in the range $10-10^3$ for reasonable values of z and it exceeds 1 for values of (M_{seed}, z) below the "amplification" boundary in Figure (1). Whether fragmentation occurs depends upon rather uncertain astrophysical details but a simple dynamical criterion (Ostriker & Cowie 1981) suggests that it happens above the "fragmentation" boundary in Figure (1). If the shell cools to the CBR temperature, the fragment mass would be in the range $10^3-10^6 M_\odot$, though it is not clear that it gets that cool (Wandel 1985).

After the redshift z_c, radiative cooling dominates and the cooling timescale may be   longer than the cosmic expansion timescale at the explosion epoch. The amplification achieved by a later redshift z is then

$$ \eta \simeq 400 \ (1+z)^{-1.2} \ \left[\frac{M_{seed}}{10^{11}M_\odot} \right]^{0.2} \left[\frac{\epsilon}{10^{-4}} \right]^{-0.2} \Omega_g \ \Omega^{-1.2} \ h^{-0.4} \tag{2} $$

(Carr & Ikeuchi 1985) where Ω_g is the gas density parameter and we have now normalized the seed mass to a galactic-scale. Although η can still be large, cooling is only possible on a cosmological time for M_{seed} less than about $10^{12}M_\odot$, as indicated by the "cooling" line in Figure (1). The associated maximum comoving shell radius is

$$ R_{shell} \simeq 3 \ (1+z)^{-0.2} \ \left[\frac{D}{4} \right]^{0.2} \left[\frac{\Omega_g}{0.1} \right]^{0.2} \Omega^{-0.6} \ h^{-0.8} \ \text{Mpc} \tag{3} $$

where D is the overdensity in the shell relative to the cosmological background (normalized to the value appropriate for an isolated shell). Since the fragment mass is likely to be comparable to a galactic scale when radiative cooling dominates (Bertschinger 1983), one has a natural

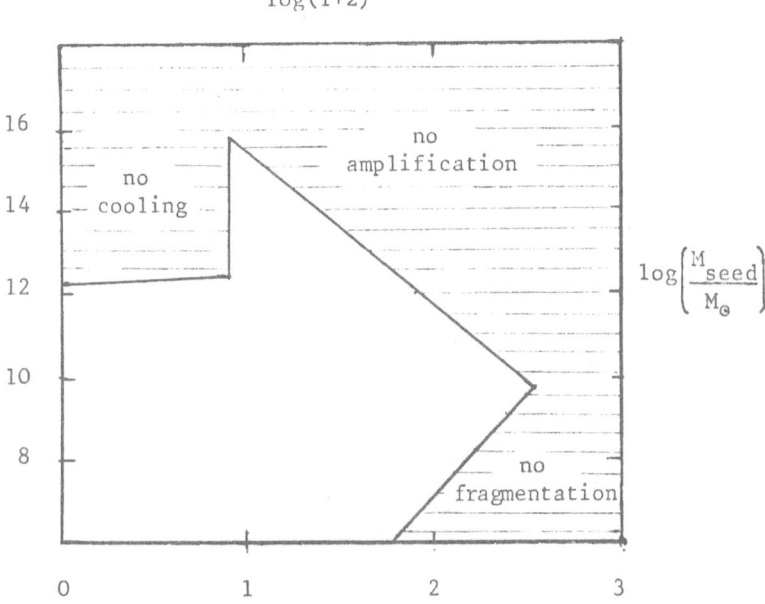

Figure (1). This shows the (M_{seed}, z) regime in which an explosive seed produces a shell which can amplify the scale of structure, as well as cool and fragment.

way of generating clusters of galaxies. However, the value of R_{shell} is less than the size associated with supeclusters or large-scale voids. To explain the observations, we must therefore assume that the explosions occur in the Compton-cooled era.

Figure (1) suggests that the largest fragmenting shells would be generated at the redshift z_c and have a mass of order $10^{16} M_\odot$. As it happens, this is about the time at which one expects the first seeds to form in many scenarios. However, one only attains the maximum shell mass in a single explosive step if the seed itself has a mass of $10^{16} M_\odot$ (viz. $\eta=1$ at the amplification boundary). The maximum shell size is only naturally attained if one invokes the "hierarchical" explosion scenario, in which the shell fragments themselves explode (Carr & Ikeuchi 1985). Equivalently, one can envisage a detonation wave (Bertschinger 1985). In either case, one can show that the shell radius asymptotes to $\sqrt{\epsilon}$ times the particle horizon size. In comoving coordinates this is

$$R_{shell} \simeq 10 \left[\frac{\epsilon}{10^{-4}} \right]^{0.5} \left[\frac{1+z}{10} \right]^{-0.5} \Omega^{-0.5} \ h^{-1} \ Mpc \qquad (4)$$

Such a bootstrap process is possible in the Compton era because the fragment mass could be in the exploding star range but it is implausible in the radiative era because galactic-scale fragments are unlikely to generate explosive energy with high efficiency. Thus the maximum comoving shell radius is probably around $10 h^{-1} Mpc$; this corresponds to putting $z=z_c$ in eqn (4).

3. THE INTERACTION BETWEEN SHELLS

If the observed large-scale structure is to be explained by explosions, it is clear that we need the shells to overlap. Thus the intershell interaction is very important. In this context, one can envisage two possible situations, depending on whether the shells fragment before or after they overlap; we will refer to these as scenarios A and B, respectively.

Scenario A has the advantage that it is easy to simulate numerically (Saarinen et al. 1987, Dekel 1988). As initial conditions, one envisages a region of the Universe containing a number of randomly placed shells: the baryons within each shell are randomly distributed around its surface as galaxies, while the baryons outside the shells remain as gas. Any non-baryonic dark matter is assumed to maintain its original (relatively smooth) distribution. The shells of galaxies initially have a small filling factor but they continue to expand (without sweeping up further gas) until they overlap, whereupon their gravitational interactions becomes important.

The important features of the Saarinen et al. simulations are as follows: (i) The initial bubbly structure tends to smear out and become frothy. (ii) The parts of the shells which overlap become bound and form elongated superclusters. (iii) The galaxy correlation function initially scales as $\varsigma_G \propto r^{-1}$ for $r < R_{shell}$ (as expected) but it then steepens with time, becoming $\varsigma_G \propto r^{-1.8}$ (as observed) after an expansion factor of about 10. (iv) In order to explain the amplitude of ς_G, the initial shell radius must be about $10h^{-1}$Mpc. (v) If there is background dark matter, its correlation function ς_{DM} is initially very small but it increases as the dark matter is accreted and $\varsigma_G/\varsigma_{DM}$ attains a value of about 5 after an expansion factor of 10. (vi) The void distribution evolves in time towards larger voids; it roughly fits the CFA data at the present epoch (Ryden & Turner 1984), although it may be hard to generate voids as large as the one in Bootes.

These considerations suggest that Scenario A has many attractive features. In particular, one has a form of biased galaxy formation which naturally explains the galaxy correlation function if galaxies form at a redshift of about 10 on shells with radius of about 10 Mpc. However, the scenario has several drawbacks. Firstly, since the shells fragment before colliding, most of the baryons in the Universe must be left as background gas. On the other hand, the filling factor of the shells cannot be too small: since the density of galaxies is at least 1% of the critical density, whereas cosmological nucleosynthesis constraints show that the total baryon density is at most 10% of the critical value (Yang et al. 1984), the filling factor must exceed 10%. One thus has a fine-tuning problem, in the sense that the shells must fragment just before they collide. Secondly, the gas which does remain outside the shells must be very hot and, in this case, it might generate too large an X-ray background.

We therefore focus on Scenario B, in which the shells do not fragment until after they have collided. This might happen rather naturally because, once the shells have collided, the extra compression would decrease the cooling timescale. In this case, one would expect most of the gas in the Universe to be compressed into sheets , with only the non-baryonic matter being left in the voids. The typical fragment mass would still be comparable to a galactic-scale but one has the added feature that each fragment is likely to become the focus of a high pressure cooling flow. Such cooling flows appear to turn gas into dark matter with high efficiency (Fabian et al. 1984), so one has a natural mechanism for producing both large-scale structure and dark baryonic galactic halos (Ashman & Carr 1987).

One would expect the fragments formed after shell collisions to exhibit characteristic streaming motions (Allen & Carr 1987): velocities of order 600 km/s on scales of up to 20 Mpc would be typical. Although this scenario is harder to simulate numerically than the first one (because it requires gas dynamics), one would anticipate that most of the qualitatively attractive features of Scenario A would be preserved providing one has $R_{shell} \simeq 10$ Mpc and $z \simeq 10$ when the shells fill the Universe. Of course, the detailed galaxy distribution which comes out of the scenario depends on the distribution of the explosive seeds. If the seeds were regularly spaced, one would expect a polygonal structure but, if they were placed irregularly or correlated, one would have a more complex galaxy distribution, with some voids being much larger than average.

4. DISCUSSION

The above considerations suggest that explosions could provide a natural explanation for many features of large-scale structure providing the shells eventually overlap and providing they do not fragment before overlapping. However, a number of important questions remain.

When do the shells overlap? If the shells are to attain the characteristic radius of 10 Mpc required to explain the observations, they almost certainly need to overlap in the Compton era. This is because one can get a much larger amplification just before the redshift z_c than after it, as illustrated by Figure (1). There are also other problems with having the shells overlap in the radiative era. In particular, the requirement that shells should be able to cool on a cosmological timescale place an upper limit of only a few Mpc on the shell radius, as indicated by eqn (3).

Where do the explosive seeds come from? To get 10 Mpc shells in a single explosive step, one needs a seed mass of order $10^{15} M_{\odot}$ if one invokes ordinary nucleosynthetic sources of explosive energy. Such large seeds are possible in principle. For example, in a baryon-dominated Universe with adiabatic fluctuations, the first objects to bind would have a mass $M_S \simeq 10^{13} \Omega_b^{-5/4} M_{\odot}$; in a neutrino-dominated Universe, the first objects would have a mass $M_\nu \simeq 10^{15} \Omega_\nu^{-2} M_{\odot}$. However, we have seen that it is more attractive to invoke the detonation wave picture because, in this case, the shell radius naturally asymptotes to about $10h^{-1}$Mpc by the end of the Compton era. A still more exotic proposal is to invoke superconducting cosmic strings since these generate explosive energy with greater efficiency than stars. This possibility is discussed elsewhere in this volume (Ostriker 1988).

Why do the shells overlap so late? At first sight, the scenario requires considerable fine-tuning since we need the shells to overlap at the very end of the Compton era in order to generate large enough structure. However, there is one way in which this could happen rather naturally. Let us assume that the smallest surviving fluctuations after decoupling have some characteristic mass-scale M. Let us also assume that the fluctuations on that scale have a Gaussian distribution with an rms value $\varepsilon_*(M)$. The average region of mass M will then bind at a redshift $z_* \simeq 10^3 \varepsilon_*(M)$ and, for $z > z_*$, the fraction of the Universe in seeds of mass M will be

$$f(z) \simeq \exp\left[-\frac{z^2}{2z_*^2}\right] \qquad (z > z_*) \qquad (5)$$

As z decreases, the filling factor of the shells will increase, both because f(z) increases and because the volume within each detonation front increases. For reasonable parameters, one can show that the filling factor will reach unity at a redshift z_1 given by

$$\frac{z_1}{z_*} \simeq \left[10 - 3\ln\left\{\frac{1+z_1}{10}\right\} - 2\ln\left\{\frac{M}{10^{12}M_\odot}\right\}\right]^{0.5} \qquad (6)$$

Thus z_1/z_* is expected to be about 3, with only a logarithmic dependence on the value of M. One can think of the ratio z_1/z_* as a "bias" factor since it specifies the fraction of the Universe in explosive seeds when the shells first overlap. We term this "supernatural bias" since it is a form of natural bias which invokes supernovae!

REFERENCES

Allen, A.J., and Carr, B.J., 1987. Preprint.
Ashman, K.M., and Carr, B.J., 1987. Preprint.
Bertschinger, E., 1983. *Astrophys.J.*, 268, 17.
Bertschinger, E., 1985. *Astrophys.J.*, 295, 1.
Carr, B.J., and Ikeuchi, S., 1985. *Mon.Not.R.astr.Soc.*, 213, 497.
de Lapparent, V., Geller, M., and Huchra, J., 1986. *Astrophys.J.*, 306, L1.
Dekel, A., 1988. This volume.
Dressler, A., Faber, S.M., Burstein, D., Davies, R.L., Lynden-Bell, D., Terlevich, R.L., and Wegner, G., 1987. *Astrophys.J.*, 313, L37.
Fabian, A.C., Nulsen, P.E.J., and Canizares, C.R., 1984. *Nature*, 310, 733.
Giovanelli, R., Haynes, M.P., and Chincarini, G.,1986. *Astrophys.J.*, 300, 77.
Ikeuchi, S., 1981. *Pub.Astron.Soc.Japan*, 33, 211.
Kirshner, R., Oemler, A., Schechter, P., and Schechtman, S.A., 1981. *Astrophys.J.*, 248, L57.
Ostriker, J.P., 1988. This volume.
Ostriker, J.P., and Cowie, L.L., 1981. *Astrophys.J.*, 243, L127.
Ryden, B.S., and Turner, E.L., 1984. *Astrophys.J.*, 287, L59.
Saarinen, S., Dekel, A., and Carr, B.J., 1987. *Nature*, 325, 598.
Tully, R.B., 1982. *Astrophys.J.*, 257, 389.
Wandel, A., 1985. *Astrophys.J.*, 294, 385.
Yang, J., Turner, M.S., Steigman, G., Schramm, D.N., and Olive, K.A., 1984. *Astrophys.J.*, 281, 493.

MODELS FOR THE EVOLUTION OF THE TWO POINT CORRELATION FUNCTION

H.M.P. Couchman and J.R. Bond
Canadian Institute for Theoretical Astrophysics,
University of Toronto,
Toronto, ON M5S 1A1,
Canada

The spatial two point galaxy correlation function, $\xi(r)$, is, at present, the most useful statistic for comparing theoretical models to observational data. We have derived an expression for the dynamical evolution of ξ for structures arising from Gaussian initial conditions under the assumption that non-linear evolution may be described by the Zel'dovich approximation. The observed angular correlation function of galaxies, $w(\theta)$, may place useful constraints on the spectrum of initial fluctuations on large scales. The angular correlation function due to the growth of structure from Gaussian fluctuations with the Cold Dark Matter spectrum is calculated. It is shown that w_g for the $\Omega = 1$ biassed CDM model exhibits a gentle falloff near the position of the break found by Groth and Peebles in the Shane–Wirtanen catalogue.

The determination of the galaxy correlation function from redshift surveys is hampered by the small samples ($\sim 10^3$ galaxies) and by the transformation from redshift space. Davis and Peebles (1983) found $\xi_g \simeq (r_0/r)^\gamma$ with $\gamma \simeq 1.8$ and $r_0 = 5.4h^{-1}$ Mpc, although the new CfA survey suggests that r_0 may be larger. The errors are too large to substantiate any deviations from a power law on scales 15–$20h^{-1}$ Mpc. The angular two point function, w_g, has the advantage that it can be estimated from large samples such as the Shane–Wirtanen catalogue ($\sim 10^6$ galaxies, with an effective depth $D_* = 360h^{-1}$ Mpc). Groth and Peebles (1986) suggest that there is a break in $w_g(\theta)$ at $\theta \sim 3°$, although this is controversial.

The large scale texture of the Universe, as revealed by the CfA redshift survey, can be reproduced at least qualitatively in Hot Dark Matter universes with moderate evolution by pancake formation and in Cold Dark Matter (CDM) models by a combination of dynamical formation of pancakes and preferential galaxy formation in clusters and pancake sites. Moderate dynamical evolution can be modeled with some success by the Zel'dovich approximation — $\mathbf{x}(\mathbf{q},t) = \mathbf{q} - D(t)\mathbf{s}(\mathbf{q})$, where $\mathbf{q}$ and $\mathbf{x}$ are the initial and final positions of a particle and $\mathbf{s}$ denotes the displacement field, related to the peculiar velocity by $\mathbf{v}_p = -a\partial\mathbf{s}/\partial t$, where $a = (1+z)^{-1}$ is the expansion factor. For an $\Omega = 1$ universe the growth factor $D(t) = a$. Although the Lagrangian mapping formally breaks down when the first caustics (e.g., pancakes) form, i.e., when the mass overdensity $1 + \delta(\mathbf{q},t) = (1 + \det(\partial s_i/\partial q_j))^{-1}$ first becomes infinite and 'shell crossing' occurs, it is expected to remain reasonably accurate until shells propagate through each other to such an extent that the texture is 'washed out'.

The two point mass correlation, $1 + \xi_Z(x) = P(x|m1,m2)$ ($x = |\mathbf{x}_1 - \mathbf{x}_2|$), is expressible as an integral over $\{\mathbf{s},\mathbf{q}\}$ assuming that $\mathbf{s}(\mathbf{q})$ is Gaussian. In general this would require a functional integral over *all* positions and velocities; an intractable problem. The Zel'dovich approximation allows an enormous

N. Kaiser and A. N. Lasenby (eds.), The Post-Recombination Universe, 263–265.
© *1988 by Kluwer Academic Publishers.*

simplification. Only pairs of points determining the correlations appear. Indeed the sums $\mathbf{r}_1 + \mathbf{r}_2$ and $\mathbf{s}_1 + \mathbf{s}_2$ integrate out leaving only the relative initial separations, $\mathbf{r}$, and velocities, $\mathbf{s}$ which enter in the form of a conditional probability for x qiven $\mathbf{q}$ and $\mathbf{s}$. The two point function can thus be written as

$$1 + \xi_Z = \int d^3\mathbf{q}\, d^3\mathbf{s}\, P(x|\mathbf{q}, \mathbf{s})\, P(\mathbf{s}|\mathbf{q}) P(\mathbf{q}) \tag{1}$$

The first probability is deterministic for the Zel'dovich map, $P(x|\mathbf{q}, \mathbf{s}) = \delta(x - |\mathbf{q} - D\mathbf{s}|)/(4\pi x^2)$. The probability $P(\mathbf{s}|\mathbf{q})$ is a Gaussian distribution derived from the initial spectrum. It is characterized by the covariance matrix $\langle s_i(q)s_j(0)\rangle = [(\xi'_W - q\xi''_W)\hat{q}_i\hat{q}_j - \xi'_W \delta_{ij}]/q$, where $\xi_W = \sum_k P_L(k)k^{-4}e^{ik\cdot q}$. The probability $P(\mathbf{q})$ is, of course, unity.

We have also derived a closed form expression for the full non-linear power spectrum which arises when fluctuations grow according to the Zel'dovich prescription:

$$P_Z(k) = e^{-k^2\sigma_s^2} \int d^3\mathbf{q}\, e^{-i\mathbf{k}\cdot\mathbf{q}} \exp[-(\mathbf{k}.\nabla)^2 \xi_W(q)], \tag{2}$$

where $\sigma_s^2 = \langle s_i^2(0)\rangle$.

For these results to be useful it is necessary to demonstrate that, for moderate non-linear evolution, the Zel'dovich approximation does not seriously underestimate the degree of dynamical evolution. The accuracy of the approximation has been assessed by calculating the non-linear power spectrum $P_D(k) \equiv \langle|\delta\rho(k)/\rho|^2\rangle = P_L + P_{D2}$ from the full dynamical equations to quadratic order $\mathcal{O}(\sigma_\rho^4)$ and comparing it with the power spectrum for the Zel'dovich approximation expanded to the same order, $P_Z = P_L + P_{Z2}$ (Bond and Couchman 1987). The results are shown in Figure 1.

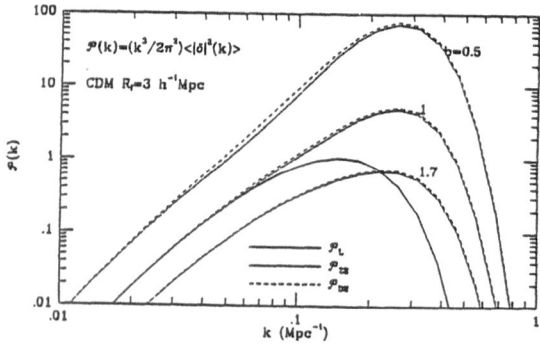

Figure 1. The power spectrum including quadratic nonlinearities derived from the full dynmical equations is shown by the dashed line; that for the Zel'dovich approximation is shown by the dotted line. The agreement is remarkable for all the biassing factors indicated, including the relatively large $b = 0.5$ ($\sigma_\rho = 2.3$) case. The linear spectrum (solid line) plotted for the $b = 1$ case shows that waves above $7h^{-1}$ Mpc evolve linearly, whilst substantial power is generated below the filtering scale, $R_f = 3h^{-1}$ Mpc.

Using the Zel'dovich map to model the two point functions thus appears to be an excellent approximation.

If galaxies trace the mass distribution, then the galaxy two point function, ξ_{gZ}, is also given by the form in equation (1). A similar result obtains for schemes in which galaxies are 'biassed' tracers of the underlying mass distribution, occurring towards the high peaks in the matter density (necessary if $\Omega = 1$ CDM models are to be viable). In this case $1 + \xi_{gZ} = P(x|g1, g2)$ is a more complicated reduction into conditional probabilities; it may be written as

$$1 + \xi_{gZ} = \int d^3\mathbf{q}\, d^3\mathbf{s}\, d\nu_{b1}\, d\nu_{b2}\, P(x|\mathbf{q}, \mathbf{s}) P(\mathbf{s}|\mathbf{q}, \nu_{b1}, \nu_{b2})$$
$$P(\mathbf{q}) P(g1|\nu_{b1}) P(g2|\nu_{b2}) P(\nu_{b1}, \nu_{b2}|\mathbf{q}) \tag{3}$$

The 'peak–background split' is used to describe the statistical clustering: $P(g|\nu_b) \propto \exp[(b-1)\sigma_{\rho b}\nu_b]$, where ν_b is the overdensity of the background field in units of the r.m.s. and b is the bias factor. The *linear* two point galaxy correlation, ξ_{gL}, is then related to the linear mass correlation, ξ_L, by $\xi_{gL} = b^2\xi_L$. Nascent galaxy peaks are defined by the density fluctuation power spectrum between the galactic scale and a larger (background) scale (which is chosen to be $3h^{-1}$ Mpc), whilst the density of galaxies is determined by the local value of the mass field smoothed on the background scale R_f.

To calculate the evolution of ξ for the Cold Dark Matter (CDM) model we cannot choose the filtering scale to be as small as galaxy scales $\sim 0.5h^{-1}$ Mpc since far too much shell crossing would have occurred. Instead we take the smallest filtering scale consistent with the validity of the Zel'dovich approximation; $R_f \simeq 3h^{-1}$ Mpc. The amplitude is found by normalizing $J_3(10h^{-1}$ Mpc$)$ to the observed value. The bias factor, b, is chosen to be 1.7, which is a typical value required by the observations for an $\Omega = 1$ universe. It is found that very little non-linear evolution occurs on scales above $\sim 7h^{-1}$ Mpc.

The limitations of the filtering on the dynamical evolution of smaller scales are overcome by attaching to ξ_{gz} a power-law as predicted by N-body simulations ($\xi \sim r^{-1.8}$ at least between $1h^{-1}$ Mpc and $7h^{-1}$ Mpc). As can be seen from Figure 2 the match between the observed power-law and the result calculated here at $7h^{-1}$ Mpc is excellent. The spatial function, ξ, and resulting angular function, w, are shown in Figure 2 for the CDM spectrum and a spectrum with extra power on large scales; CDM+X. The latter spectrum with extra power is designed to reproduce cluster-cluster and cluster-galaxy correlation functions which extend as power-laws beyond ~ 20–$30h^{-1}$ Mpc (which the standard CDM spectrum fails to do) as may be suggested by some observations.

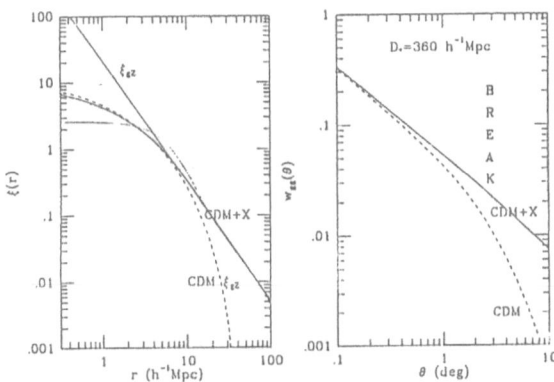

Figure 2. The function ξ_{gZ} with small scale extension for the biassed CDM and CDM+X models together with the corresponding angular correlations, $w(\theta)$. The depth, D_*, corresponds to that of the Shane–Wirtanen catalogue. In (a), the dotted line is the correlation function as calculated in linear theory for CDM+X. The confirmation of the 3° break of Groth and Peebles therefore supports the standard CDM spectrum and appears to contradict the observed cluster-galaxy and cluster-cluster correlation functions. ($\Omega = 1$, $h = 0.5$.)

REFERENCES

Bond, J.R. and Couchman, H.M.P., 1987. In: *Proceedings of the Second Canadian Conference on General Relativity and Relativistic Astrophysics*: eds. C. Dyer, A. Coley; World Scientific.

Davis, M. and Peebles, P.J.E., 1983. *Astrophys. J.*, **267**, 465.

Groth, E.J. and Peebles, P.J.E., 1986. *Astrophys. J.*, **310**, 499.

SOME POSSIBLE MECHANISMS FOR THE FORMATION OF LARGE SCALE STRUCTURES IN THE UNIVERSE

Ruth A. Daly
Institute of Astronomy
Madingley Road
Cambridge CB3 0HA
England

ABSTRACT. Three mechanisms for the formation of large voids are discussed. These have been considered elsewhere, however, since these were presented we have considered them in more detail which lead to a few modifications: these are presented here.

1. INTRODUCTION

The relative positions of luminous galaxies define extremely large voids with diameters up to ~ 50 h^{-1} Mpc (Joeveer and Einasto 1978; Kirshner et al. 1981) which appear to be roughly spherical in shape (de Lapparent, Geller and Huchra 1986). In the context of standard models for galaxy formation where the dark matter gravitationally grows to form potentials into which the baryons fall, some biasing mechanism must be invoked to allow for the existence of these voids; however, de Lapparent, Geller and Huchra (1986) note that the remarkably sharp edges of the voids indicates that hydrodynamic processes were important in their formation. One possibility is that the dark matter plays more than a passive, purely gravitational role. The dark matter may play a dynamical role which could drastically effect the type and location of the luminous galaxies which form as the universe evolves, and could lead to the formation of large, roughly spherical voids. For example, the dark matter may be unstable and decay, setting hydrodynamic processes into motion.

The formation of large voids is intimately connected with galaxy formation because the voids are defined by the relative positions of the luminous galaxies. There are some numerical coincidences suggesting a connection between the formation of structures in the universe and the origin of the diffuse x-ray background (Daly and Turner, 1987). It is possible that the major key to understanding the formation of galaxies and large scale structures is an understanding of the origin of the diffuse x-ray background. If this background is truly produced by thermal bremsstrahlung emission, characterized by the extremely high temperature of ~ $4 \times 10^8(1 + z_{emit})$ K (where z_{emit} is the redshift of emission), then it should be a major consideration in any model for evolution of structures in the universe, since it must originate at cosmological distances (Schwartz, Murray and Gursky 1976). Indeed, as the microwave background radiation is the key to the early universe, so the x-ray background may be the key to the epoch of the formation of structures in the universe.

In this vein, we have discussed in detail the possibility that a new class of object may have existed, which are referred to as condensates (Daly 1987a,b,1988). The condensates would form at redshifts of about three to five from the gravitational collapse of the matter within the comoving volumes corresponding to the voids. They are massive (~ $10^{15}-10^{16}$ $M_\odot$), having virial temperatures ~ temperature characterizing the diffuse x-ray background, and the thermal bremsstrahlung emission from the gas in these potentials could produce the x-ray background. The subsequent evolution of

267

N. Kaiser and A. N. Lasenby (eds.), The Post-Recombination Universe, 267–269.

these objects would lead to the formation of the voids. Three possible evolutionary tracks are discussed. The first involves unstable dark matter. The second involves stable but annihilating dark matter. The third involves stable dark matter. Below, each of these is discussed in turn.

1.1 Condensates Disrupt Due To Unstable Dark Matter

A model in which the condensates disrupt due to unstable dark matter is presented by Daly (1987a,b). In this picture the condensates form, the thermal bremsstrahlung emission from the gas in these potentials produces the x-ray background, the dark matter then decays into relativistic secondary particles which free stream from the potential, causing the baryons to go into free expansion. The expanding gas collides with the ambient gas triggering galaxy formation and the motion of the shock system as a unit produces the voids. It was pointed out to me at this workshop by Nick Kaiser that I neglected the fact that when the matter ceased expanding with the Hubble flow, it has an overdensity of about 5.5 the mean density at that time. The consequences of including this factor are: (1) the free fall time for the matter is shorter and (2) the object has a smaller final radius, so that the number density of gas and dark matter in the condensate is higher. Including this factor does not significantly change any of the results presented earlier; factors of 1.4 or 1.5 creep into the calculations. We find that the virial temperature of the gas is

$$T_v \simeq 4 \times 10^8 \, (1 + z_{form}) \, (\Omega(t_{form})/1.0) \, (R_0/12.5 \; h^{-1} \; Mpc)^2 \; K$$

where R_0 is the comoving radius of the volume from which the matter collapsed set at a redshift of zero and $\Omega(t_{form})$ is the ratio of the average mass-energy density in the universe to the critical density at the time the object forms, when the age of the universe is t_{form}, corresponding to a redshift z_{form}. The integrated thermal bremsstrahlung emission matches that observed for $\Omega_B \simeq 0.04 \, (fC_L)^{-1/2}$, where Ω_B is the ratio of the mass density in baryons to the critical density of the universe, C_L is the clumping factor for the gas in the condensate and f is the volume filling factor of condensates with $R_0 \simeq 12.5 \; h^{-1} \; Mpc$. The expansion phase of the gas leads to a relationship between the redshift corresponding to the lifetime of the unstable particle, z_d, and the redshift at which the expanding gas reaches the edge of the comoving volume from which the matter originally collapsed z_R: $(1 + z_d) \simeq 2.7(1 + z_R)$, where we have assumed that the condensate forms when the age of the universe is equal to the lifetime of the unstable particle (however, the results are insensitive to this choice). So, if the matter ceases expanding with the Hubble flow at a redshift $(1 + z)$ of 7.5 [10] the object forms at a $(1 + z)$ of 5.7 [7.6], the gas goes into free expansion at a $(1 + z)$ of 4.3 [5.7] and reaches the edge of the comoving volume from which it collapsed at a $(1 + z)$ of 2.1 [2.8]. Fluctuations in the temperature characterizing the microwave background radiation due to the Sunyaev-Zeldovich effect while the condensates are producing the x-ray background is slightly larger then that given previously and the angular size of each object is slightly smaller; these figures will be detailed elsewhere. All of these modifications are very minor.

1.2 Condensates Disrupt Due To Annihilating Dark Matter

Recently, we have analyzed the annihilating dark matter model in detail. The idea is that, subsequent to the formation of the condensate, the dark matter in the core of the condensate annihilates, producing relativistic secondary particles which free stream causing the potential to become more shallow. The question is, do the outermost shells of gas go into free expansion as a result of this change in the potential?

To evaluate this question we assume that a constant number density core forms which satisfies the equation of hydrostatic equilibrium. The initial annihilation timescale must exceed the free fall timescale in order that the core (and condensate) may form (this is necessary so that the x-ray background may be produced). As the annihilation proceeds the core expands, causing the annihilation rate to decrease. We find that, irrespective of the temperature gradient beyond the core, if the initial sound crossing time to the edge of the condensate is less than the characteristic expansion time so that hydrostatic equilibrium is achieved, hydrostatic equilibrium will be maintained for all time; the expansion will be quasistatic. Hence, the outermost shells of gas will never go into free expansion. Unless there is some mechanism which will cause the core to collapse as the annihilation proceeds, a model in which the unstable dark matter is replaced by annihilating dark matter may be ruled out.

1.3   Condensates Evolve Into Low Luminosity Sources

The third possibility is that galaxies and clusters of galaxies form directly from the gravitational growth of primordial density perturbations in the regions exterior to the voids. The matter within the voids gravitationally collapses to form bound, virialized objects (condensates) which produce the diffuse x-ray background. These must evolve into low luminosity x-ray sources so that the x-ray background is not dominated by low redshift objects and is not overproduced. This will occur if the object is expanding (for example, due to the annihilation discussed above) or if the ratio of the core radius to the total radius of the object is increasing rapidly enough with time.

2.   CONCLUSIONS

The extremely large sizes of the voids (Joeveer and Einasto 1978; Kirshner *et al.* 1981) and their rough spherical shape and sharply defined edges (de Lapparent, Geller and Huchra 1986) suggest that some event occurred which inhibited galaxy formation in these regions (and/or triggered galaxy formation in the regions exterior to the voids), and this event may have been hydrodynamic in nature. One possibility is that this event has left its signature at extremely high energies by heating the baryons to a temperature $\sim 4 \times 10^8 \, (1 + z_{emit}) \, K$ which then produce the x-ray background via thermal bremsstrahlung radiation. If the x-ray background is produced by thermal bremsstrahlung emission, as opposed to resulting from the combined emission of non-thermal sources such as quasars, the origin of this background should be a major consideration in any model for evolution of structure in the universe.

ACKNOWLEDGEMENTS

I would like to thank Alan P. Marscher, Sheldon L. Glashow and Martin J. Rees for helpful discussions. This work was supported by NASA's Graduate Student Researchers Program, NSF Grants AST-8315556 and AST-8516549 (A.P. Marscher, P.I.), and NSF Fellowship Award RCD-8751127.

REFERENCES

Daly, R.A. 1987a, Ph.D. Thesis, Boston University.
Daly, R.A. 1987b, *Ap. J.*, **322**, in press.
Daly, R.A. 1988, IAU Symposium No. 130, *Evolution of Large Scale Structure in the Universe*, (Dordrecht: D. Reidel), in press.
Daly, R.A., and Turner, E.L. 1987, in preparation.
de Lapparent, V., Geller, M. and Huchra, J. 1986, *Ap. J. (Letters)*, **302**, L1.
Joeveer, M. and Einasto, J. 1978, IAU Symposium No. 79, *The Large Scale Structure of the Universe*, (Dordrecht: D. Reidel), 241.
Kirshner, R.P., Oemler, A., Jr., Schechter, P.L., and Shectman, S.A. 1981, *Ap. J. (Letters)*, **248**, L57.
Schwartz, D.A., Murray, S.S., and Gursky, H. 1976, *Ap. J.*, **204**, 315.

EVOLUTION OF QUASAR CLUSTERING

A. Iovino and P. Shaver
European Southern Observatory
Karl-Schwarzschild-Str. 2
D-8046 Garching bei München
Federal Republic of Germany

SUMMARY

We have combined three different samples, totalling 371 quasars, to study the evolution of their clustering properties with cosmic time. Clustering is detected at low redshifts ($z<1.5$) on scales less than $10 \ h^{-1}$ Mpc at the 4.7σ level, but not at high redshifts ($z>1.5$). The difference between these two results is significant at the 2.2σ level.

SAMPLES USED

The samples used are the following: Boyle (1986): 171 UVX quasars, area 4 sq. deg., mag. lim. B=20.9; Crampton et al. (1987): 125 grens quasars, area 5.2 sq. deg., mag. lim. 20.5; Barbieri et al. (1987): 80 UVX quasars, area 10 sq. deg., mag. lim. B=19.5.
    The redshift distribution of the three samples is quite smooth (figure 1), and the surface density distribution is essentially constant over the fields, except for the Barbieri et al. sample, where there is some indication of mild ($\lesssim 25\%$) vignetting.

METHOD

The correlation function method has been used to analyze the samples. We have done many random simulations for each of the different samples, exactly reproducing both the redshift distribution and the angular distribution of the real data. These samples have then been used to evaluate ξ according to the formulae:

$$\xi(r) = N(r)_{obs} \ / \ N(r)_{rand} - 1 \ ,$$

$$\Delta\xi = \sqrt{(1+\xi) \ / \ N_{rand}}$$

(Peebles, 1980).

N. Kaiser and A. N. Lasenby (eds.), The Post-Recombination Universe, 271–274.

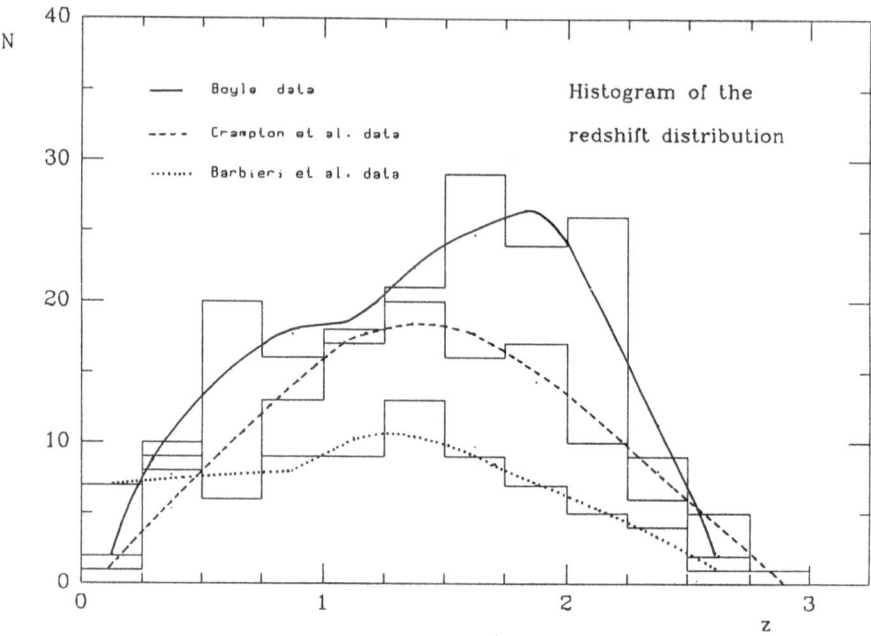

Figure 1 - Redshift distribution of the three samples used. The lines
indicate the smoothing adopted.

CONCLUSIONS

The results of our analysis are illustrated in figures 2 and 3 and in
Table 1.

Table 1

# of pairs with sep. < 10 h^{-1} Mpc	$z < 1.5$	$z > 1.5$
EXPECTED	4.7	4
OBSERVED	15	5
	4.7σ	0.5σ

We have found evidence for clustering of quasars at low

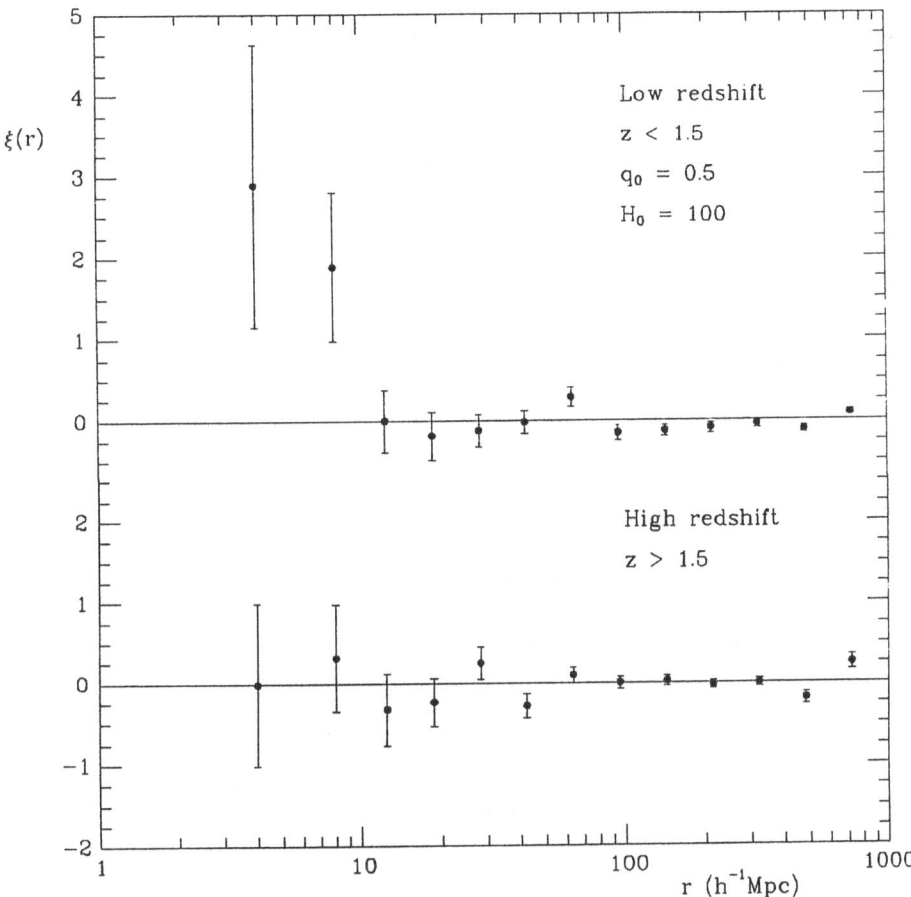

Figure 2 - Quasar two-point correlation function at low (z < 1.5) and
high (z > 1.5) redshifts, from the three samples used.

redshifts, up to scales of 10 h^{-1} Mpc. By contrast there is no
indication of clustering in the high redshift bin. Thus, there is
evidence for the evolution of clustering.

The amplitude of the correlation function for quasars at low
redshifts is similar to that expected for clusters. This is consistent
with a picture in which quasars are located in rich environments,
where they may perhaps be triggered by interactions.

If the clustering of quasars really reflects that of clusters,
than its evolution may really be that of large-scale structure of the
universe.

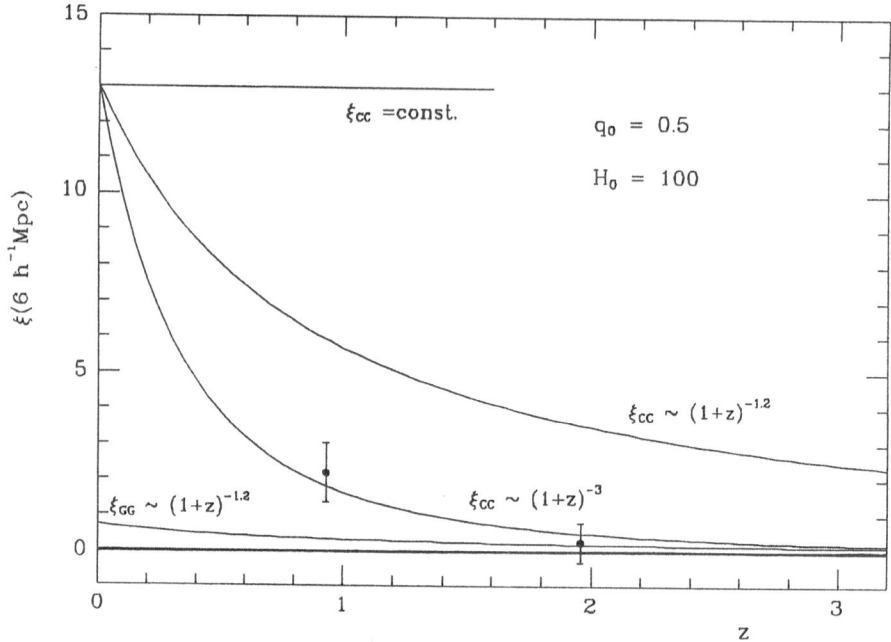

Figure 3 - Amplitude of the quasar correlation function at 6 h^{-1} Mpc
as a function of redshift, from the three samples used. The
smooth curves correspond to possible forms of evolution of
the correlation function for galaxies (GG) and clusters
(CC): comoving (ξ = constant), stable clustering ($\xi \sim$
$(1+z)^{-1.2}$), and collapsing ($\xi \sim (1+z^{-3})$).

REFERENCES

Barbieri, C., et al., 1987, A&A, to be submitted.
Boyle, B.J., 1986, Ph.D. Thesis, University of Durham.
Crampton, D., et al., 1987, Ap.J., **314**, 129.
Peebles, P.J.E., 1980, The Large Scale Structure of the Universe,
Princeton Series in Physics.

FROM 2-D TO 3-D BY MAXIMUM ENTROPY METHOD

Ofer Lahav and Donald Lynden-Bell
Institute of Astronomy, Madingley Road, Cambridge CB3 0HA, UK

Stephen F. Gull
Mullard Radio Astronomy Observatory, Cavendish Laboratory, Madingley Road, Cambridge CB3 0HE, UK

ABSTRACT. We present a method of estimating distances to clusters of galaxies from two-dimensional catalogues by using the "Maximum Entropy Method".

The three-dimensional distribution of galaxies is a key observation for our understanding of cosmology. Magnitude limited catalogues give the angular position of galaxies almost over the entire celestial sphere. Distances to galaxies are mainly estimated from their redshifts. However, only a fraction of the sky has been surveyed so far in a complete way. Furthermore, the distribution of galaxies as seen in redshift surveys is distorted due to local gravitational fields.

On the other hand, the angular diameters (or magnitudes) of galaxies can be used as distance indicators as well. The mapping from 2-D to 3-D can be done by using a diameter function (or a luminosity function), which is based on a redshift survey from a section of the sky.

The diameter function can be viewed as an a priori statistical knowledge of the distribution of galaxy metric diameters, or as an unfocused lens, which casuses blurring of an image. This suggests the application of "Maximum Entropy Method" to our problem. For review of the Maximum Entropy Method see for example Gull & Skilling (1984).

The problem is formulated as follows. The number of galaxies with a metric diameter D in a volume element d^3r is:

$$dN = n(\mathbf{r})/n_b \, d^3r \, \phi(D)dD \qquad 1$$

where n(r) is the 'true' number density of galaxies at position $\mathbf{r}$, n_b is the mean number density of galaxies in the universe and $\phi(D)dD$ is the diameter function. We assume that within a narrow cone n(r) = n(r) and then express $N(\geq \theta)$, the number of galaxies greater than a certain angular diameter θ. In a discrete form we write the relation as:

$$N(\geq \theta_k) = \sum_i n_i P_{ik} \qquad 2$$

N. Kaiser and A. N. Lasenby (eds.), The Post-Recombination Universe, 275–276.

276

where n_i is the density at the $i - th$ distance bin and P_{ik} is our "point spread function", which is a function of the diameter function and Galactic obscuration. We express the deviations of the measurements from the predictions (2) in terms of χ^2 statistics. We require that χ^2 to be smaller than a certain value.

The **entropy** of the image is expressed as :

$$S = \sum_i [n_i - n_b - n_i \, log(n_i/n_b)] \qquad 3$$

The procedure now is to maximaize the entropy (3) under the χ^2 constraint.

As an example we have applied the method to galaxies from the UGC catalogue which are within 6° of Virgo's centre. We have used a diameter function which has been derived from the CfA redshift survey (Lahav, Rowan-Robinson & Lynden-Bell 1987) and the Maximum Entropy algorithm of Skilling & Bryan (1984). The results are shown in Figure 1 by plotting the number density profile in each cone versus distance. The dashed line indicates the mean number density of galaxies in the universe (the a priori model). The algorithm identifies Virgo at a distance of about 1200 km/sec, in a good agreement with its known distance. Currently the method is insufficiently developed to determine the amplitudes and breadths of the density bumps.

We intend to develop the method further by improving the distance indicator, by making use of the partial redshift information and by modifying the algorithm. The method can be also applied to new deep catalogues, e.g. the Cambridge APM survey.

REFERENCES

Gull, S.F. & Skilling, J. 1984. *IEE proceedings (F)*, **131**, 646.
Lahav, O., Rowan-Robinson, M. & Lynden-Bell, D., 1987. *preprint.*
Skilling, J. & Bryan, R.K. 1984. *Mon. Not. R. astr. Soc.*, **211**, 111.

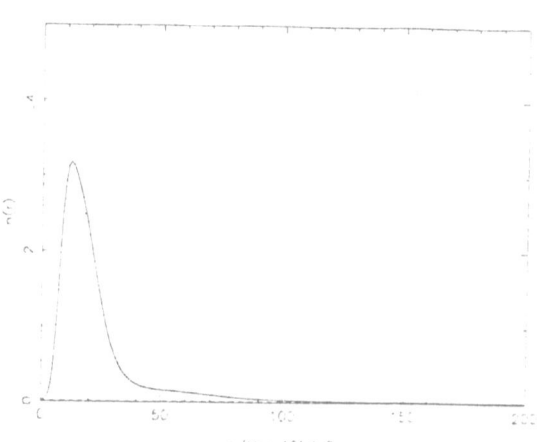

Figure 1. The radial density profile in a cone
of 6 degrees centred on Virgo.

CLUSTER-GALAXY CROSS-CORRELATIONS

Per B. Lilje and G. Efstathiou
Institute of Astronomy
Madingley Road
Cambridge CB3 0HA
England

ABSTRACT. The cross-correlations between Abell clusters and the $10' \times 10'$ Lick galaxy counts are analysed. We correct for galactic obscuration, check the effects of large-scale gradients in the Lick counts, use redshifts for Abell clusters to compute the spatial cross-correlation function, and use modern determinations of the galaxy luminosity function to compute the amplitude of the spatial function. For $r \lesssim 20\,h^{-1}$Mpc, our results are well described by the power-law $\xi_{cg} \approx (r/8.8\,h^{-1}\text{Mpc})^{-2.2}$. On scales $1\,h^{-1}$Mpc $\lesssim r \lesssim 10\,h^{-1}$Mpc, these results are found to be in surprisingly good agreement with theoretical predictions for ξ_{cg} in the $\Omega = 1$, biased CDM model. On other scales there are some minor discrepancies.

1. INTRODUCTION

The galaxy and rich cluster correlation functions provide simple and convenient tests of models for galaxy formation. They set especially sharp constraints on scales $r \gtrsim 10\,h^{-1}$Mpc where the density fluctuations must be nearly linear. However, the most frequently applied statistic, the galaxy two-point correlation function ξ_{gg}, has a very low amplitude on these scales and the reliability of the results are not yet satisfactory. Fortunately, the rich cluster two-point correlation function, ξ_{cc}, (eg., Bahcall & Soneira 1983) and the cluster-galaxy cross correlation function, ξ_{cg}, (eg., Seldner & Peebles 1977a,b) have higher amplitudes than ξ_{gg} and are therefore easier to measure out to larger separations. Bahcall & Soneira found that ξ_{cc} is positive out to $150\,h^{-1}$Mpc. Seldner & Peebles found that the cross correlation of rich Abell clusters (Abell 1958) and the $10' \times 10'$ Lick galaxy counts (Shane & Wirtanen 1967; Seldner et al. 1977) was well described by $\xi_{cg}(r) = (r/7\,h^{-1}\text{Mpc})^{-2.5} + (r/12\,h^{-1}\text{Mpc})^{-1.7}$ for $0.5\,h^{-1}$Mpc $< r < 40\,h^{-1}$Mpc. But Bahcall & Soneira's result was based on a sample of only 104 clusters. The reliability of the determination of the amplitude of ξ_{cc} has also recently been questioned (eg., Sutherland, this conference).

The so called biased cold dark matter (CDM) model has been found to be in good agreement with several properties of the structure and clustering of galaxies (eg., Davis et al. 1985; White et al. 1987). However, it has been shown (White et al. 1987) that the CDM model is incompatible with positive correlation functions extending to $r \gtrsim 40\,h^{-1}$Mpc implied by the quoted results for ξ_{cc} and ξ_{cg}. Due to the mentioned uncertainties with ξ_{cc}, the strongest constraints come from ξ_{cg}. This statistic has an important advantage over ξ_{cc} since the huge size of the Lick catalogue allows a reasonably accurate determination of ξ_{cg} from the relatively small number of Abell clusters.

Here we reevaluate ξ_{cg} using modern data. Especially, we take into account the measured redshifts of clusters by evaluating the cross-correlation function of "projected distance", $\sigma = y\theta$, where θ is the angular separation between a galaxy and the centre of a rich cluster at distance y

N. Kaiser and A. N. Lasenby (eds.), The Post-Recombination Universe, 277–279.

from the observer. This cross-correlation function, $w_{cg}(\sigma)$, is related to the spatial function $\xi_{cg}(r)$ by an expression which is simple to invert numerically. With this method, $\xi_{cg}(r)$ can be evaluated from the measured $w_{cg}(\sigma)$ without assuming any special functional form for ξ_{cg}. This inversion is of course dependent on the assumed luminosity function for the galaxies.

2. THE CROSS-CORRELATION FUNCTIONS AND THE CDM MODELS

Redshifts for the rich Abell clusters are from the catalogue of Struble & Rood (1987). Our analysis is based on the 204 clusters in that catalogue with $R \geq 1$, $|b| \geq 40°$, $\delta \geq -22°.5$ and $r \leq 400\,h^{-1}$Mpc. As our galaxy sample we use the Lick counts reduced to $10' \times 10'$ cells multiplied with the correction factors of Seldner et al. (1977). We also correct the counts for atmospheric extinction and the gross effects of galactic obscuration. Groth & Peebles (1986) have shown that the plate matching procedure of Seldner et al. generates artificial large-scale gradients of similar amplitude to the gradients seen in the counts. We remove such large-scale gradients on scales $\gtrsim 20°$ by fitting smooth surfaces to projections of each galactic hemisphere.

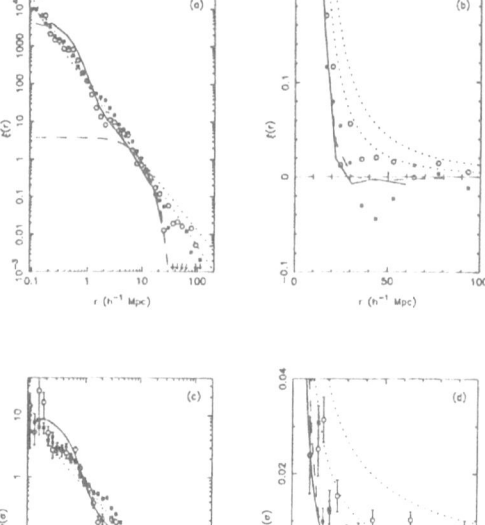

The cross-correlation function of projected distance is computed from the formula $w_{cg}(\sigma) = dN_D / dN_R - 1$, where dN_D represents the number of cluster-galaxy pairs within a fixed σ ring and dN_R is estimated by cross-correlating the cluster sample with a "random" catalogue. This catalogue is constructed by dividing the Lick counts into a large number of intervals in galactic latitude, and randomising the cell counts inside each interval. This procedure corrects for any residual large-scale gradients which only are dependent on galactic latitude. The results for the smoothed sample is shown in figure 1c,d. They are well approximated by the single power-law $w_{cg}(\sigma) = 0.83\sigma^{-1.2}$ for $0.2\,h^{-1}$Mpc $< \sigma < 20\,h^{-1}$Mpc. At larger projected separations the signal disappears in the noise. The unsmoothed sample gives the same results on small scales. However it gives a somewhat higher amplitude on scales $5\,h^{-1}$Mpc $< \sigma < 30\,h^{-1}$Mpc, and a weak anticorrelation on scales larger than that. Large-scale gradients therefore limit the accuracy of our analysis on scales $\gtrsim 20\,h^{-1}$Mpc. The determination of ξ_{cg} from $w_{cg}(\sigma)$ is dependent upon the chosen luminosity function. However, one can check the luminosity function by computing $w_{cg}(\sigma)$ for clusters in different distance intervals and compare the estimates of ξ_{cg}'s from them.

Figure 1. The cross-correlation functions and the CDM results. Figure (a,b) show ξ_{cg}, figure (c,d) show $w_{cg}(\sigma)$. The estimates of the correlation functions for the northern galactic hemisphere are shown by the filled symbols, while the open symbols show the results for the south. The lower of the two dotted lines show the model $\xi_{cg}(r) = (r/8.8\,h^{-1}\text{Mpc})^{-2.2}$. The solid lines show the CDM N-body results while the dashed lines show the theoretical estimate.

We have tried three different forms of luminosity functions with eleven different sets of parameters. The best consistency is given by a Schechter (1976) function with $\alpha = -1.1$ and $M_*(B_T) = -19.68$ (for $h = 1$). This is in excellent agreement with a an analysis of five redshift surveys made by Efstathiou, Ellis & Peterson (1987). The spatial cross-correlation function computed with this luminosity function is shown in figure 1a,b. It is well fitted by the power-law $\xi_{cg}(r) = (r/8.8\,h^{-1}\mathrm{Mpc})^{-2.2}$ on scales $0.2\,h^{-1}\mathrm{Mpc} < r < 20\,h^{-1}\mathrm{Mpc}$, although this power-law slightly overestimates the amplitude on scale $> 10\,h^{-1}\mathrm{Mpc}$. Our result is not compatible with the result of Seldner & Peebles which is quoted in section 1. However in calculating their result, they used a luminosity function which is not compatible with modern data. Using their luminosity function, we do not get agreement between the determinations of ξ_{cg} from different distance intervals. Seldner & Peebles also did not apply any corrections for galactic obscuration or large-scale gradients in the Lick-counts.

We have determined ξ_{cg} for the biased CDM model for $h = 0.5$ using N-body simulations identical to those of White et al. (1987). Each simulation corresponds to a cubical volume of box length 280 Mpc containing 32768 particles. The cross-correlation function obtained from 22 simulations is shown as the solid line in figure 1. The results show a clear two-component structure with a transition at ~ 3 Mpc. At separation > 30 Mpc, ξ_{cg} falls rapidly and crosses zero at $r \sim$ 45 Mpc. Due to the limited size of the volume, there could be problems with the results for the largest separations. Therefore we have computed a theoretical estimate of ξ_{cg} from the theory of the statistics of Gaussian density fields described by Bardeen et al. (1986). This result is shown as the dashed line in figure 1, and is in good agreement with the N-body result at large separations. Comparing these results with the data, we see that the biased CDM model predicts too high an amplitude, by a factor of ~ 2, on scales $\sim 0.4\,h^{-1}\mathrm{Mpc}$. On scales $\sim 10\,h^{-1}\mathrm{Mpc}$ the CDM model drops below the observational results by a factor of ~ 1.5–2. On scales $> 20\,h^{-1}\mathrm{Mpc}$ the CDM model predicts $\xi_{cg} \approx 0$, which in our judgement is compatible with the observational results.

3. CONCLUSIONS

We have shown the following main points: a) The amplitude of ξ_{cg} on scales $\lesssim 20\,h^{-1}\mathrm{Mpc}$ inferred by Seldner & Peebles is larger than expected given modern determinations of the galaxy luminosity function. b) From $w_{cg}(\sigma)$, we infer a ξ_{cg} which is well fitted by a single power-law and which is not significantly different from zero on scales $\gtrsim 20\,h^{-1}\mathrm{Mpc}$. c) Large-scale gradients in the Lick counts restrict the accuracy with which one can determine the amplitude of ξ_{cg} at large separations. d) The cross-correlation function does not rule out the biased CDM model, although there are some discrepancies. A more complete description of this analysis is given by Lilje & Efstathiou (1987).

REFERENCES

Abell, G.O. 1958. *Astrophys. J. Suppl.*, **3**, 211.
Bahcall, N.A. & Soneira, R.M. 1983. *Astrophys. J.*, **270**, 20.
Bardeen, J.M, Bond, J.R., Kaiser, N. & Szalay, A.S. 1986. *Astrophys. J.*, **304**, 15.
Davis, M., Efstathiou, G., Frenk, C.S. & White, S.D.M. 1985. *Astrophys. J.*, **292**, 371.
Efstathiou, G., Ellis, R.S. & Peterson, B.A. 1987. *preprint*.
Groth, E.J. & Peebles, P.J.E. 1986. *Astrophys. J.*, **310**, 507.
Lilje, P.B. & Efstathiou, G. 1987. *Mon. Not. R. astr. Soc.*, in press.
Schechter, P.L. 1976. *Astrophys. J.*, **203**, 297.
Seldner, M. & Peebles, P.J.E. 1977a. *Astrophys. J.*, **214**, L1.
Seldner, M. & Peebles, P.J.E. 1977b. *Astrophys. J.*, **215**, 703.
Seldner, M., Siebers, B., Groth, E.J. & Peebles, P.J.E. 1977. *Astron. J.*, **82**, 249.
Shane, C.D. & Wirtanen, C.A. 1967. *Publ. Lick. Obs.*, **22**, part 1.
Struble, M.F. & Rood, H.J. 1987. *Astrophys. J. Suppl.*, **63**, 555.
White, S.D.M., Frenk, C.S., Davis, M. & Efstathiou, G. 1987. *Astrophys. J.*, **313**, 505.

CLUSTER CORRELATIONS IN COLD DARK MATTER SCENARIOS

E. Martínez - González and J.L. Sanz
Departamento de Física Moderna
Universidad de Cantabria
39005-Santander, Spain

ABSTRACT. The cluster correlation function   is shown to be a strong test of biased scenarios. The standard cold dark matter models, either with adiabatic or isocurvature fluctuations, fit the observed ξ_c in the range 10-50 h^{-1} Mpc only marginally (i.e. for weak bias). On The other hand there is a wide range of the parameters Ω and h which allow to fit the observed galaxy - cluster cross correlation function.

1. INTRODUCTION

Observations of rich clusters of galaxies (Bahcall and Soneira 1983, Klypin and Kopylov 1983, Postman et al. 1986)lead to a spatial correlation function consistent with a power law $\xi_c=(rh/25\ \text{Mpc})^{-1.8}$ in the range $rh\in[5,100]$Mpc. On the other hand,it has been claimed in the literature (Bardeen et al. 1986, Bond 1987) that the adiabatic CDM spectrum may lack of sufficient power on large scales within the biasing scheme to account for ξ_c , but as far as we know no exhaustive analysis has been done. We are interested to test both the idea of bias and the CDM scenarios.

2. THE CLUSTER AUTOCORRELATION FUNCTION

Kaiser (1984) introduced the idea of "bias" as regards rich clusters of galaxies. The basic assumption is that such structures form only at high density regions with the matter distribution represented by a Gaussian random field at the initial time, which will be transfered until the present time through the standard linear model. We shall assume that the Gaussian character is preserved above a certain scale $\sim$10Mpc, which is supported by N-body simulations.

   The autocorrelation function ξ_c of the regions, with characteristic scale R, lyng above the threshold $\delta \equiv \nu\sigma$ is given by the expression for the bivariate Gaussian

$$\xi_c(r,R,\nu)=-1+2\pi^{-1/2}\left[\text{erfc}\left(2^{-1/2}\nu\right)\right]^{-2}\int_{2^{-1/2}\nu}^{\infty}dx\ e^{-x^2}\ \text{erfc}\left[\frac{2^{-1/2}\nu - zx}{(1-z^2)^{1/2}}\right]\qquad(1)$$

N. Kaiser and A. N. Lasenby (eds.), The Post-Recombination Universe, 281–283.

where $z(r,R)$ is the autocovariance function, i.e. $z(r,R) = \xi_o(r)/\xi_o(0)$ and $\xi_o(r)$ is the correlation of the matter distribution Gaussian filtered on the comoving scale R.

Given the number density of clusters $n_c \simeq 6 \times 10^{-6} \, h^3 \, Mpc^{-3}$, we have a relationship between the filtering scale R and the threshold ν (Vanmarcke 1983)

$$n_c = (2\pi^3 \, 3^{3/2})^{-1} \left[erfc \left(2^{-1/2} \nu \right) \right]^{-2} e^{-\frac{3}{2}\nu^2} (\sigma_1 \sigma_o^{-1})^3 \tag{2}$$

where $\sigma_1^2 \equiv \int dk \, k^4 \, P(k) \, e^{-k^2 R^2}$, $\sigma_o^2 \equiv \int dk \, k^2 \, P(k) \, e^{-k^2 R^2}$. Now, if we assume a cluster mass range $M_c \in (0.3, 4.5) \times 10^{15} \, h^{-1} M_\odot$, corresponding to $R_c \in (4,10)(h\Omega)^{-1/3} \, Mpc$ we are able to calculate the range of values of ν. This is the only free parameter apart from the two cosmological ones h and Ω.

We have calculated ξ_c for the values of Ω and h between (0.5,1) and we have considered the whole range of allowed values of the filtering scale. Firstly, we have selected only those cases which satisfy the following criterion: a positive ξ_c until 50h^{-1}Mpc and a corelation length $r_o \in (16,30) \, h^{-1} \, Mpc$, in agreement with the observations. For the adiabatic spectrum, none of the cases considered fit the observed ξ_c (we find too steep slopes and anticorrelations appearing at scales smaller than 50 h^{-1}Mpc), except for the marginal cases h=1, Ω=0.5 and $R_c \simeq 12h^{-1}$Mpc, h=0.5, Ω=0.5 and $R_c \simeq 10.7h^{-1}$ Mpc, respectively. On the other hand, isocurvature fluctuations give ξ_c with no anticorrelations below 50h^{-1} Mpc but again with too steep slopes. The only marginal case surviving corresponds to h=1, $\Omega = 0.5$ and R=10.4 h^{-1} Mpc. All these results will be published elsewhere (E.Martínez-González and J.L. Sanz 1987).

3. THE GALAXY-CLUSTER CROSS CORRELATION

The cross correlation function of the regions, with characteristic scales R_1 and R_2, lyng above the thresholds $\delta_1 \equiv \nu_1 \sigma(R_1)$ and $\delta_2 \equiv \nu_2 \sigma(R_2)$, respectively, ξ_{gc} is given by a generalization of equation (1) (see J.L. Sanz and E. Martínez-González 1988).

As we did for the clusters, we now consider for the number density of galaxies $n_g \simeq 2 \times 10^{-2} h^3 \, Mpc^{-3}$ and a galaxy mass range $M_g \in (0.3,9) \times 10^{11} \, h^{-1} \, M_\odot$. The corresponding range for $R_g \in (0.2,0.6)(h\Omega)^{-1/3} \, Mpc$. The analogous criterion for ξ_{gc} is: a positive cross correlation below 40h^{-1} Mpc and a correlation length in the interval $(3,20) \, h^{-1} \, Mpc$. Contrary to the previous case, there are many cases that fit the observed ξ_{gc}, but the two marginal cases that were selected for adiabatic fluctuations are clearly eliminated. In the isocurvature case none of the selected cases are eliminated.

4. COMMENTS

We have found negative results for biased COM scenarios as regards cluster corelations. We have exhaustively analized all the cases for h and Ω in the range (0.5,1) using filtering scales and thresholds which correspond to the observed masses ranges and number densities for clusters and galaxies. It is found that ξ_{gc} can be fit to the observa - tions for certain values of the parameters whereas ξ_c is fit only for marginal cases, because of problems with anticorrelation and steep slopes.

REFERENCES

Bahcall, N.A. and Soneira, R.M. 1983, Ap. J. **270**, 20.
Bardeen, J.M., Bond, J.R., Kaiser, N. and Szalay, A.S. 1986, Ap.J. **304**, 15.
Bond, J.R., In Proceedings of the Theoretical Workshop on "Cosmology and Particle Physics", I. Hinchliffe ed., World Scientific 1987 p. 22.
Kaiser, N. 1984, Ap.J. (Letters) **284**, L9.
Klypin, A.A. and Kopylov, A.I. 1983, Soviet Astr. Letters, **9**, 41.
Martínez-González, E. and Sanz, J.L. 1987, Ap.J. in press.
Postman, M. Geller, M.J. and Huchra, P. 1986, Ap.J. **91**, 1267.
Sanz, J.L. and Martínez-González, E., "Cluster correlations for scale-free spectra" in Proceedings I.A.U. Symposium no. 130 "The structure of the Universe", Balantonfured, Hungria, Reidel 1988, in press.
Vanmarcke, E.M. 1983, "Random fields: Analysis and Synthesis", MIT Press, Cambridge, Massachusetts.

COSMIC STRINGS AND THE LARGE-SCALE STRUCTURE

Albert Stebbins
NATO/Fermilab Astrophysics Center
FNAL   MS 209
Box 500
Batavia, IL   60510
USA

Abstract. *A possible problem for cosmic string models of galaxy formation is presented. If very large voids are common and if loop fragmentation is not much more efficient than presently believed then it may be impossible for string scenarios to produce the observed large-scale structure with $\Omega_0 = 1$ and without strong environmental biasing.*

In order for cosmic string theories to be viable they must be able to explain the observed large-scale galaxy distribution. There are some indications that the distribution of Abell clusters is predicted by string theories[1], but the theory must predict all aspects of the large-scale structure to succeed. Until our understanding of the details of the evolution of the string network increases we should be wary of results from computer simulation 'black boxes'. Hopefully a second generation of string simulations will bolster our confidence in the results obtained even if they don't provide a clear theoretical understanding of things like loop correlations. Here I would like to explore some aspects of the large-scale structure which may be studied even with very little knowledge of the dynamics of strings. In particular the requirement that string evolution obey causality puts severe constraints on the type of large-scale structure that strings can produce.

Recently large regions devoid of galaxies have been discovered in a complete magnitude-limited galaxy redshift survey[2]. This was not predicted by those developing the string theory of galaxy formation. Here we will examine whether these voids are inconsistent with this theory. If galaxies form around cosmic string loops then these voids must correspond to voids in the loop distribution unless there is some environmental bias which prevents the matter accreted around loops in the voids from forming observable galaxies. For the purposes of this paper I shall assume that no such biasing mechanism exists. It has been suggested[3,4] that cosmic string wakes could produce structures not unlike the sheets of galaxies observed. This still would not explain what has happened to the condensations around loops which should still be in the voids. One reason that voids in the spatial distribution of loops seems implausible is that the loops which seed galaxies are formed at very early times when the comoving horizon size was very small. It seems unlikely that the apparent scale length of the large-scale galaxy distribution is much larger than the largest possible coherence length of the of the loops which seed them. It is true that loops do move away from their place of origin but if the initial loop velocities are not too large ($v_i < 0.2c$) they do not move as far as a comoving horizon size

N. Kaiser and A. N. Lasenby (eds.), The Post-Recombination Universe, 285–287.

before they start to accrete. In any case, I would expect loop motions to randomly move loops into a void rather than to coherently move loops out. Also the peculiar velocities expected in string scenarios are not sufficient to sweep galaxies from large voids[5].

One may calculate using a spherical infall model[6] the amount of mass within the shell just beginning to fall back onto a loop-seeded condensation today in the CDM model. Expressing the comoving horizon size when loops were formed in terms of this mass we obtain

$$\lambda_H = 6.0\, h_{50}^{-1} \text{Mpc} \left(\frac{M_T}{10^{12} M_\odot} \right)^{\frac{1}{2}} \frac{\sqrt{0.72\Omega_0 + 0.28}}{\Omega_0 \sqrt{\mu_6 \xi_{0.1}}} \qquad \begin{matrix} \mu_6 \equiv 10^6 G\mu/c^2 \\ \xi_{0.1} \equiv \xi/0.1 \end{matrix} \qquad (1)$$

for loops formed in the radiation era ($M_T \ll 1.6 \times 10^{14} M_\odot \xi_{0.1} \mu_6 h_{50}^{-3}$). The parameter ξ is the length of a loop (M_l/μ) measured in units of the horizon size (c/H) at the time it was produced. I shall assume that ξ is monochromatic, although in reality we expect that at any given time there is a distribution in the masses of loops produced. The value of $\xi \approx 0.1$ is really just a guess and is somewhat lower than one would estimate from the results presented in ref. [7]. The fiducial values chosen are meant to be reasonable values for loops which seed L^* galaxies. Loops which seed less luminous galaxies would have formed when the comoving horizon size was even smaller. The reason why this quantity is of interest is because loop production is statistically independent at points separated by more than this distance. As suggested above, the positions of L^* galaxy-seeding loops are only correlated on distances considerably smaller than the size of observed voids. Can we show that this is inconsistent?

Divide up the universe into a lattice of cubes of length λ_H on a side. The probability that any such cube has no galaxies brighter than a given magnitude in it I denote by P_H. Now consider a larger cube, call it V, containing N^3 smaller cubes. A simple estimate of the probability that the larger cube is empty would be $P_V = P_H^{N^3}$. However this is not obviously true since all the small cubes in the big cube are not statistically independent. In particular, the loop production in adjacent cubes could be correlated. Now consider a sparse cubic lattice of small cubes (λ_H on a side) that has twice the period of the original lattice (i.e. $2\lambda_H$). This sparse cubic lattice only fills 1/8 of the total volume but since each point in a cube of the sparse lattice is at least a distance λ_H from any point in any other cube on the lattice, the loop production in each cube of the sparse lattice is statistically independent. Combining this statistical independence with the fact that a necessary but not sufficient condition for V to be empty is that all small cubes on the sparse lattice contained within V be empty, we have the strict inequality $P_V < P_H^{(N/2)^3}$, at least if N is even. Let $P(\lambda)$ be the probability that a randomly chosen cube of length λ on a side is empty of bright galaxies. The above argument suggests that

$$\lambda_H > \kappa \lambda_V \left(\frac{\ln P(\lambda_H)}{\ln P(\lambda_V)} \right)^{1/3} \qquad \kappa = \frac{1}{2}. \qquad (2)$$

I have only proved this for λ_V an even multiple of λ_H but I shall proceed heuristically and assume that equation (2) holds for $\lambda_V \geq \lambda_H$. Note that if we had stuck with the original naive estimate of P_V the only difference would be that κ would become unity. Probably a stronger inequality (larger κ) can be proven, but we have no space to pursue this here.

It would be very difficult to determine $P(\lambda)$ for a given string theory but one can empirically determine $P(\lambda)$ for the actual distribution of galaxies. If we had uniformly surveyed a large enough volume of space one could determine the actual $P(\lambda)$ by randomly laying down cubes of various sizes on the surveyed part of the universe and seeing for each sized cube what fraction has no bright galaxies in it. From this empirically determined $P(\lambda)$ one can then place a lower bound on the distance over which galaxy positions are correlated. To do this divide up the λ_H-λ_V plane into regions in which equation (2) is satisfied and regions in which it is not. The largest value of λ_H for which the inequality is violated places a lower bound on the actual value of λ_H. To prove the validity of this procedure one must use the fact that $P(\lambda)$ is a non-increasing function of λ, which one can show.

To avoid confusion we stress that the probability that a random cube is empty can be quite small even though voids of that size or somewhat larger are common. This is because if the cube size is comparable to the void size then it is unlikely that a randomly placed cube would not stick out of a void. It is only in the limit of the cubes being much smaller than the voids that the probability of the cubes being empty approaches the filling factor of voids. Given the recent observations it would not be unreasonable to expect that the probability that a $30\,h_{50}^{-1}$Mpc cube is empty is 0.001 and the probability that a $6\,h_{50}^{-1}$Mpc cube is empty is 0.2. Equation (2) then tells us that $\lambda_H > 3.5\,h_{50}^{-1}$Mpc. While according to equation (1) this would not pose any problems for $10^{12}M_\odot$ galaxies it would pose a problem for $3 \times 10^{11}M_\odot$ galaxies. Present observations are close to being inconsistent with a model using our fiducial parameters and future observations or more refined statistical tests may pose real problems.

Note that the left-hand-side of (2) depends only very weakly on the probabilities. Thus even with fairly poor statistics one can set fairly firm lower limits on the distance over which galaxies are correlated. Also this method can be used even if some loops seed underluminous galaxies which we do not see. We only require that the factors which determine the probability of forming a 'failed galaxy' are not correlated over larger distances than the loops. Instead of using the probability that some region is empty of bright galaxies we could just as well have used the probability that everywhere in the region the 'local galaxy density' is less than some fixed fraction of the mean. Thus we could put constraints on the statistics of underdense regions as well empty regions. Some difficulties in this analysis come from uncertainties in determining the mass of the loop which seeded a galaxy. If very large-scale structure is indeed present then string models with low Ω_0, or low μ, or increased fragmentation (low ξ) are favored. Finally we note that even if the large-scale structure is consistent with causality we may find that loop production just doesn't produce the structure we observe.

References

[1] N. Turok 1985, *Phys. Rev. Lett.* **55**, 1801.

[2] V. de Lapparent, M. Geller, and J. Huchra 1986, *Ap. J., Lett.* **302**, L1.

[3] T. Vachaspati 1986, *Phys. Rev. Lett.* **57**, 1655.

[4] A. Stebbins, S. Veeraraghavan, R. Brandenberger, J. Silk, and N. Turok 1987, *Ap. J.* **321**, 607.

[5] P. Shellard, R. Brandenberger, N. Kaiser, and N. Turok 1987, *Phys. Rev. D* **36**, 335.

[6] A. Stebbins 1986, *Ap. J. Lett.* **303**, L21.

[7] A. Albrecht and N. Turok 1985, *Phys. Rev. Lett.* **54**, 1868.

THE 3-D DISTRIBUTION OF ABELL CLUSTERS

Will Sutherland
Institute of Astronomy, Cambridge CB3 0HA, England.

Summary
The Struble & Rood catalogue of all measured Abell cluster redshifts is analysed, with correction for the redshift incompleteness in the $D = 5 - 6$ subsamples, and clear evidence is found for selection effects in the Abell catalogue. When these are compensated for, a much reduced cluster correlation amplitude is indicated, with

$$\xi_{cc}(r) \approx \left(\frac{r}{14 \; h^{-1} Mpc} \right)^{-1.8} ,$$

and an upper limit $\sim 800 \, km/s$ is placed on cluster peculiar velocities.

1. Introduction
Bahcall & Soneira (1983, hereafter BS) analysed a complete sample of $D \leq 4$ Abell clusters, and claimed that the observed angular correlations are entirely due to intrinsic clustering in redshift space, with $\xi_{cc}(r) \approx (r/25 \; h^{-1} Mpc)^{-1.8}$. They also noted that ξ_{cc} is elongated in the line-of-sight direction and thus deduced large peculiar velocities $\sim 2000 \, km/s$ between clusters (BS 1983, Bahcall, Soneira & Burgett 1986). These results are in clear disagreement with most theoretical models. However, their sample contains only 104 redshifts and thus is subject to considerable uncertainties (Ling, Frenk & Barrow 1986).

The Struble & Rood (1987) catalogue lists 588 Abell cluster redshifts; while it is subject to angular selection biases, the redshift information is effectively unbiased since one cannot tell *a priori* whether cluster pairs at small angular separations are really associated in redshift.

2. The Redshift Correlation Function
This is calculated by making histograms of redshift difference $|z_1 - z_2|$ for pairs of clusters of given D in a given range of angular separation $\theta_1 \leq \theta \leq \theta_2$, ie $N_{\theta}^c(\Delta z)$, and for all pairs, $N_{\theta}^R(\Delta z)$; the latter is then normalised using the angular correlation function of the appropriate subset of the whole Abell catalogue, so that the totals $N_{\theta}^c/N_{\theta}^R = 1 + w_A(\theta)$, and then the redshift correlation function is defined by

$$\xi(\Delta z) = \frac{N_{\theta}^c}{N_{\theta}^R} - 1.$$

Results are shown in Fig. 1; although noisy, $\xi(\Delta z)$ clearly remains positive out to large Δz, in contradiction to the peculiar velocity model. Similar results are found

N. Kaiser and A. N. Lasenby (eds.), The Post-Recombination Universe, 289–291.

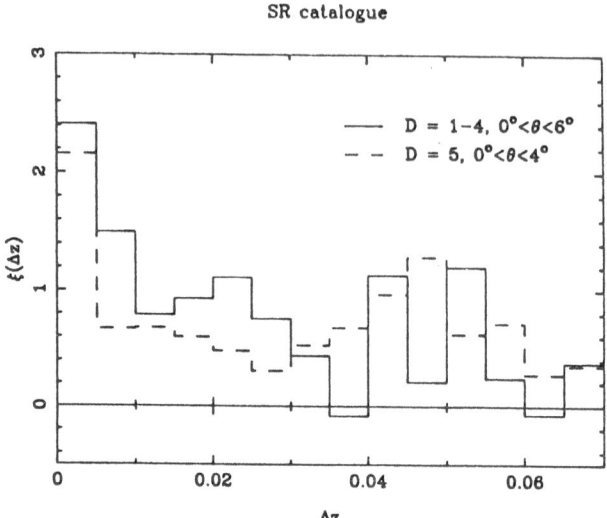

Figure 1:   The redshift correlation function is shown for the (nearly com plete) D = 1-4, $R \geq 0$ sample and the $D = 5$ sample, for angl corresponding to projected separation $r_p \lesssim 20\ h^{-1} Mpc$.

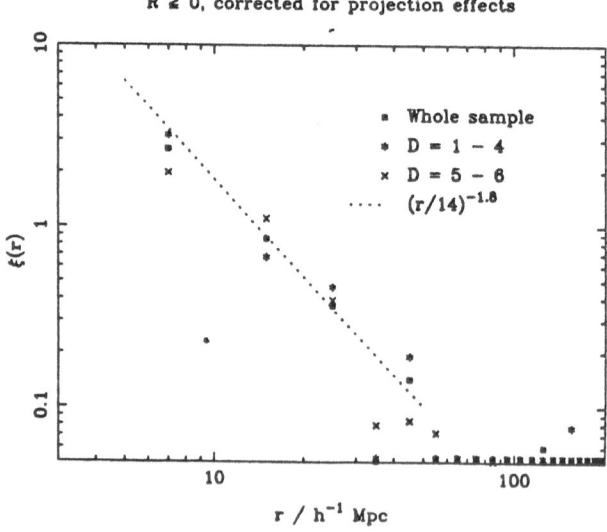

Figure 2:   The spatial correlation function $\xi(r)$ for the $R \geq 0$ sample, co rected for projection effects as described in Section 3. The $\xi(r) \approx (r/14\ h^{-1} Mpc)^{-1.8}$ is shown.

for Bahcall & Soneira's complete sample; although they note that the excess *number* of pairs $N_\theta^c - N_\theta^R$ falls with increasing Δz, on dividing by the random distribution we find a "tail" of positive $\xi(\Delta z)$. This means that not enough of the small-angle cluster pairs are really associated in redshift to account for the observed angular correlations; thus line-of-sight selection effects exist in the Abell catalogue.

Such effects have been modelled by Kaiser (1987); the basic principle of the model is that since the cluster-galaxy correlation function is positive out to large separations (Lilje & Efstathiou 1987), an Abell cluster tends to be surrounded by a large "halo" of galaxies. This will enhance the apparent richness of other clusters along a nearby line of sight, and hence enhance their probability of selection in the Abell catalogue, since the number of clusters selected is very sensitive to the lower richness cutoff.

Plate matching errors are another factor which may contribute to such effects, since Abell made no magnitude corrections between plates. This could cause significant enhancement of $w(\theta)$ for $\theta < 6°$.

3. Compensation for projection effects.

The projection effects described above clearly cause the observed ξ_{cc} to be greater than the "intrinsic" ξ_{cc} which would be measured for a cluster sample with no projection effects. We can correct for this with the method of "normalisation to large scales" used by Shaver (1985 & this volume) in the analysis of quasar clustering.

In practice, we produce a histogram of cluster-cluster pairs in both projected separation r_p and line-of-sight separation r_z, and repeat this for cluster-random pairs. Then we normalise each "column" of constant r_p so that $\overline{\xi(r_z, r_p)} = 0$ at large r_z, then rebin in r to give the intrinsic $\xi_{cc}(r)$. The result is shown in Fig. 2; ξ still approximately obeys an $r^{-1.8}$ power law, but the correlation length r_0 is now reduced to $r_0 \approx 14 \ h^{-1} Mpc$, with approximate 95% confidence interval $11 < r_0 < 18 \ h^{-1} Mpc$. No correlations are found beyond $50 \ h^{-1} Mpc$ (this result is confirmed by redshift shuffling calculations).

Examining the corrected ξ as a function of r_z and r_p, only marginal elongation in the z-direction is found. We find that peculiar velocities are not required, and an upper limit $\sim 800 \ km/s$ is placed on cluster peculiar velocities.

These conclusions indicate that the standard cold dark matter model (eg White *et al*, 1987) should not be excluded.

These results will be published in full in MNRAS, mid-1988.

References
Abell, G.O., 1958. *Astrophys. J. Suppl.*, **3**, 211.
Bahcall, N.A. & Soneira, R.M., 1983. *Astrophys. J.*, **270**, 20.
Bahcall, N.A., Soneira, R.M. & Burgett, W., 1986. *Astrophys. J.*, **311**, 15.
Hauser, M.G. & Peebles, P.J.E., 1973. *Astrophys. J.*, **185**, 757.
Kaiser, N., 1987. Unpublished work.
Lilje, P.B. & Efstathiou, G.P., 1987. *Mon. Not. R. astr. Soc.*, in press.
Ling, E.N., Frenk, C.S. & Barrow, J., 1986. *Mon. Not. R. astr. Soc.*, **223**, 21P.
Shaver, P., 1985. *Astron. Astrophys.*, **136**, L9.
Struble, M.F. & Rood, H.J., 1987. *Astrophys. J. Suppl.*, **63**, 543.
White, S., Frenk, C.S., Davis, M., & Efstathiou, G.P., 1987. *Astrophys. J.*, **313**, 505.

QUASAR PAIRS, CLUSTERING AND ITS EVOLUTION

Xingfen Zhu and Yaoquan Chu
Institute for Astrophysics, University of Bonn, F. R. G.
The Center for Astrophysics, University of Science and Technology of China, Hefei, China.

Do clusters of quasars exist? Does quasar clustering depend on the cosmological time? These are some of the most interesting questions in current research on quasars. In this paper we would like to summarize our statistical research on the distribution of quasars. The aim of our study mainly concerns to search for large scale structure in the universe and its evolution.

1) Quasar Pairs
   Dr. Shaver (1984) suggested that clustering is expected to show up amongst pairs or groups of quasars which are close to each other in both redshift and position on the sky. By contrast,pairs which have large differences in either redshift or position are expected to be physically unassociated with each other. The search for clustering can then be done by comparing these two groups. With the Veron Catalogue (1984), he revealed possible clustering of quasars .
   Recently we apply the same method to the Hewitt and Burbidge Catalogue (1987),and also discover evidence of quasar clustering (see Fig.1).

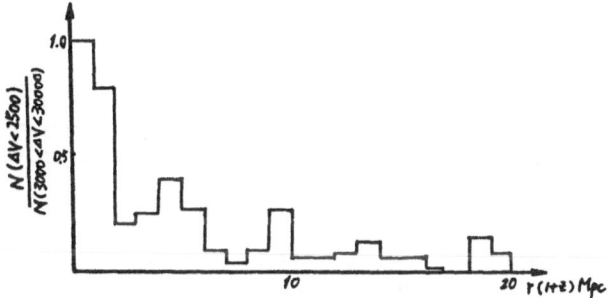

Fig.1   Ratio of the number of quasar pairs with small redshift difference to the number with large redshift difference, against projected linear separation

It clearly shows an excess of quasar pairs with small separation and

N. Kaiser and A. N. Lasenby (eds.), The Post-Recombination Universe, 293–295.
© 1988 by Kluwer Academic Publishers.

redshift difference. If there were no clustering, the ratio in Fig.1 should be constant as a function of separation difference. The problem of distinguishing between gravitational lens and physical pair cannot easily be solved in individual case, particularly in the case that two quasars are seperated only few arcseconds. However, in our sample there are only few quasar pairs in which the gravitational lens interpretation is now reasonably established ,namely, 0957+561, 2016+112. Other three quasar pairs 1146+111, 1634+267 and 2345+006, which have no obvious visible lens objects, have been claimed to be physical pairs with small seperation (Bahcall, et. al. 1986). So the contamination due to lenses seems unlikely to be statistically serious in our analyses.

2) Clustering of Quasars from Homogenious Survey
In the early stage, the statistical results from different researchers were sometimes not consistent with each other. Some of them were even contradictory. For example, with the same quasar sample in the CTIO survey, Arp (1980) found some evidence of quasar grouping, while Osmer(1981) and Webster (1982) got a negative result. Chu and Zhu (1983) analysed the three-dimension distribution of quasars in two 5 × 5° survey done by Savage and Bolton (1979), and found weak clustering in one field (02h, -50°) but no clustering in another field(22h,-18°).
Lately more attention has been paid to the space distribution of quasars , yet the results for quasar clustring are still inconsistent: Boyle et.al. (1986) and Shanks et.al. (1987) find evidence for clustering in their UVX sample. He, Chu et. al. (1986), Kundth and Sargent (1986), Crampton et. al. (1987) and Clowes et. al. (1987) find no clustering amongest quasars from their optical-prism or grens survey.
The differences among these results are mainly due to the sam-ples,i.e., the numbers of quasars in these early surveys are not large enough to do statistical analysis with high significance. Meanwhile the criteria of clustering in different statistical methods maybe not equivalent and various selection effects may play important role in different surveys.
In spite of the difference in different statistical researches, a conclusion seems to already be acceptable: the clustering of quasars is rather week, at least weeker than that of galaxies.

3) Evolution of Quasar Clustering
After carefully comparing all the previous statistical results, we find a very interesting fact that all results from the samples contai-ning UVX quasars(or quasar candidates) always show stronger clustering than others (see, Chu and Fang 1986). As we know that the redshifts of quasars found by objective prism and grism are generally higher than that of quasars found by UVX methods. So, the stronger clustering in UVX samples imply that the strength of quasar clustering depends on the redshifts of quasars, namely, the clustering is evolving: the larger the redshifts, the weaker the clustering.
Fang, Chu and Zhu (1985) reanalysed clustering of quasars at diffe-

rent redshift ranges in the Savage-Bolton sample, which consists of both quasars identified by objective prism technique and UVB two colour method. We divided whole sample into two groups : Z>2 and Z<2, applied the nearest neighbor test and found apparent clustering at 96% significant level for quasars with Z<2 in both fields and no clustering with Z>2 at all. The main results are listed in Table 1, were N is the number of quasars, $\langle D_< \rangle$ denotes the sample's mean of nearest neighbor separations, $D_<^*$ and $\bar{\sigma}$ the mean and standard deviation from Monte Carlo sample respectively. $1-P(\delta)$ is the probability of clustering.

Table 1: Nearest neighbor test for the Savage-Bolton sample

Redshift	N	Quasar data $\langle D_< \rangle$ Mpc	Monte Carlo data $D_<^*$ Mpc	$\bar{\sigma}$	δ	$1-P(\delta)$
(02h ,-50°) field						
Z< 2	62	141.7	159.0	79.6	-1.72	96%
Z> 2	48	201.0	205.9	83.2	-0.40	66%
(22h ,-18°) field						
Z< 2	57	146.7	165.8	77.9	-1.84	97%
Z> 2	26	207.1	193.0	75.9	> 0	---

Iovino and Shaver (1988) analyse three deep quasar samples and find strong clustering for quasars with Z <1.5 and no clustering for quasars with Z > 1.5. Using the quasar pairs from Veron catalogue, Kruszewski (1986) also find strong redshift depence of quasar clustering.

In one word, a common result in the study on quasar clustering is that the clustering exists in the distribution of Z< 2 quasars, but not of Z> 2 quasars.

REFERENCES
Arp,H. 1980. Ap.J. vol.239, 463. Bahcall, J. N., Bahcall, N. A., and Schneider, D. P., 1986, Nature, vol.323, 515.
Boyle, B. J., et. al., 1986, in "Observational Cosmology", IAU Symp. 124, ed. Hewitt, A., et. al., 643.
Chu, Y.,and Zhu, X., 1983, Ap. J., vol.267, 4.
Chu, Y., and Fang, L., 1986,in "Observational Cosmology", IAU Symp. 124, ed. Hewitt, A., et. al.,
Clowes, R. G., Iovino, A. and Shaver, P., 1987, MNRAS, vol.227, 921.
Crampton,D., Cowley, A., P. and Hartwick, F. D. A., 1987, Ap.J. vol.314, 129
Fang, L., Chu, Y. and Zhu, X., 1985, Ap. S. S., vol. 115, 99.
He, X., Chu, Y., et.al.,1986, Acta Astronomia Sinica, vol.27,No.2, 144.
Hewitt, A. and Burbidge, G. 1987, Ap. J. Suppl. vol. 63, No.1.
Iovino, and Shaver, P. A., 1988, in IAU symp. 130, Hungary.
Kunth, D., and Sargent, W. W., 1986, A. J., vol.91, 761.
Kruszewski, A., 1986, Princeton Observatory Preprint.
Osmer,P. S., 1981, Ap. J., vol. 247, 761.
Savage, A., and Bolton, J. G., 1979, MNRAS, vol.188, 99.
Shanks, T., Fong, R., et. al. 1987, MNRAS, vol. 227, 739.
Shaver, P. A., 1984, A. Ap., vol.136, L9.

GRAVITATIONAL LENSES

S. Refsdal, R. Kayser
Universität Hamburg, Hamburger Sternwarte
Gojenbergsweg 112, 2o5o Hamburg 8o
F. R. G.

ABSTRACT. After some introductory comments on basic gravitational lens theory, we discuss briefly the probability of lensing and the present state of the determination of H_O and the mass of the deflector in gravitational lens systems. Then follows a more detailed discussion on astrophysical applications of micro-lensing.

1. INTRODUCTION

Several good reviews on the theory of gravitational lensing have been published recently (Gott [1], Canizares [2], Blandford [3]). The observations of gravitationally lensed quasars have been reviewed in [2] and [3]. Since then Surdej et al. [4] have found a new lens system, UM673 with two images separated by 2.2". We here concentrate mainly on new theoretical developments, particularly on micro-lensing which offers many interesting astrophysical applications. Of most interest are the possibilities of determining:

   a) the size and the one dimensional luminosity profile of very compact and distant sources (quasars),

   b) the transverse velocity of distant sources (deviations from the Hubble flow), and

   c) masses and number density of the micro-lenses (stars, black holes, brown dwarfs).

2. BASIC THEORY

The physical basis for gravitational lensing is the gravitational deflection of electromagnetic waves. From general relativity it follows that the deflection of a ray passing a mass M at distance r is given by

$$\alpha = \frac{4\,G}{c^2}\frac{M}{r} \qquad (\alpha \ll 1) \qquad (1)$$

This deflection has been verified to 2o % accuracy my measuring star

N. Kaiser and A. N. Lasenby (eds.), The Post-Recombination Universe, 297–309.

298

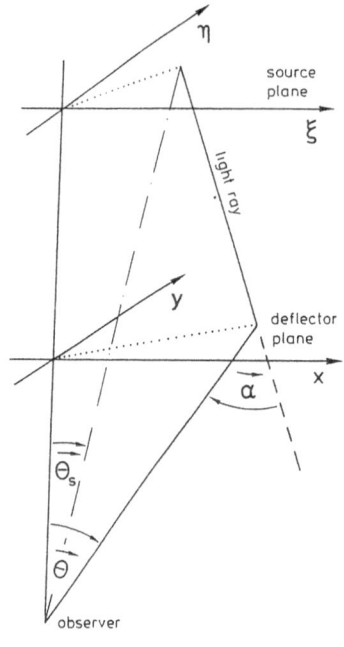

Figure 1.
Gravitational lens geometry

positions during solar eclipses and recently to 1 % accuracy by radio-
interferometric measurements on quasars. For an extended mass distribution the total deflection is obtained by superposing the effect of
all mass elements. Gravitational lensing is essentially geometrical
optics of thin achromatic lenses with small deflections. Introducing
a source plane with coordinates $\vec{\zeta} = (\xi, \eta)$, and a deflector plane with
coordinates $\vec{z} = (x, y)$, both perpendicular to the line source-observer,
we can describe the mapping from the deflector plane onto the source
plane by the lens equation (see Fig.1).

$$\vec{\zeta} = \vec{\zeta}(\vec{z}) = \frac{D_s}{D_d} \vec{z} - D_{ds} \vec{\alpha}(\vec{z}) \qquad , \ |\vec{\alpha}| \ll 1 \qquad (2a)$$

where D_d, D_s, and D_{ds} are apparent size distances. (In some
cases lens situations with several deflector planes must be considered
[5]) Introducing the angular position $\vec{\theta}_s$ of the source and the angular
position $\vec{\theta}$ of the image, the lens equation can also be written in the
form

$$\vec{\theta}_s = \vec{\theta}_s(\vec{\theta}) = \vec{\theta} - \frac{D_{ds}}{D_s} \vec{\alpha}(\vec{\theta}) \qquad , \ |\vec{\alpha}| \ll 1 \qquad (2b)$$

The lens equation describes a reversed ray tracing, since the mapping
is in the direction opposite to the light propagation. A very similar
mapping is obtained from the deflector plane onto the observer plane
(for a fixed source point).

The most interesting cases of lensing occur when the inversion of the lens equation is multivalued so that several images can be seen of a single source. Usually no global formulation of the inversion of Eq.(2) exists and the numerical problems are not trivial. Considerable progress in dealing numerically with gravitational lensing has recently been achieved by Kayser and Schramm [6, 7].

An important property of gravitational lensing is the conservation of surface brightness [8]. The total amplification A_T is then obviously due to the change in the solid angle covered by the image(s) relative to that of the unlensed source

$$A_T(\vec{\theta}_s) = \sum_i A(\vec{\theta}_i) = \sum_i \left| \frac{\partial(\theta_{s,x}, \theta_{s,y})}{\partial(\theta_x, \theta_y)} \right|^{-1}_{\vec{\theta}=\vec{\theta}_i} \tag{3}$$

3. SIMPLE EXAMPLES

a) "Point mass" lens

The point mass lens (see Fig.1) is the classical gravitational lens. Due to the axial symmetry it is essentially a one dimensional problem, and the lens equation can be written in the form

$$\theta^2 - \theta_s \theta - \theta_o^2 = 0 \tag{4a}$$

where

$$\theta_o = 2 c^{-1} (G M D_{ds}/D_d D_s)^{1/2} \tag{5a}$$

is the angular radius of the luminous Einstein ring which is seen by exact alignment $(\theta_s = 0)$ [9]. For cosmological distances $(D_d \approx D_s \approx 10^9 \text{ pc})$ we get

$$\theta_o \approx 1'' (M/10^{11} M_\odot)^{1/2} \tag{6}$$

i.e. some arcseconds for a massive galaxy and some micro-arcseconds for a solar mass star. θ_o gives the typical angular scale of the lensing ($\sim$ distance between images). The amplification depends only on θ_s/θ_o and increases with decreasing θ_s/θ_o. Significant amplification (somewhat arbitrarily defined as total amplification $A_T := A_1 + A_2 > 1.34$) occurs for $\theta_s < \theta_o$, see next section.

b) Uniform disk lens

A uniform circular disk of matter acts as a normal converging lens for light rays normal to the disk. The focal length of this lens is

$$f = c^2/4\pi G \sigma \tag{7}$$

where σ is the surface mass density. The amplification is given by

$A = (1 - \sigma/\sigma_{crit})^{-2}$ where

$$\sigma_{crit} = c^2 D_s/(4\pi G D_d D_{ds}) \tag{8}$$

is the critical surface mass density. For $\sigma = \sigma_{crit}$ the beam is focused onto the observer with infinite amplification (point source, geometrical optics). For cosmological distances σ_{crit} is about 1 g/cm^2, roughly equal to the average surface density in the central parts of massive galaxies. For extended deflectors one must usually have $\sigma \geq \sigma_{crit}$ over at least a part of the deflector in order to create multiple images.

c) Uniform disk plus a point mass

When a uniform disk with $\sigma < \sigma_{crit}$ is added to a point mass the lens equation (4a) is unchanged, whereas for $\sigma > \sigma_{crit}$ the lens equation can be written as [1o]

$$\theta^2 - \theta\theta_s + \theta_0^2 = o \;. \tag{4b}$$

It is important to note however that θ_s is now the position of the source seen through the disk and that θ_0 changes to

$$\theta_0 = 2\, c^{-1}\, (GMD_{ds}/(D_d D_s \, | 1 - \sigma/\sigma_{crit} | \,))^{1/2} \tag{5b}$$

For $\sigma < \sigma_{crit}$ θ_0 is again the angular radius of the Einstein ring. For $\sigma > \sigma_{crit}$, however, there is no Einstein ring, and there is no real solution of Eq.(4b) for $\theta_s < 2\,\theta_0$ which means that no image is seen through the disk for $\theta_s < 2\,\theta_0$ [11].

4. SIMPLE PROBABILITY CONSIDERATIONS

We have seen that a point mass lens causes a total amplification larger than 1.34 for $\theta_s < \theta_0$ (point source). The Einstein ring can therefore be considered as a kind of cross section for gravitational lensing. For randomly located point sources at distance D_s the probability of signi-ficant lensing ($A_T > 1.34$) caused by a single deflector is

$$P = \frac{\pi\theta_0^2}{4\pi} = \frac{D_{ds}}{D_s}\frac{G}{c^2}\frac{M}{D_d} = \frac{D_{ds}}{D_s}\frac{\phi}{c^2} \tag{9}$$

where ϕ is the gravitational potential of the deflector at the observer. We see that the probability is linear in ϕ so that Eq.(9) is also valid for several deflectors acting independently of each other, which is the case for small optical depth. With a constant density of deflectors (static universe) we then get by taking an appropriate average of D_{ds} for the probability P (or optical depth τ) [12]

$$P = \tau = <\frac{D_{ds}}{D_s}>\frac{\phi_T}{c^2} = \frac{1}{3}\frac{\phi_T}{c^2} \tag{1o}$$

where ϕ_T is the potential of all possible deflectors $(D_d < D_s)$. From this simple result it is clear that stars in our galaxy have an extremely small optical depth for lensing $(\phi_T/c^2 \approx 10^{-6})$.

For very distant sources the situation is much more promising. It is then easy to show that the average optical depth is given by [13]

$$\tau \approx \frac{\pi}{4} \varepsilon \Omega \, z^2 \qquad (z < 1) \qquad (11)$$

where z is the redshift of the source, Ω the density parameter and ε the mass fraction which is in lenses. We see that the optical depth can be large for $z \gtrsim 1$ if $\varepsilon \Omega$ is not too small.

A simple relation exists between τ and D_c, the reduction of luminosity (in magnitudes) for a light bundle propagating far from all deflectors (empty light cone) [14]. From Eq.(27) in [15] we find for small z that $D_c = 0.54 \, \varepsilon \Omega \, z^2$ so that

$$\tau = 1.44 \, D_c \qquad (D_c \ll 1) \qquad (12)$$

D_c is a good parameter to use since it relates directly to the statistical properties of the lensing [14, 16, 17].

The probability for high amplifications is of particular interest, and it is easy to show that the asymptotic behaviour of the probability P for amplification larger than A (A >> 1) goes as $P \sim A^{-2}$ for a point source [15, 3]. For extended sources there is however a maximum amplification depending on the size of the source and the lensing parameters [15, 18].

5. H_0 AND MASS DETERMINATION

The determination of H_0 and the mass of the deflector by gravitational lensing is intimately connected with the determination of the time delay Δt between the images, since H_0^{-1} and M are both proportional to Δt [19]. Three methods have been used in order to determine Δt theoretically, the wavefront method [19, 20, 21, 22], a method based on splitting up the time delay in a geometrical part and a potential part [23, 24, 25, 26, 27, 28] and a method based on time delay surfaces and Fermat's principle [29].

The most extensive monitoring of a gravitationally lensed quasar has been carried out on the double quasar QSO o957+561 A, B. Tentative time delays of 1.57 years [30] and 1.02 years [3] have been reported, and very recently 1.2 years [32]. By continued monitoring one will hopefully be able to agree upon a value of Δt within some years.

The main uncertainty by the determination of H_0 (for a given Δt) comes from the badly known mass distribution in the deflector. For QSO o957+561 A, B the largest uncertainty relates to the influence from the galaxy cluster which surrounds the deflecting galaxy. Since the gravitational field of the cluster has a much larger scale than that of the galaxy, one can model the cluster field by a power expansion [26, 28, 33]. The approach of Borgeest and Refsdal [26] is best suited for systems with a symmetric galaxy and small dog-leg angle (which seems to

be the case for QSO o957+561 A, B). Since the cluster helps focusing
the light they could give an upper limit of H_o, or a lower limit of H_o^{-1}

$$H_o^{-1} > 5 \cdot 10^9 \, \Delta t \qquad (13)$$

With $H_o^{-1} < 25 \cdot 10^9$ years we then get an upper limit of $\Delta t < 5$ years. In
a more general and complicated analysis which included the luminosity
ratio as model parameters, Falco et al. [28] came to a similar result.
They also showed that there are inherent restrictions as to which
characteristics can be determined from lens observations. It is for
instance not possible to get a (point) estimate of the Hubble constant
without additional information on the lens (i.e. virial mass of the
galaxy or virial mass of the cluster).

The use of a luminosity ratio as model parameter is a bit risky,
due to its sensitivity to local inhomogeneities (micro-lensing by stars
or globular clusters). In the case of QSO o957+561 A, B the risk seems
however to be reasonably small since the ($\approx$ same) luminosity ratio has
been measured in optical continuum, in optical emission lines and in
radio. Since the radio source is too large to be micro-lensed by stars,
only globular clusters (or black holes with $M \gtrsim 10^6 \, M_o$) seems possible,
however not very probable since they comprise a very small fraction
of the total mass in the galaxy.

Alcock and Anderson [34] have argued that since multiple images
by gravitational lensing are rare, the conditions involved in their
formation are unusual. For instance, the distribution of matter along
the light rays from the observer through the deflector to the quasar
may be very different from mean conditions. If, apart from the lens
itself, there is an excess of matter along the light rays ($\gamma > 1$ in
[34]) which helps focusing the light, one will get a too high value
of H_o, as would also be the case if the cluster helps focusing the
light [26]. An effective negative mass density along the light rays
($\gamma < o$, H_o too small) due to tidal divergences (Weyl focusing) is
unlikely, since it requires a distribution of large and very compact
masses in the universe, and very special geometrical situations. In
addition comes the fact that larger galaxy masses are required for
$\gamma < o$. We therefore conclude that $\gamma = o$ is the smallest realistic
value of γ, but that it is difficult to set an upper limit on γ. This
means however that the lower limit of H_o^{-1} in Eq.(13) is still valid
since the possibility $\gamma = o$ is included in the cosmological correction
factor [22].

Mass determinations by the gravitational lens effect are more
accurate than the determination of H_o since there is no influence from
the cosmological model [2o] or from large scale gravitational fields
such as a cluster surrounding the deflecting galaxy or an excess of
matter along the light rays [27]. For the double quasar QSO o957+561
A, B the mass inside an angular radius of 3.1" is [27]

$$M = 7.5 \cdot 10^{11} M_\odot \, \Delta t \, (yrs) \pm 2o \, \% \qquad (14)$$

6. GRAVITATIONAL MICRO-LENSING

Stars or other compact objects close to the light path in an intervening galaxy can act as "micro-lenses". This can lead to a splitting-up of a "macro-image" of a quasar into several "micro-images" [35] and to large variations in brightness due to transverse motions. The separations between the micro-images are of the order micro-arcseconds for solar mass stars and cosmological distances, i.e. by far not resolvable with methods available today. However, the brightness variations in the (unresolved) macro-images, which can be several magnitudes, are easily observable.

Following the work of Chang and Refsdal [35] micro-lensing has been investigated by a number of authors [36, 37, 11, 38, 1o, 39] . A simple way to demonstrate the effects of micro-lensing is by a so-called ray plot diagram [39] which is a mapping of a grid of points in the de-flector plane onto the source plane (observer plane). An example of such a diagram is shown in Fig.2. The randomly distributed star field used for the ray plot in Fig.2 corresponds to an optical depth for micro-lensing equal to o.4 [4o], which means that the Einstein rings of the deflectors cover a fraction o.4 of the sky and that the smoothed out surface density of the deflectors is o.4 $\cdot$ σ_{crit}. This case could possibly correspond to image B in QSO o957+561 A, B.

In the source plane the point density (each point represents a light ray traced back from the observer through the deflector to the source plane) is proportional to the flux of a macro-image of a point source. For extended sources with constant surface brightness the amplification factor is simply proportional to the number of points covered by the source.

In our calculations [39] we have neglected the relative motion of the stars, so that a ray plot does not change with time. However, due to transverse motions the source will move across the ray plot, causing a variation in brightness. It is immediately clear that the effects of micro-lensing are larger and can occur faster for the smaller sources. This is also seen from the light curves shown in Fig.2 which are obtained by moving sources of different sizes with constant velocity along the line η = o. The length unit in the ray plot diagram, ζ_0 , is the radius of the Einstein ring projected into the source plane (ob-server plane). For cosmological distances ($D_d \approx D_{ds} \approx 1o^9$ pc and stars of mass M) the length unit is

$$\zeta_0 \approx o.o2 \ (M/M_\odot)^{1/2} \ pc \qquad (15)$$

and the time unit Δt_0 for moving one length unit is

$$\Delta t_0 = \frac{\zeta_0}{V} \approx 2o \ years \ (M/M_\odot)^{1/2} \ V/1ooo \ km \ s^{-1} \qquad (16)$$

where V is the effective transversal velocity, see Table 3 and Appendix B in [39].

Whereas intrinsic changes in the luminosity shows up with a time delay in the different macro-images, micro-lensing effects are com-pletely independent in the different macro-images. Micro-lensing may

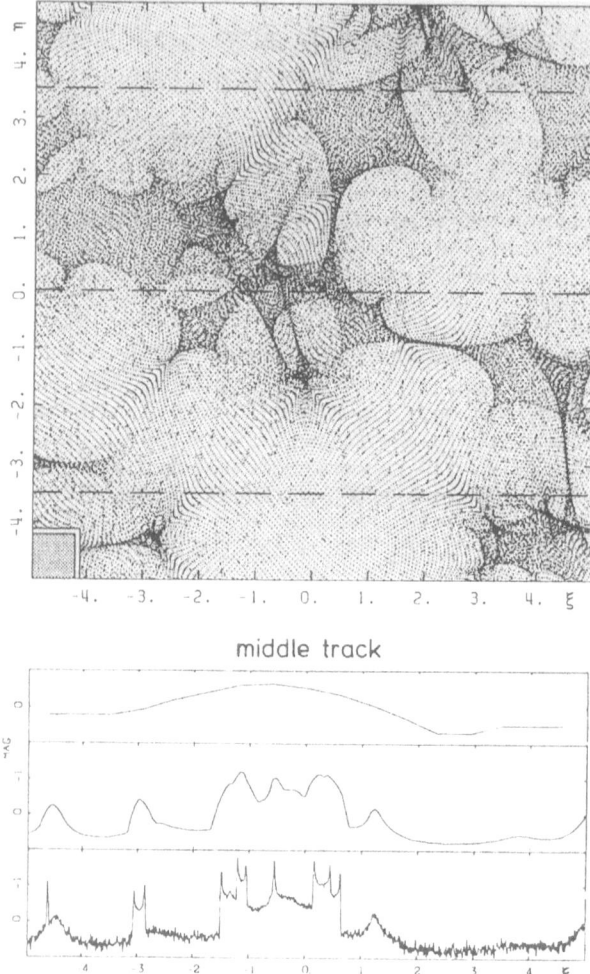

middle track

Figure 2. Ray plot diagram: A projection of a regular grid of rays in the deflector plane into the source plane (observer plane). The total amplification of a point source is proportional to the point density. The small square in the lower left corner of the diagram shows the point density for the corresponding smoothed out galaxy. Below: Amplification in magnitudes relative to that produced by the smeared out galaxy as a function of source center position for the middle track and for three different values of the source radius (o.875, o.o972 and o.o1o8 in normalized units ζ_0; the largest radius at the top). By replacing ξ with $\xi \zeta_0 v^{-1} = \xi \Delta t_0$ we get the corresponding time depending light curve.

therefore complicate the determination of time delays from observed light curves.

Gravitational deflection of electromagnetic waves is independent of wavelength, gravitational lenses are therefore achromatic. However, since the amplification by micro-lensing depends on the source size indirect chromatic effects can occur [39, 41]. Spectral components, which are produced by regions of different extensions (e.g. continuum source and broad line region) can be differently amplified by micro-lensing. Since the macro-images of a multiple quasar are seen through different parts of the galaxy, they are influenced by different micro-lenses, thus differences in their spectra can show up (e.g. different equivalent width and profiles of emission lines). Such differences are therefore not an argument against the lens hypothesis for a multiple quasar.

If the intrinsic polarization varies over the source it is also clear that micro-lensing can cause variations in the observed polarization, and in particular differences in the polarization between macro-images.

a) High amplification events

Of special interest are the so-called high amplification events (HAE), which appear as typical asymmetric peaks in the light curves of compact sources. A HAE happens when the observer crosses a caustic, a line with infinite amplification for a point source (in the approximation of geometrical optics). The caustics are easily identified on the ray plot diagram as discontinuities in the point density. They separate regions with different number of images. By the crossing of a caustic the number of micro-images changes by two. Thus, the crossing of a caustic, or a HAE, is an eclipse-like phenomenon. The source appears in two new micro-images (or two micro-images disappear) whereas the other images scarcely change. It is possible to retrieve the intrinsic one-dimensional brightness profile $P(\xi)$ of the source in a similar way as by a normal eclipse (e.g. eclipsing binaries), by analyzing the lightcurve during a HAE. In contrast to a normal eclipse, however, we must take into account the increasing amplification towards the caustic ($A \sim d^{-1/2}$, where d is the distance to the caustic) [35, 42]. As discussed in more detail in another contribution in this volume [43], see also [44], $P(\xi)$ can be expressed as the solution of an Abelian integral equation. By simultaneously measuring the light curves for several spectral components, we can obtain information on the profiles and relative sizes of the different source regions. A scaling of the profile, i.e. the determination of the source size is possible if V_T, the transversal velocity between source and critical curve, can be determined, see below. The timescale of a HAE for a source with radius R_S is

$$\Delta t_E = 2 R_S / V_T . \qquad (17)$$

The method indicated above can be tested very nicely by using simulated lightcurves [44]. Usually profiles can be retrieved for sources with $R_S < 0.1 \zeta_0$ which is about 10^{-3} pc for solar mass stars and cosmological distances.

b) Parallax effect

From the ray plot diagram we see that micro-lensing causes brightness
gradients in the observer plane, so that a brightness difference δm
is observed by two observers a distance δr apart [45]

$$\delta m = \text{grad } m \cdot \vec{\delta r} \quad . \tag{18}$$

During a HAE the value of grad m can reach values up to about $0.01\ AU^{-1}$
(typical case, e.g. QSO o957+561 and $R_s \approx 10^{-3}$ pc). For two observers
A and B, 1 AU or more apart, it should then be possible to observe the
time lag δt_{AB} in the lightcurves. The transverse velocity of the observers (for negligible relative velocity) perpendicular to the
critical curve is then given by

$$v_T = r_{AB} \sin \beta / \delta t_{AB} \leq r_{AB}/\delta t_{AB} \quad . \tag{19}$$

r_{AB} is the projected distance between the observers perpendicular to
the line of sight, and β is the angle between $\vec{r}_{AB}$ and the critical
curve. Since β is unknown we get an upper limit of v_T. A point value
of v_T can be obtained either if the relative velocities are not negligible or if there are 3 observers [44]. The corresponding velocity V_T
in the source plane, with time measured by the observer, is then

$$V_T = -v_T D_{ds}/(D_d (1 + z_d)) \tag{20}$$

and we can thus scale the source profile P(t) and determine the size of
the source.

We note that a determination of these transversal velocities has a
great value in itself, since it gives important indications for the
deviations from the Hubble flow.

The method described above, using space-borne observers $\lesssim 1$ AU away
from the earth, seems a realistic possibility since it should be possible to determine time lags in the light curves for the brightest quasars
(m $\lesssim$ 17) with a rather small telescope during a HAE (aperture 10 to
20 cm).

For larger baselines (~ 100 AU for M $\approx$ M$_\odot$) a significant fraction
of distant quasars should show measurable brightness differences
($\Delta m \gtrsim 0.01$), even without a HAE. Such measurements could therefore give
very valuable information on the distribution of masses in the range
10^{-4} M$_\odot$ to 10^2 M$_\odot$ in the universe. With even larger baselines (see
below) or more accurate observations masses up to 10^4 M$_\odot$ or more should
be possible to detect. This is very close to the present limit of about
10^5 M$_\odot$ which can be detected by direct image splitting by means of VLBI
observations.

Spectacularly large baselines would be offered by projects like
the astrometry probe TAU ("thousand astronomical units") recently proposed by JPL [46].

An additional important aspect of space-borne observations
over large baselines is the possibility to distinguish between intrinsic variability and micro-lensing variability, since no parallax

effect will show up for intrinsic variability alone.

c) Candidates for micro-lensing

The most promising candidates for micro-lensing are the known gravitationally lensed quasars with two or more macro-images. Since we roughly know how massive the lens is due to its ability to produce multiple images, one usually finds that at least one image has an optical depth for micro-lensing about equal to or larger than the mass fraction ε which is in "stars". A rough estimate of the frequency of HAE has been made in [39] : A HAE should occur about every third year for the multiple quasars known in 1985.

An additional reason to focus our interest on multiple quasars is that variations due to micro-lensing are easier to distinguish from intrinsic variations in those systems than in single quasars.

For single quasars any sign of intervening mass increases the probability of micro-lensing. Such signs can be galaxies close to the light path, absorption line systems or a very high absolute luminosity.

Due to the important information which can be obtained from micro-lensing, strong efforts should be made in order to prove the existence of this effect, and to measure HAE lightcurves with high accuracy. This can only be achieved by regular monitoring of the best candidates. In order to notice a HAE in time, each candidate should be observed, if possible, once a week and at least once a month. Such an ambitious project obviously requires international cooperation.

d) Observational indications for micro-lensing

R.E. Schild [32] has reported a tentative time delay of 1.2 years for the double quasar QSO o957+561 A, B, based on 7 years of observations. He finds however that the lightcurves cannot be explained by intrinsic variations alone. This argument holds even if his value for the time delay is not correct, since the theoretical upper limit for Δt is 5 years ($H_o > 4o$ km s^{-1} Mpc^{-1}). The only explanation seems to be that micro-lensing is contributing to the variations in at least one of the components.

Another example is the multiple quasar 1115+o8o. Its two brightest components, A_1 and A_2 are separated by o.55". Theoretical calculations lead to a time delay of a few days between A_1 and A_2 . Foy et al. [47] have observed a change in the brightness difference of nearly one magnitude between March and May 1984. Since QSO 1115+o8o shows no sign of variability on a timescale of days this cannot be explained by intrinsic variations, micro-lensing seems to be the only possible explanation. It is interesting in this connection to note that A_1 and A_2 are probably strongly amplified by macro-lensing, this increases the probability for strong micro-lensing [11].

References

1. Gott III,J.R.: 1987, Proc. IAU Symp. 117, 219
2. Canizares,C.R.: 1987, Proc. IAU Symp. 124, 729
3. Blandford,R.D., Kochanek,C.S.:1987, Proc. 13th Jerusalem Winter
        School, in press
4. Surdej,J., Magain,P., Swings,J.-P., Borgeest,U., Courvoisier,T.J.-L.
        Kayser,R., Kellermann,K.I., Kühr,H., Refsdal,S.: 1987,
        Nature 329, 695
5. Schneider,P., Borgeest,U.: 1988, submitted to Astron.Astrophys.
6. Schramm,T., Kayser,R.: 1987, Astron.Astrophys. 174, 361
7. Kayser,R., Schramm,T.: 1988, Astron.Astrophys., in press
8. Etherington,I.M.H.: 1933, Phil.Mag. 15, 761
9. Refsdal,S.: 1964, Mon.Not.Roy.Astr.Soc. 128, 295
1o. Paczynski,B.: 1986, Astrophys.J. 3o1, 5o3
11. Chang,K., Refsdal,S.: 1984, Astron.Astrophys. 132, 168
12. Refsdal,S.: 1965, Proc.Int.Conf.Rel.Theor. Gravitation, London
13. Press,W.H., Gunn,J.E.: 1973, Astrophys.J. 185, 397
14. Zel'dovich,Ya.B.: 1964, Soviet Astr. 8, 13
15. Refsdal,S.: 197o, Astrophys.J. 159, 357
16. Kayser,R., Refsdal,S.: 1988, preprint, submitted to Astron.Astrophys
17. Schneider,P., Weiss,A.: 1988, preprint MPA 295, submitted to
        Astrophys.J.
18. Schneider,P.: 1987, Astrophys.J. (Lett.) 316, L7
19. Refsdal,S.: 1964, Mon.Not.Roy.Astr.Soc. 128, 3o7
2o. Refsdal,S.: 1966, Mon.Not.Roy.Astr.Soc. 132, 1o1
21. Chang,K., Refsdal,S.: 1976, Proc.IAU Coll. 37, 369
22. Kayser,R., Refsdal,S.: 1983, Astron.Astrophys. 128, 156
23. Cooke,J.H., Kantowski,R.: 1975, Astrophys.J. (Lett.) 195, L11
24. Young,P., Gunn,J.E., Kristian,J., Oke,J.B., Westphal,J.A.: 198o,
        Astrophys.J. 241, 5o7
25. Young,P., Gunn,J.E., Kristian,J., Oke,J.B., Westphal,J.A.: 1981,
        Astrophys.J. 244, 736
26. Borgeest,U., Refsdal,S.: 1984, Astron.Astrophys.141, 318
27. Borgeest,U., 1986, Astrophys.J. 3o9, 467
28. Falco,E.E., Gorenstein,M.V., Shapiro,I.I.: 1985, Astrophys.J.
        (Lett.) 289, L1
29. Schneider,P.: 1985, Astron.Astrophys. 143, 413
3o. Florentin-Nielsen,R.: 1984, Astron.Astrophys. 138, L19
31. Schild,R.E., Cholfin,B.: 1986, Astrophys.J. 3oo, 2o9
32. Schild,R.E.: 1987, priv. communication
33. Gorenstein,M.V., Falco,E.E., Shapiro,I.I.: 1988, Astrophys.J. in
        press
34. Alcock,C., Anderson,N.: 1986, Astrophys.J. 3o2, 43
35. Chang,K., Refsdal,S.: 1979, Nature 282, 561
36. Gott III,J.R.: 1981, Astrophys.J. 243, 14o
37. Young,P.: 1981, Astrophys.J. 244, 756
38. Subramanian,K., Chitre,S.M., Narasimha,D.: 1985, Astrophys.J. 289,37
39. Kayser,R., Refsdal,S., Stabell,R.: 1986, Astron.Astrophys. 166, 36
4o. Vietri,M., Ostriker,J.P.: 1983, Astrophys.J. 267, 488
41. Ostriker,J.P., Vietri,M.: 1985, Nature 318, 446

42. Chang,K.: 1984, Astron.Astrophys. 13o, 157
43. Grieger,B., Kayser,R., Refsdal,S.: 1988, this volume
44. Grieger,B., Kayser,R., Refsdal,S.: 1988, Astron.Astrophys. in press
45. Grieger,B., Kayser,R., Refsdal,S.: 1986, Nature 324, 126
46. Meinel,A.B., Meinel,M.P.: 1986, Bull.Am.Astr.Soc. 18, 1o12
47. Foy,R., Bonneau,D., Blazît,A.: 1985, Astron.Astrophys. 149, L13

GRAVITATIONAL MICRO-LENSING AS A CLUE TO QUASAR STRUCTURE

B. Grieger, R. Kayser, S. Refsdal
University of Hamburg
Hamburger Sternwarte
Gojenbergsweg 112
2o5o Hamburg 8o / F. R. G.

ABSTRACT. We discuss the possibilities of determining the one-dimensional intrinsic luminosity profile of a quasar which is affected by gravitational micro-lensing (lensing of stars in a distant galaxy). When the
source crosses a critical curve (Jacobian of the lens mapping vanishes,
i.e. anti-caustic) due to relative transverse motion, a typical asymmetric peak occurs in the lightcurve, a so-called high amplification
event (HAE). It corresponds to the appearance (or disappearance) of
two bright micro-images with a separation of some microarcseconds, which
is not resolvable. However the brightness changes during a HAE are
easily observable. Since it is an eclipse-like effect, it gives the
possibility of calculating the one-dimensional source profile from the
lightcurve. We present the method and discuss the results of its application on simulated lightcurves. It should be possible to obtain
information on the structure of quasars on a scale smaller than o.oo1
pc for solar mass stars.

1. THEORETICAL FOUNDATIONS

The amplification of a point source near a critical curve is given by
[1,2]

$$A(d) = A_o + \begin{cases} k \cdot d^{-1/2} & \text{for } d > o \\ o & \text{for } d < o \end{cases} \qquad (1)$$

with A_o being the amplification of the other images, k a constant of
the order of unity, and d the distance from the critical curve in
normalized units. For the general validity of this expression see [7]
(this volume). By using Eq.(1) for the integration of the amplification
over the source we have to make the following assumptions:
    1. The expression is valid in the region of interest, which means
that the source radius is small compared to the radius of curvature of
the critical curve.
    2. The amplification of the other images is constant in the region
of interest, which means that the source radius is small compared to
the typical distance between critical curves.
    The radius of curvature and the typical distance between critical

311

N. Kaiser and A. N. Lasenby (eds.), The Post-Recombination Universe, 311–313.

curves are both of the order of the normalized unit, so we assume the source radius to be small compared to the normalized unit ζ_o , which is of the order of 10^{-2} pc for typical lensing situations [3].

2. THE ABELIAN INTEGRAL EQUATION

We consider an extended source with a surface brightness $I(\xi,\eta)$ being crossed by a critical curve which is parallel to the η-axis. Since in a small region close to the critical curve the amplification depends only on ξ , we can integrate $I(\xi,\eta)$ over η and define a one-dimensional source profile $P(\xi)$. In order to obtain the lightcurve which is generated by the crossing of the critical curve we have to integrate $P(\xi)$ weighted with the amplification factor, see Eq.1, and so we get the flux [5]

$$F(\xi_c) = k \int_{\xi_{min}}^{\xi_c} \frac{P(\xi)}{\sqrt{\xi_c-\xi}} \, d\xi + F_o \quad . \tag{2}$$

This is the well-known Abelian integral equation, which has the analytical solution

$$P(\xi) = \frac{1}{\pi k} \frac{d}{d\xi} \int_{\xi_{min}}^{\xi} \frac{F(\xi_c) - F_o}{\sqrt{\xi - \xi_c}} \, d\xi_c \quad . \tag{3}$$

3. HOW TO OBTAIN THE SOURCE PROFILE

In order to obtain a numerical solution of Eq.(2), there are several possibilities for a discret formulation. A simple and accurate method is to assume the source profile to be given by a polygonian

$$P(\xi) = P_{j-1} + s_j(\xi-\xi_{j-1}) \quad \text{for} \quad \xi_{j-1} < \xi < \xi_j \quad .$$

Then we can solve the integral in Eq.(2) and derive an algorithm to obtain all P_i at the points ξ_i where the flux F_i is measured.
The points P_i of the polygonian source profile are given by

$$P_o = o \quad , \quad P_1 = \frac{3}{4} \frac{F_1 - F_o}{\sqrt{\xi_1 - \xi_o}} \quad ,$$

$$P_i = \frac{3}{4} \frac{(F_i - F_o)\sum_{j=1}^{i-1} f_{ij}}{\sqrt{\xi_i - \xi_{i-1}}} - \frac{1}{2} P_{i-1} \quad \text{for} \quad i > 1 \quad , \tag{4}$$

with f_{ij} defined by

$$f_{ij} = 2(P_{j-1} + s_j(\xi_i - \xi_{j-1})) \, ((\xi_i - \xi_{j-1})^{\frac{1}{2}} - (\xi_i - \xi_j)^{\frac{1}{2}}) -$$

$$- \frac{2}{3} s_j \, ((\xi_i - \xi_{j-1})^{\frac{3}{2}} - (\xi_i - \xi_j)^{\frac{3}{2}}) \quad . \tag{5}$$

We have tested the method on simulated lightcurves for different source
profiles [5],[6] (this volume). The source profile can be retrieved up
to a source radius of the order of o.1 normalized units [3], which is
about 10^{-3} pc ($\approx 10^{-7}$ arcsec !) for solar mass stars and cosmological
distances. There is no lower limit of resolution.

4. PROBLEMS

In some cases the retrieved source profile does not exactly fall down
to zero again. This is caused by a too large distance from the critical
curve or a variation of the other micro-images (A_0 or k not constant,
compare Eq.(1)). We can estimate it by an interpolation between the
parts of the source profile we guess to be zero, and then obtain a
corrected profile by subtraction.

A general feature of deconvolution problems is noise amplification.
Since an observed lightcurve may have much larger noise than our simulated lightcurves, the retrieved source profile can be completely distorted. We are at the moment investigating mathematical methods to
stabilize the solution (e.g. the maximum entropy method).

The intrinsic variability of a quasar could be indistinguishable
from variability due to gravitational micro-lensing. However there are
two situations in which we can make a distinction:

First, if we have two macro-images and know the time-delay between
them, a difference of the two "synchronized" and appropriately scaled
lightcurves cannot be intrinsic.

The second situation requires a second observer more than about 1
A.U. away. In this case the variation caused by micro-lensing is time-
shifted relative to the earth-bound observer due to a parallax-effect
[4] and therefore distinguishable from intrinsic variation, which is
not affected.

5. CONCLUSION

The observation of a high amplification event offers a unique possibility to obtain extremely high resolution "images" of quasars. In particular it should be possible to distinguish between different quasar models,
for instance the models of spinars and magnetoids and the model of a
black hole with an accretion disk.

We therefore propose very frequent monitoring of those quasars
which are good candidates for gravitational micro-lensing.

REFERENCES:

[1] Chang,K., Refsdal,S.: 1979, Nature 282, 561
[2] Chang,K., Refsdal,S.: 1984, Astron.Astrophys. 132, 168
[3] Kayser,R., Refsdal,S., Stabell,R.: 1986, Astron.Astrophys. 166, 36
[4] Grieger,B., Kayser,R., Refsdal,S.: 1986, Nature 324, 126
[5] Grieger,B., Kayser,R., Refsdal,S.: 1987, Astron.Astrophys. in press
[6] Refsdal,S., Kayser,R.: 1988, this volume
[7] Schramm,T.: 1988, this volume

GRAVITATIONAL LENSING BY CLUSTERS OF GALAXIES

Israel Kovner,
Institute of Astronomy, Cambridge, England, and
Physics Department, Weizmann Institute, Rehovot 76100, Israel

ABSTRACT: *I examine the gravitational lens hypothesis for Giant Luminous Arcs. In particular: possible associated selection effects and statistics, ways in which such arcs can be formed on a large scale and small scale details (demonstrated by graphics), tests of the hypothesis, astrophysical applications and predictions if lensing is confirmed.*

A cluster at redshift z_l can split images of a point source at z_s, if its core surface mass density is $\Sigma > \Sigma_{min}(z_l, z_s)$. The largest estimates of Σ are close to the smallest values of Σ_{min} for high redshift rich and compact clusters (HRRC) at optimal redshifts[1] (this may be not a coincidence[2]). Apparently, most HRRC, even at optimal z_l, cannot split images. A small fraction of HRRC, at the tail of their distribution in Σ/Σ_{min}, may either split images on their own or give an unusually strong assistance to a galaxy. Such assistance is a kind of 'marginal' lensing[1] which has a small probability unless there are selection effects working against ordinary lensing. One such is the magnification bias[3,4], another is the density bias[1]. There may be yet another bias to favor marginal lensing of high redshift galaxies (HRG) by HRRC well able to split images, as follows.

The HRG are numerous but it is difficult to identify lensed galaxies among hundreds of cluster galaxies. Merging pairs of galactic images are conspicuous but look like physically interacting pairs[1] (a similar question arose in ref. 5). Three merging images can form a more conspicuous arc-like structure. As an example of statistics, let us assume magnifications of $\sim 2 \ mag$ for twin images and $\sim 4 \ mag$ for arcs, a very deep survey in $J-$band down to $J \leq 23$, number counts[6] $N(< J) = 4.5 \cdot 10^{0.4(J-24)} \ arcmin^{-2}$ valid to $J = 27$, that most galaxies of $J < 25$ are HRG[7] of $z_s \gtrsim 1$, a typical angular radius $\sim 1''$ for a galaxy, $\sim 50''$ of source plane caustics (SPC) in a compact HRRC for a typical galaxy redshift. Then ~ 0.3 twin galaxtic images are expected to saddle the image plane critical curve (IPC) and there is a probability $\sim 0.3 - 0.7$ for a galaxy to be close enough to one of the ≥ 4 cusps on the SPC, to form 3 close or merging images.

Large structures attract attention, as it happened with the Giant Luminous Arcs[8] (GLA), which are also remarkably circular, thin and regular. Paczyński[9] proposed that these arcs may be lensed HRG (redshifts are still unknown). A necessary configuration can occur in compact clusters, if a (sub)cluster has a sufficiently close to circular segment of IPC, and a few tens of such arcs in the whole sky would be consistent with what is known about clusters and HRG[10].

A thin lens is a mapping $\vec{r} \rightarrow \vec{s} = \vec{r} - \nabla\Phi(\vec{r})$, where $\vec{r}$ and $\vec{s}$ are, respectively, image and source angular positions on the sky. Let us assume that Φ is dominated by one or two smooth 'halos' of typical deflection angle $\Delta\theta_c$, with IPC containing close to circular stretches mapped to cusps. An extended source in the vicinity of such a cusp(s) produces a large arc in the vicinity of the IPC, broken or continuous, fig. 1.

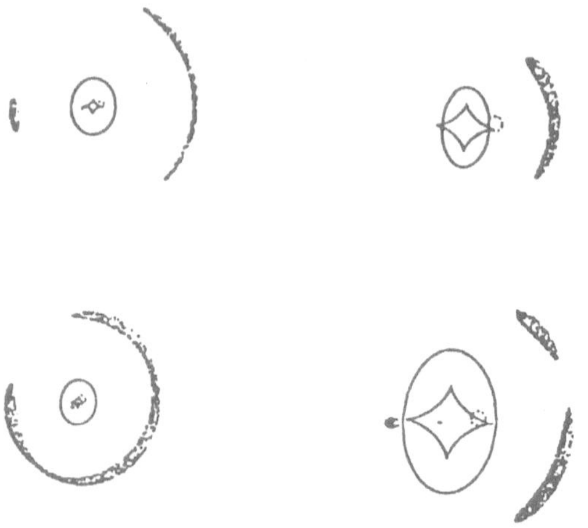

Figure 1: Arcs formed by smooth halos of elliptic potentials of different ellipticities. Solid lines are SPC, dashed circles are source boundaries.

Figure 2: 'Fingers' and 'filaments': perturbations of an IPC and of a galaxy image, respectively, by relatively 'heavy' galaxies superposed on a smooth halo, in a model resembling A370. 'Wheels' represent galaxies, crosses show centers of halos, long dashed and solid lines are IPC and SPC, respectively.

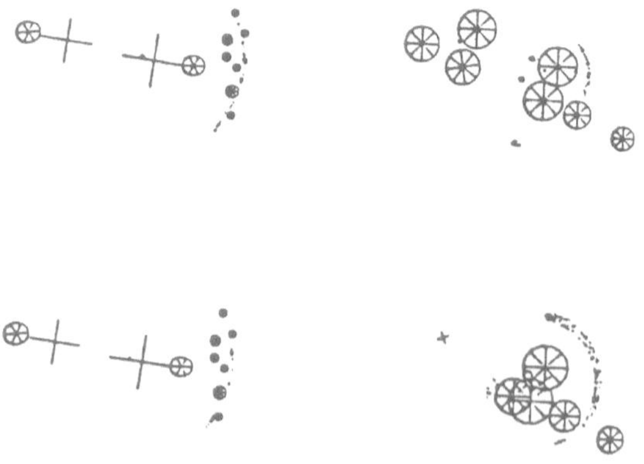

Figure 3: Double halo lens models resembling A370 (left) and 2242-02 (right), with arcs produced by galaxies in small solid circles. 'Wheels' represent galaxies, crosses show halo centers.

The arc thickness is the size of the source, magnified by ~ 2. The source galaxy does not have to be near a cusp to form an arc, however covering both branches of a cusp or a higher singularity can ensure the large extent and continuity of an arc[11]. For instance, in the models for 2242-02 described below, the galaxy happens to be near a (may be unformed yet) swallow-tail.

If GLA are lenses, they give evidence for compactness of the (sub)cluster cores, and an estimate of the velocity dispersion σ_v (line of sight). Measuring σ_v is a test of the lensing hypothesis, though it may happen to be not decisive: the arc sizes are just right for rich clusters, e.g. A370 has $z_l = 0.37$ and its arc radius is $\Theta \approx 15''$, so that, say, for $z_s = 1, \sigma_v \sim 850(\Theta/15'')^{1/2}$ km s^{-1}. The smooth part is perturbed by galactic fields of typical $\Delta\theta_g \ll \Delta\theta_c$. A galaxy near an arc causes a distortion $\Delta\Theta/\Theta \sim \Delta\theta_g/\Delta\theta_c$. Observational bounds on the distortions may bear constraints on the galaxies, in particular on their tidal radii. Fig. 2 shows strong distortions of an IPC 'fingers', and of an image of a large galaxy, 'filaments'. In the numerical models resembling A370 and 2242-02, fig. 3, I assumed the following: (a) the smooth part is composed of two halos with elliptic potentials of form $\Phi = \Delta\theta_c \left[b^2 + x^2(1-e)^2 + y^2(1+e)^2\right]^{1/2}$; (b) halo positions are related to concentrations of galaxies; (c) galactic potentials are either of the form above, with $\Delta\theta_c$ replaced by $\Delta\theta_g$, or $\Phi = (\Delta\theta_g^2/2)\ln(b^2 + r^2)$; (d) $\Delta\theta_g$ are derived from a luminosity - velocity relation, $L = C_l\sigma^4$, where L is the area of a galaxy image in ref. 7, and C_l is taken the same for all galaxies; (d) $\sigma_v = 250$ km/s for the giant ellipticals; (e) $z_s = 1, \Omega = 1$.

Tests of lensing hypothesis: In the absence of arc redshifts, direct tests can be σ_v, a thrice repeated (once in reverse) substucture along the arc of the part of the galaxy inside the caustic (if it is clumpy or has a color gradient), stratification of the substructure along the arc, the high probability of another arc to be found in a sufficiently deep survey.

Derivable information if lensing confirmed: Core compactness. Galactic tidal radii. Properties of the lensed galaxy (via mapping $\vec{r} \to \vec{s}$).

Predictions: There should be arcs in other HRRC, some of which should consist of several segments. In some cases arcs may be formed by two or more galaxies, in general at different redshifts, so that arc radii and centers are different in the same cluster.

Acknowledgements: this work was made possible by computer graphics of VAX at the Institute of Astronomy, by kind invitation to IoA by M. Rees, and by SERC grant SG/C 22555. I also thank R.D. Blandford for a friendly challenge.

References
1. Kovner, I., *Ap. J.*, **321**, 686 (1987).
2. Rees, M., private communication (1987).
3. Turner, E.L., Ostriker, J.P., & Gott J.R., *Ap. J.*, **284**, 1, (1984).
4. Narayan, R., Blandford, R., & Nityananda, R., *Nature*, **310**, 112 (1984).
5. Cowie, L.L., and Hu, E.M., *Ap. J. (Letters)*, **318**, L33 (1987).
6. Tyson, J.A. & Jarvis, J.F., *Astrophys. J. (Letters)*, **230**, L153 (1979).
7. Pritchett, C. & Kline, M.I., *Astron. J.*, **86**, 1859 (1981).
8. Soucail, G., Fort, B., Mellier, Y. & Picat, J.P., *Astr. Ap. (Letters)*, **172**, 414 (1987); and Lynds, R. & Petrosian, V., *Bull. Am. astr. Soc.*, **18**, 1014 (1986).
9. Paczyński, B.P., *Nature*, **325**, 572 (1987).
10. Kovner, I., *Nature*, **327**, 193 (1987).
11. Blandford, R.D., & Kovner, I., in preparation.

PROPERTIES OF GRAVITATIONAL LENS MAPPINGS

Thomas G. Schramm
Hamburger Sternwarte
Gojenbergsweg 112
2o5o Hamburg 8o
F. R. G.

ABSTRACT. The vector formulation of the gravitational lens theory is used to apply catastrophe theory and elimination theory to obtain expressions for the caustics and the distribution of cusps.

1. INTRODUCTION

Light of distant sources is deflected by masses lying near the light-path. If the extension of the mass distribution (deflector) is small compared with the distances deflector-source and deflector-observer, the total deflection can be described similar to the ray-approximation of the theory of thin lenses of geometrical optics. The total vector deflection angle is then proportional to the gradient of the logarithmic potential of the mass distribution projected onto the deflector plane DP perpendicular to the line observer-source. Introducing a source plane SP, parallel to DP, containing the source and taking the geometry into account one yields the Gravitational Lens Mapping GLM which expresses the SP-coordinates as a function of the DP-coordinates for a fixed observer. For details of the physical implications, the geometry and notation see [7] and [3], here we show some techniques in handling the obtained formulations of gravitational lens scenarios.

2. THE GRAVITATIONAL LENS MAPPING

From the gravitational lens theory we obtain a surjective mapping from DP to SP with coordinates $\vec{z} = (x,y)$ and $\vec{\zeta} = (\xi,\eta)$:

$$\text{GLM: DP} \to \text{SP}; \ \vec{z} \to \vec{\zeta}; \ \text{or implicit in components:}$$
$$\xi - \xi(x,y) = 0 \ ; \ \eta - \eta(x,y) = 0 \tag{1}$$

Since our map is a gradient (Lagrange)-mapping a scalar formulation is always possible and leads to the so-called Time Delay Function TDF [1]. For our purposes, however, the vector formulation is easier to use and gives more insight from a geometrical point of view.

N. Kaiser and A. N. Lasenby (eds.), The Post-Recombination Universe, 319–321.

2.1. Rayshooting Methods (Rayplots)

The simplest method in examining the GLM of a given model is just to map (shoot) a set of points of DP into SP using the GLM of Eq. (1) and then to analyse the distribution $I(\xi,\eta)$ of points in SP. If the points are chosen on a grid, one gets interesting insights in the geometrical and topological properties of the mapping. If one connects the grid-points to a net, it becomes clear that it is possible to interpret the point distribution in SP as the projection of a folded and possibly self-intersected surface. Extended sources can be simulated by interpreting the GLM as a mapping from DP to the observer plane OP for a fixed source element and then overlay all rayplots for each source element which nicely shows that the structures (obtained for point sources) are washed out with increasing size of the source. For large extended sources the best results are obtained, if the source elements and the deflector points are taken randomized. The distribution $I(\xi,\eta)$ corresponds to the intensity distribution in OP for a fixed source and is proportional to the sum of the inverse Jacobians $J^{-1}(x,y)_i$ of the GLM for each image i of any source point seen from OP in DP. Some rayplots can e.g. be found in [4]. These rayplots show sharp structures called caustics, which can be associated with very high intensity in OP. Caustics are well-known from optics and correspond mathematically to the image $D(\xi,\eta) = GLM(J = 0)$ of the vanishing Jacobian $J(x,y) = 0$. In the next sections we show how to express the properties of the GLM near its singular structure and derive an analytic formulation for D under certain assumptions.

2.2. The Local Structure of the GLM

In order to understand the variation of the intensity in OP and therefore the behaviour of the caustics, we apply elements of the catastrophe theory CT. In its general form it describes the singular structure of smooth mappings $R^n \rightarrow R^m$ [2]. The scalar version of CT for mappings (as TDF) $R^n \rightarrow R$ is easier to handle [6], [1], but both have the disadvantage of using somewhat unfamiliar mathematical objects like orbits, determinacy etc. Fortunately there is a full theory for the GLM which is very simple to apply, the Whitney-theory for mappings $R^2 \rightarrow R^2$ [10]. We shortly summarize the main points: The caustics of generic GLMs consist only of smooth curves (folds) and isolated cusps on the folds. The GLM is everywhere regular (R) except on folds (F) and cusps (C) and local equivalent to one of the following forms:

$$
\begin{array}{llll}
R: & J(x,y) \neq 0 & \xi = x \;; & \eta=y & (2R) \\
F: & J(x,y) = 0; \; \vec{dT}(GLM) \neq 0 & \xi = x^2 \;; & \eta=y & (2F) \\
C: & J(x,y) = 0; \; \vec{dT}(GLM) = 0; \; \vec{dT}(\vec{dT}(GLM)) = 0; & \xi = xy-x^3 \;; & \eta=y & (2C_1) \\
\multicolumn{4}{l}{\vec{dT}(GLM) = 0 \iff P_1(x,y,\vec{p}) = 0; \; P_2(x,y,\vec{p}) = 0; \; \vec{p} = (p_1, p_2 \dots)} \\
\multicolumn{4}{r}{(2C_2)}
\end{array}
$$

where the vector $\vec{dT}(GLM)$ is the directional derivative in the direction tangential to $J = 0$. From Eq. (2F) it is easy to see that near folds the intensity is inverse proportional to the square root of the distance from the fold in SP. Eq. ($2C_2$) gives us two equations depending only on the DP-coordinates and the model parameters $\vec{p}$ for the cusp points. If P_1 and P_2 are polynomials the resultant $R = R(x,\vec{p}) = 0$ can be built to eliminate e.g. y to obtain a single polynomial depending only on x and the parameters $\vec{p}$ [8]. Interpreting $R(x,\vec{p}) = 0$ as a hypersurface in the $(x,\vec{p})$-space it is easy to see that the number of cusps changes, if R and the derivative of R with respect to x vanish simultanuously. Again the resultant can be built and a single polynomial $D(\vec{p}) = 0$ remains, which devides the model parameter space in areas of constant cusp numbers. $D(\vec{p})$ is simply the discriminant of R.

2.3. A Global Caustic Equation

If Eq. (1) is polynomial (if not, an expansion could eventually bring it into the right form), we can use the resultant theory again to obtain a closed form for the caustic. So we obtain the resultant of Eq. (1) $R(x,\xi,\eta) = 0$. The degree of R in x gives us the maximum number of images of a given point source. As we mentioned in Chap.2.1. we interpret R as a hypersurface in (x,ξ,η)-space and similar to Chap.2.2. the number of solutions alter where R has an extremal value with respect to x, i.e. where the discriminant $D(\xi,\eta)$ of R vanishes, which is therefore an implicit equation for the caustic in dependence of the SP-coordinates.

3. CONCLUDING REMARKS

We showed that the Whitney-theory is a useful tool in determining the local structure of the GLM. In the case of polynomial GLM we obtain closed forms for caustics and the model dependence of cusps. However, if done by hand, the computation of resultants can be rather extensive and should therefore be done by using software-tools for symbolic computation like REDUCE. Nevertheless there are remaining problems in finding global properties of lens models, e.g. the image construction for given sources or the source reconstruction for observed images. Some of these problems are treated in [5] and [9].

References

[1] Blandford,R.D., Narayan,R.: 1986, Astrophys.J. 310, 568
[2] Gibson,C.G.: 1978, Singular points of smooth mappings
[3] Grieger,B.: This volume
[4] Kayser,R., Refsdal,S., Stabell,R.: 1986, Astron.Astrophys. 166, 36
[5] Kayser,R., Schramm,T.: 1987, Astron.Astrophys. in press
[6] Poston,T., Steward,I.: 1978, Catastrophe Theory and its Appl.
[7] Refsdal,S., Kayser,R.: This volume
[8] Salmon: 1877, Algebra der linearen Transformationen
[9] Schramm,T., Kayser,R.: 1986, Astron.Astrophys. 174, 361
[1o] Whitney,H.: 1955, Annals of Mathem. 62, No. 3, 374

THE FORMATION OF ELLIPTICAL GALAXIES

Bernard J.T. Jones,
NORDITA,
Blegdamsvej, 17,
Copenhagen, Denmark 2100

ABSTRACT. We review theories for the formation of galaxies, concentrating on those aspects of the observational data that might provide clues to the problem. Data samples have grown rapidly in the last few years and a number of key issues have become clarified such as the role of rotation in elliptical galaxies and the existence of more than one family of objects.

A key issue is why there should be two apparently quite distinct classes of galaxy: the S and E type, and why there are two distinct classes of E: the "normal" and the "dwarf" E. I shall present arguments that a normal elliptical is a galaxy that for a variety of reasons was unable to form a disk. The accumulated material simply went into the spheroidal system rather than into an organized disk.

1. OBSERVATIONAL MILESTONES

Not so long ago elliptical galaxies were regarded as highly relaxed stellar systems whose degree of flattening was determined by rotation. The model was of a stellar version of the Maclaurin spheroid. In the past decade, we have divested ourselves of that simplistic notion. In this article I shall review the reasons for that change of viewpoint and present some of the theoretical ideas we have about the formation of galaxies.

The development of the study of elliptical galaxies has been well reviewed (see for example Binney, Kormendy and White, 1982). It is appropriate here to merely recall the key observational milestones that have lead to our current understanding:

1. Hubble's sequence Sm Sc Sb Sa S0 E7 $\cdots$ E0

2. A universal light profile (Hubble or de Vaucouleurs law)

3. Rotation of Ellipticals small

N. Kaiser and A. N. Lasenby (eds.), The Post-Recombination Universe, 323–345.

4. There are two sequences of E's: Normal and Dwarfs

5. Normal E's are a two-parameter family

6. Small E's are rotationally supported

7. E's are found predominately in dense environments

8. E's inside and outside clusters are indistinguishable

9. S's are surrounded by extensive massive halos

10. There is no evidence for dark matter in ellipticals

Some of these statements may be regarded as slightly contentious. The "fact" that ellipticals inside and outside of clusters are indistinguishable depends on somewhat limited surveys containing only a few hundred elliptical galaxies. It may be fairer to say that there are no manifest differences.

The key issue is to distinguish clearly the "normal" ellipticals from the "dwarf" ellipticals. In what follows, when I talk of "Elliptical Galaxies" I shall be referring to the "normal" family of elliptical galaxies, as defined by Kormendy (1985, 1986). (The idea seems to have originated with Wirth and Gallagher, 1984).

2. GENERAL RELATIONSHIPS: CHEMISTRY AND DYNAMICS

The observational data is systematized via a set of relationships between the global properties of galaxies. Many of these relationships have been known or at least hinted at for a number of years, but it is with the study of large samples of elliptical galaxies by Djorgovski and Davis (1987), Dressler et al. (1987), Davies et al. (1987), Burstein et al. (1987), that a concrete picture has emerged.

For Elliptical galaxies the most interesting relationships are

1. Faber Jackson relationship

2. Deviations from Faber Jackson significant (2nd Parameter)

3. 2nd parameter is surface brightness

4. In normal E's surface brightness correlates with luminosity

5. Velocity dispersion correlates with metallicity

6. $\mathcal{M}/\mathcal{L}$ correlates with $\mathcal{L}$ and is "normal"

(We use $\mathcal{M}$ to denote mass and $\mathcal{L}$ to denote luminosity in some appropriate waveband. M will denote absolute magnitude). In essence, the elliptical galaxies are characterised by three parameters: luminosity, velocity dispersion and metallicity. Examination of the distribution of the Ellipticals in this space reveals that they lie on a relatively narrow two dimensional surface.

There is a parallel set of relationships between the properties of spiral galaxies, the most important of which are perhaps

1. The Fisher-Tully relationship

2. The infrared colour magnitude relationship

There are other more contentious relationships in disk galaxy samples that could be of importance:

1. The constancy of the central surface brightness of disks

2. A metallicity absolute magnitude relationship

The link between the Elliptical types and the Spirals is probably through the bulge to disk ratio. I shall discuss that later since it forms a central part of the most promising theories for the existence of two distinct morphological classes of galaxy.

3. KEY THEORETICAL NOTIONS

There are a few key theoretical ideas underlying our understanding of elliptical galaxies:

1. Violent relaxation responsible for profiles and shapes

2. Models supported by velocity anisotropy

3. Galaxies are probably triaxial (twisting isophotes)

The universality of the luminosity profiles has long been recognized as being due to relaxation processes, the most popular idea being that Lynden-Bell's "violent relaxation" is the main mechanisim. (For recent discussions on this subject see Tremaine et al. (1986), Madsen (1987) and Shu (1987)). Given suitable initial conditions (see discussion on numerical models below) the relaxation towards a universal density profile is certainly seen in the numerical simulations of van Albada (1982) and Quinn et al. (1986). The closest we have to an explanation of how this relaxed state is achieved is given by White and

Narayan (1987) who argue that the final state is only a local maximum entropy state, not a global maximum entropy state. To get this result they apply the constraint that the state have a power law density profile. Interestingly, it turns out that the power law needed to give a state of local maximum entropy is just the observed power law in density.

The fact that, in general, ellipticals owe their non-spherical shapes to the anisotropy of the stellar distribution function rather than rotation is of importance (Binney, et al. 1982). Whatever the relaxation processes were that lead to the universal profile, they were clearly not efficient enough to isotropize the velocity distribution function, even in the central parts of the galaxies.

It is a simple step to assert that the distribution function could be triaxial and hence that ellipticals are essentially triaxial systems. Observational verification of this comes from the twisting of the isophotes. (See, for example, Jedrzejewski et al. 1987).

4. INTERESTING IDEAS ON GALAXY FORMATION

Peebles' first paper on galaxy formation in the hot big bang theory appeared even before the paper in Astrophysical Journal Letters announcing the discovery of the microwave background radiation. That paper is interesting because it set out clearly the idea that galaxy formation is a hierachical process with the force of gravity acting on a succession of scales selected from a random distribution of density fluctuations. With the exception of the "pancake" scenarios, this view of a gravitational hierachy is still with us, the major difference in the modern version being the key role played by hypothetical non- baryonic material. The "monolithic collapse" scenarios, in which a quasi spherical galactic mass goes into gravitational collapse and forms stars, are regarded as oversimplifications.

Some of the ideas that have been discussed in recent years can be summarized in the following list:

1. Galaxies are built in a gravitational hierarchy

2. Non baryonic material plays a significant role

3. Galaxies form in the collapse of large scale structures

4. E's are consequence of mergers

5. E's and S's discriminated by early cooling processes

6. Galaxies are formed at local maxima of density field

7. E's were 3 σ perturbations, S's 2 σ

8. Angular momentum is a consequence of tidal torques

9. Where does the other family of Ellipticals come from?

The list is not supposed to be in any way exhaustive. In the remaining sections I shall focus on a selection of these ideas that have been the subject of relatively recent investigations. So I shall not discuss either mergers of galaxies or the role of nonbaryonic material in establishing a clustering hierachy.

4.1. The spectrum of density fluctuations

The "Cold Dark Matter" (CDM) model of the universe makes rather precise predictions about the shape of the spectrum of primordial potential wells on all scales from dwarf galaxies up to great clusters and perhaps superclusters. The free parameters are the normalization of the amplitude of the spectrum, and the shape of the initial (Planck time) spectrum. In principle, it is the task of a theory of the early universe to explain these (see the review of Brandenburger, 1986). Since this has not been achieved the assumption is usually made that the power spectrum has the Harrison– Zel'dovich power law form, $\mathcal{P}(k) = \langle |\delta_k{}^2| \rangle \sim k$ and its amplitude is fixed by demanding agreement with observational data on some specific scale. It should be remembered that there is nothing sacred about the Harrison-Zel'dovich spectrum: it could be multiplied by logarithmic terms, or it could be that the spectrum has a completely different slope for reasons that we do not as yet understand.

The physical processes taking place before recombination modify the spectrum of fluctuations on the smaller scales. In the CDM picture, the smallest scales have their amplitudes frozen when they enter the horizon and the spectrum of amplitudes there is: $\langle |\delta_k^2| \rangle \sim k^{-3}$. A $\mathcal{P}(k) \sim k^n$ spectrum gives rise to mass fluctuations $\delta \mathcal{M}/\mathcal{M}$ whose variance scales with mass $\mathcal{M}$ as $\mathcal{M}^{-1/2-n/6}$. Thus, in the CDM picture, on large (ie. cluster) scales the mass fluctuations scale as $\mathcal{M}^{-2/3}$ and are independent of mass on smaller (galactic) scales. The independence of fluctuation amplitude on galaxy–like scales means that density fluctuations condense out of the universe at times that depend only on the local density enhancement, not on the mass. Objects that are 3σ fluctuations therefore collapse before objects that come from 2σ fluctuations, regardless of mass.

It has been suggested that the Harrison–Zel'dovich spectrum may not have sufficient power on large scales to explain very large scale phenomena such as cosmic voids, cluster–cluster correlations or reported large scale cosmic streaming motions (Dressler et al. 1987). The point is addressed in terms of the numerical simulations of CDM dominated universes by White et al. (1987), and in terms of possible modifications of the spectrum by Bond (1987) and by Bardeen, Bond and Efstathiou (1987).

4.2. Galaxies are formed at local maxima of density field

Our models of galaxy formation are almost entirely controlled by gravitational forces,

328

and so refer to the matter that contributes most to the fluctuations in the gravitational potential. This is not necessarily the luminous matter that we see in stars and galaxies. Since galaxy formation is a complex process it is desirable to relate, as simply as possible, the luminous objects that form to some aspect of the initial density perturbations.

A recent and currently popular ansatz is to say that luminous systems formed only where there were relatively high local maxima in a suitable spatial average of the density field just after the universe recombined. The idea seems to have originated with Kaiser (1984). The two key points here are that the maximum should somehow be a strong one (there should be a threshold value of some sort), and that the density fluctuation field should be averaged. (Blumenthal et al., 1984; Bardeen et al., 1985; Davis et al., 1985; Frenk et al., 1985).

The idea of averaging is a logical one: if in a random density field you wish to isolate structures on a given length scale, you do that by spatially convolving (windowing) the density field with an appropriate filter that removes the structure on the smaller scales that are not of interest. For lack of any obvious alternative, a Gaussian spatial filter is at least as good as anything else.

The idea of imposing a threshold is simply to stop most of the universe forming into luminous systems. Indeed, the threshold should depend on the windowing scale in such a way as to reproduce the observed luminosity function and the variations in mass to light ratio with scale. This idea is referred to as "threshold biasing".

Ultimately, the post hoc assignment of the threshold is to be replaced by an astrophysically motivated theory involving gas dynamics and star formation. For discussion of some specific mechanisms see Rees (1985), Silk (1985), Couchman and Rees (1986), Couchman (1987b) and the review by Dekel and Rees (1987).

4.3. E's were 3 σ perturbations, S's 2 σ

The higher binding energy of ellipticals suggested that they formed earlier than disk galaxies. They are also less numerous. A simple model is that E's result from the higher amplitude fluctuations than the S's. It is possible to adjust the two thresholds so that, for each mass scale, the ratio of E to S is correct, and so that the overall mass (luminosity) function is as observed. The two functions describing the E threshold and the S threshold as a function of mass scale would come from some detailed theory of the galaxy formation process (see for example, Blumenthal et al., 1984).

Such an approach has the advantage that it gives a ready explanation for the variation of morphological type with environment: the fluctuation amplitudes are boosted wherever there is an incipient cluster of galaxies.

4.4. Angular momentum is a consequence of tidal torques

In the hierachical models for galaxy formation, there is a substantial amount of angular momentum generated through tidal interactions between neighboring proto-galaxies. The magnitude of the effect is that typically the induced rotational velocities are 5 - 10 % of the random velocities. The spread in this quantity is substantial.

What appears from recent numerical experiments (Barnes and Efstathiou, 1987) is that the rotation parameter $\lambda \sim J E^{1/2} G^{-1} M^{5/2}$ is more or less independent of the spectrum of fluctuations (because most of the transfer of angular momentum takes place when the objects on a given scale are just separating out from the background universe). It also appears that it is independent of the amplitude of the fluctuation. This is because high amplitude fluctuations collapse faster than their less prominent counterparts, allowing less time for the transfer of angular momentum.

In non-hierachical theories for the formation of large scale structures, the situation is less clear. There will be tidal forces between neighboring protogalaxies, but nobody has evaluated their net effect. It was hinted out long ago that galaxy rotation could be generated in complex shock fronts through a complex interaction of pressure and temperature gradients (not unlike the process which generates weather!). The pressure forces that govern the fragmentation of the shock front could also drive rotational motions.

5. GENERAL QUESTIONS

Here is a list of some interesting questions relating to the formation of galaxies in general, and elliptical galaxies in particular.

1. Why is there a surface brightness - luminosity relationship?

2. Has dissipation played a role in E formation?

3. When did E's form?

4. How are dark and luminous matter segregated?

5. How does environment come in?

6. Why are smallest galaxies closest to rotational support?

7. Why are voids empty?

8. What is the luminosity (mass) function?

9. What is an $\mathcal{L}^*$ galaxy?

10. Why are normal ellipticals a two-parameter family?

11. Is 2nd parameter environment dependent?

These are the questions that any effective theory of galaxy formation should purport to answer, and conversely data relating to these questions will constrain theories of galaxy formation.

5.1. Why is there a surface brightness - luminosity relationship?

Traditionally, we infer a relationship between mass and surface mass density from an assumed form of the initial (recombination) spectrum of fluctuations. This inference generally assumes that there is no dissipation, as would be appropriate for a system dominated by cold collisionless particles.

This is translated to an observed luminosity - surface brightness relationship by making assumptions about mass to light ratios. Since we know how mass to light ratios in normal ellipticals vary with light, this step is not unreasonable. If we adopt $(\mathcal{M}/\mathcal{L}) \sim \mathcal{L}^{1/4}$, we find that on galactic scales arising from an $n = -3$ spectrum of density fluctuations at recombination, the surface brightness I would vary only weakly with luminosity: $I \sim \mathcal{L}^{1/6}$. This is to be contrasted with the observed relationship $I \sim \mathcal{L}^{-1}$.

This says that the smaller galaxies have higher surface brightness than would be expected on the basis of an $n = -3$ spectrum for the density fluctuations at recombination. So either there has been dissipation, the amount of dissipation being greatest in the smallest galaxies, or the original spectrum was closer to $n = 0$.

As we shall see below there is some suggestion from the rotational properties of normal ellipticals that dissipation has varied systematically with scale: the smaller (higher surface brightness) E galaxies are closer to being rotationally supported. However, it is difficult to see why the degree of dissipation should depend in this way on mass when the $n = -3$ spectrum of fluctuations produces potential wells that all have similar density.

5.2. Has dissipation played a role in E formation?

The answer is almost certainly "yes" since we do see ellipticals that are close to being rotationally supported. It also appears that the highest surface brightness ellipticals are the ones that are closest to being rotationally supported (Wyse and Jones, 1984).

An important related question is why ellipticals that are close to rotational support have not formed thin disks like the spirals did. This could be understood if these ellipticals formed early enough to have formed stars before the gas impacted to form a disk, a process that would have been helped if the dominant cooling mechanism at the time was Compton cooling. This would tie in with various ideas on the origin of the bulge of disk galaxies (Vedel, Hesselbjerg and Jones 1987).

It has often been remarked that the power law form of the two point galaxy-galaxy correlation function extends to very small scales, and that the internal density of elliptical galaxies is considerably higher than would be implied by extrapolating the two point function to smaller scales still. Thus if ellipticals formed through a simple clustering hierarchy, dissipation is needed to boost the density to the observed values.

5.3. When did E's form?

The old argument says that ellipticals must have formed early on in order to achieve their present degree of binding. This is predicated on the assumption that there has been little dissipation during the formation of the galaxy (which may be correct, at least for the largest ellipticals). The estimate of the formation redshift has to be based on an observed density within some specific radius like the half-mass radius.

It was perhaps Gunn (1982) who first suggested that elliptical galaxies formed so early that Compton cooling dominated the gas thermodynamics. As far as I can see, this is still a plausible scheme, and it even allows a clear cut mechanism for deciding why a given protogalactic cloud should have become an elliptical or a disk system. It does not provide any obvious explanantion as to why ellipticals should prefer denser (cluster) environments unless the amplitudes of the cluster-sized perturbations were large enough to locally influence the mean epoch of galaxy formation. That would put the formation epochs of elliptical galaxies and the richer clusters at around the same redshift ($z \sim 5-10$). There is no compelling reason to doubt that a cluster like like the Coma cluster could have formed that early, though it is not a view that accords well with the N-body simulations of galaxy clustering.

The cold dark matter simulations suggest that galaxy formation was relatively recent ($z \sim 1-2$). This comes as a consequence of normalizing the fluctuation spectrum at large scales so as to give the correct peculiar velocities on those scales. The assumption that the primordial fluctuation spectrum was of the Harrison-Zel'dovich form then determines the amplitudes on galaxy scales. Making the overall spectrum steeper would make galaxies form earlier relative to clusters.

5.4. How are dark and luminous matter segregated?

In spirals like our own Galaxy, the inner regions are not dominated by dark matter, whereas the outer regions are. This segregation was probably achieved through the dissipation in the baryonic component that was necessary to create the rotationally supported baryon dominated disk system.

In slowly rotating ellipticals there is likewise no evidence for substantial amounts of dark matter in the inner luminous regions. But the fact that they are only slowly rotating presumably indicates that the gas from which the stars formed has not dissipated much. How then was the segregation achieved?

The fallacy in this argument is the assumption that slow rotation is synonymous with lack of dissipation. In a disk galaxy the dissipation has taken place in an organized and undisturbed way, allowing the gas to collapse to a plane before forming the disk stars. Moreover, the accretion of material onto the disk subsequent to its formation must be low enough so as not to disturb it.

5.5. How does environment come in?

The variation of morphological type with environment is manifest. The threshold biassing scheme for determining galaxy morphology gives a straightforward basis for this phenomenon, but is relatively devoid of physics until there is a theory for the threshold phenomenon itself. There have been two recent attempts to put some physics into morphological type segregation by Efstathiou and Lahav (1988) and by Vedel, Hesselbjerg and Jones (1987). The ideas are based on the fact that the gas cooling and star formation timescales are density dependent. (The cooling timescale would be independent of density if galaxy formation took place so early that Compton cooling was the dominant process).

5.6. Why are smallest galaxies closest to rotational support?

In the standard cold dark matter scheme the spectrum of density fluctuations on galaxy scales is rather flat, and so galaxies on all scales condense out almost simultaneously. On such a scheme it is therefore difficult to see why there should be any systematic scale dependent effects such as may be observed in the rotation of the E's.

The systematic change of galaxy properties with mass (luminosity) may be most easily interpreted if the primordial spectrum had more power on the smallest scales than suggested by a Harrison-Zel'dovich spectrum. The smallest galaxies would then be the oldest and would be presumed to have dissipated a greater fraction of their binding energy.

It may, however, be misleading to focus on the trend with mass. There is a possibly stronger trend with surface brightness. By virtue of the fact that surface brightness and luminosity are moderately well correlated in normal ellipticals, any correlation of rotational properties with surface brightness would also show up as a correlation with luminosity. The correlation of rotational properties with surface brightness indicates that galaxies have undergone differing degrees of dissipation.

5.7. Why are voids empty?

The emptiness of the voids is quite striking. The almost total lack of galaxies in the voids is most easily understood if galaxy formation is in some way dependent on the density fluctuation amplitudes (biassed). If galaxy formation were a simple threshold phenomenon, then the presence of the void would pull fluctuations that would otherwise form galaxies below this threshold value. If we believe in addition that galactic morphology depends on the actual value of the fluctuation amplitude, this says that ellipticals in voids are extremely rare.

Recent observational surveys by Bothun et al. (1986) and Thuan et al. (1987) show that the voids do not contain any dwarf galaxies either. This is important since some scenarios for biassed galaxy formation suggest that dwarfs should be good tracers of the comic mass density (Dekel and Silk, 1986; but see also the more recent article of Silk, Wyse and Shields, 1987, for a counter argument).

5.8. What is the luminosity (mass) function?

One of the most important data is the observed luminosity function for various kinds of galaxies. From this we would hope to deduce some kind of mass function, since theories and simulations of the galaxy formation process make direct statements about masses and rather indirect statements about luminosities.

There is some uncertainty as to whether there is a truly univeral luminosity function in the sense that all samples of the universe, whether taken in rich clusters or elsewhere, yield the same functional form. Given the strong environmental segregation of galaxy types, such universality would be surprising unless the luminosity functions of the different morphological types were also the same.

It has indeed been asserted on the basis of the Harvard CfA survey, that the luminosity function is independent of both environment and morphological type. This statement is not uncontested: the luminosity functions of "normal" and "dwarf" ellipticals are quite different and the Giovanelli-Haynes survey of the Perseus-Pisces region apparently shows variations in luminosity function even among the brighter galaxies.

One of the problems lies in sorting out what are the effects of intrinsic colour variations among different types of galaxy, since they would create differences among luminosity functions determined in differing wave bands. It is well known for example that the colours of disk galaxies correlate with their absolute magnitude, whereas there is no such correlation among ellipticals.

It is probably safest to refer to luminosity function measured in some near infrared band (like the H- or K-bands) where there is little obscuration from dust and little contribution from the light of young stars. It is also necessary to restrict the discussion of luminosity function to the normal ellipticals and treat the dwarf ellipticals separately. As emphasised by Bingelli (1986), the luminosity function is manifestly not a Schechter function for these.

Attempts to derive a luminosity function from a galaxy formation theory must incorporate some assumptions about how luminous material is related to the dark material. In the numerical experiments this is done by imposing a bias on the initial conditions: only objects coming from perturbations that exceed some threshold are to be counted as luminous objects.

Theoretical treatments begin with an asumption to the effect that only the maxima of the random density field lying above some pre-assigned threshold give rise to galaxies. It is then a matter of statistics to count the numbers of "objects" defined in this way. In principle, one could choose a different threshold at each mass scale so as to get the observed luminosity function and the argument would then revolve about the motivation for choosing such a threshold. If the result could be achieved with one scale independent threshold, that might be more interesting.

5.9. What is an $\mathcal{L}^*$ galaxy?

Taken at face value, the luminosity function, and by implication the mass function, has a characteristic "knee" at a luminosity referred to as $\mathcal{L}^*$. For objects substantially brighter than $\mathcal{L}^*$, the luminosity function is more or less exponential, whereas for fainter objects it is a power law.

If we look at what Bingelli (1986) says about luminosity functions, the relevance of $\mathcal{L}^*$ becomes questionable. One might instead ask why there appears to be a mimimum luminosity for the normal ellipticals. There are few (if any) normal ellipticals (as opposed to low surface brightness dwarfs) fainter than $M_B = -17$ ($H_0 = 50$ km/sec/Mpc).

There are no such scales embedded within the cold dark matter spectrum, so these scales presumably reflect the physics of the galaxy formation process. The key idea is probably the fact that internal winds generated during the forming elliptical galaxies can drive the gas out of the galaxy (Arimoto and Yoshii, 1987). If the escape velocity is low enough this mass loss affects the structure of the galaxy, leading to a dwarf elliptical.

5.10. Why are normal ellipticals a two-parameter family?

The general trend of properties of ellipticals with luminosity (mass) is a strong one, but the deviations relative to these trends are often significant. This means that there is another parameter besides the mass that determines the properties of the final galaxy.

If perturbation amplitudes were precisely related to the underlying mass scale, then the galaxies would form a strict one dimensional sequence parametrized by the mass. The fact that there is a distribution of amplitudes at a given mass allows for some spread in properties about the mean trends. If this were the only factor, we would then identify the second parameter as being related to the binding energy of the object, and this would manifest itself observationally as the surface brightness.

Rotation may also play a role. The spread in the rotation parameter, λ, is very large and the statistical distribution of λ itself seems to be almost independent of mass (or anything else). It is hard to believe that this spread would not influence galaxy properties and so play the role of a second (or third) parameter.

The relative roles played by variations in the rotation parameter and by variations in initial amplitudes is discussed by Vedel, Hesselbjerg and Jones (1987) who indeed find that both factors can play a key role in determining the eventual galaxy morphology. The rotation parameter is not affected by biassing, so alone it cannot be responsible for environment dependent variations in morphology. The conclusion is that the boosting of the amplitude of a galaxy scale fluctuation in a proto–cluster can cause substantial changes in start formation history that manifest themselves as differing galaxy morphology.

5.11. Is 2nd parameter environment dependent?

If the second parameter is related to the spread in the tidally induced rotation parameter, lambda, then this second parameter is almost surely independent of environment. This follows as a consequence of the insensitivity of the lambda distribution to mass, spectrum or anything else.

On the other hand, we see that the environmental influences, whatever they are, can affect the relative populations of S and E galaxies. It would be somewhat surprising if the same influences did not cause observable differences between ellipticals of a given mass in different environments.

In the threshold biassing theory the transformation among morphological types is achieved by changing the background level against which the perturbations are located. This changes the relative number of peaks above the thresholds for S and E galaxy formation. The theory is so simple that it makes no statement about the whether the E's so formed would be any different from the E's formed with a different background level. Some extra information about star formation and gas dynamics has to be added. In a more realistic theory, this would come mainly from the way in which the gas is accreted into the galaxy during the galaxy building process.

5.12. Where do the dwarf Ellipticals come from?

We have concentrated on the so-called "normal" elliptical galaxies. This leaves open the question as to what is the origin of the other sequence of ellipticals (the "dwarf" ellipticals, for want of a better term). Some members of this class are quite well studied: notably M32 and the classical dwarf ellipticals like the Fornax dwarf. The dwarf ellipticals in the Virgo cluster have been studied by Bingelli et al. (1985), and the dwarfs in the Fornax cluster by Caldwell (1987) and by Caldwell and Bothun (1987).

What do we know about these objects? They are mostly found in the vicinity of larger galaxies, or at least in clusters. They have a luminosity function that increases exponentially towards fainter magnitudes with no sign of cutoff. The better studied examples show evidence of recent star formation. There is a correlation between their metallicity and their luminosity, and some members (the very small ones) appear to be very metal poor. Some of them have a retinue of globular clusters. There may be two types of "dwarf" - nucleated and non-nucleated. They appear to have exponential density profiles.

Are they the relict building blocks of larger galaxies, the pieces that never made it into larger systems? If so, why are they found in abundance in galaxy clusters? Did the galaxy building process stop when the cluster was formed leaving a lot of debris around?

Alternatively, are they lately formed galaxies? The evidence for recent star formation in objects of such low density may point to a relatively recent origin. But surely such a galaxy would have looked very different when those stars formed. Was it that last burst of star formation that removed the gas, causing the galaxy to expand?

It would seem important to know whether these objects are gravitationally bound. A large $\mathcal{M}/\mathcal{L}$ may indicate a lot of dark matter, but it could equally indicate a system that has undergone considerable free expansion.

Kormendy (1986) presents a case that the dwarf ellipticals are in fact dwarf spirals and irregulars that have had their gas stripped by ram pressure. In that case the interpretation of the luminosity function becomes quite complex: the dwarf ellipticals are part of the spectrum of galaxies, but they have faded due to the loss of gas and the associated star formation. The standard argument against this scenario is the observation by Hunter and Gallagher (1985) that many dwarf E's in fact have higher surface brightness than typical dwarf irregulars. Moreover, one wonders whether it works in an environment like the Local Group where the only source of ram pressure stripping would be passage through the disk of the Galaxy.

Yoshii and Arimoto (1987) present models based on the argument that smaller protogalaxies (incipient dwarfs) can be stripped of their gas by winds during the star formation process. The mass loss transforms the density profile. Silk, Wyse and Shields (1987) go somewhat further and argue that it is the accretion of previously expelled gas that transforms a dwarf E into a dwarf Irr. Such "mini cooling flows" occur mainly in forming groups of galaxies, explaining the tendency for the dwarf Irr galaxies to be found preferentially near brighter galaxies.

6. NUMERICAL MODELS FOR E/S FORMATION

Modelling of elliptical galaxy formation started with the classical work of Larson who attempted to combine both stellar and gas dynamic processes in an axially symmetric monolithic collapse model. Since then, and until very recently, most of the effort has gone into the dynamical (collisionless) aspects of the collapse. The major kinds of simulation are as follows:

1. Larson-type monolithic (gaseous) collapse

2. Dissipationless collapse of inhomogeneous systems

3. Fragments of cosmological N-body models

4. Collapse of gas into massive halos

5. Numerical models with gas and stars

In the inhomogeneous collisionless particle models, a major problem seems to be that of getting high enough central densities. Either the initial angular momentum has to be very low, or there is a need to invoke dissipative (gas-dynamic) processes. Since there is in any case no evidence for any dark matter in the central (dense) parts of ellipticals, these collisionless particle models may be of limited relevance to the inner regions.

The collisionless particle models of van Albada (1982) and McGlynn (1984) do show the relaxation towards a smooth power-law type of profile. However, the simulations of Quinn, Salmon and Zurek (1986) show profiles that depend in detail on the nature of the initial conditions via the spectrum of inhomogeneities with which they start their simulations. Whether any simulations have enough particles to fully model violent relaxation is uncertain.

Since dissipation has obviously played a role in the formation of spirals, and probably in the formation of the smaller ellipticals, it is important to go beyond the dissipationless collapse models.

The new models of Carlberg(1988) and Evrard(1987) combining stellar dynamics with some kind of particle based gasdynamics seem most promising, though the major concern here is the uncertainties inherent in the numerical techniques.

A major problem comes in modelling the dissipation since in simulations that tends to be dominated by the numerical scheme used to handle the gas rather than by any physical rationale. It might be added that we have no real idea what the main mechanism for dissipation would be. The simplest models would be predicated on the assumption that most of the dissipation occurs in shocks that arise during the collision of differently moving streams of gas. That would probably be inadequate to describe the formation of a disk.

7. SOME TECHNICAL PROBLEM

The current trend is to think that galaxies only formed where the local density fluctuations were somewhat above the rms value for the density field averaged over galactic scales. This "threshold biassing" is a simple ansatz that excludes most of the baryonic material in the universe from partaking in galaxy formation. So by tuning the threshold we can make the Ω_{lum} of the luminous matter substantially smaller than the Ω_b of the baryonic component of the universe. A refinement of the idea is that different thresholds lead to different morphological types. Note that the bias is applied to the initial conditions where it is believed that the random process driving the density fluctuations was Gaussian. The action of gravity aggregates material into larger scale inhomogeneities and creates a non-Gaussian random density field.

This simple averaging and thresholding hypothesis has the advantage of providing a model for the sites of galaxy and cluster formation that is amenable to some analytic calculation and simple numerical experiments to evaluate quantities that are not directly calculable. The key examples of this are the papers of Kaiser (1984), Politzer and Wise (1984), Peacock and Heavens (1985), Bardeen et al. (1986), Couchman (1987a), and Bertschinger (1987) which deal with Gaussian random fields. The extension to the statistics of thresholded non Gaussian fields has been done by Matarrese, Lucchin and Bonometto (1986) and Grinstein and Wise (1986).

Unfortunately, the interpretation of the calculations is not always straightforward. The major problems arise out of the fact that the windowed distributions do not properly reflect what is going on. There are correlations introduced by the windowing procedure that extend out to many window radii, and these could have important effects on our assessment of the statistics.

7.1. Calculating the mass function

The prediction of an observationally acceptable mass function is regarded as an important point in galaxy formation theories. Even putting aside the problems arising out of the fact that we observe light rather than mass, we encounter severe technical problems when trying to predict a mass spectrum.

It is a straightforward matter to calculate the density of peaks of the windowed power spectrum (Bardeen et al., 1986), but that does not tell how much mass the peaks will accrete. If we argue that each mass scale corresponds to a given selection of window scale, we could compute the number of peaks as a function of window scale, given a threshold that depends in a given way on window scale. But even that would not yield a proper mass function since it is likely that several objects counted at one window radius would lie inside a single object at a larger radius.

The mass function as a function of window radius and threshold has been correctly calculated by Martinez–Gonzalez and Sanz (1988). Their technique is to discuss the density distribution along straight lines penetrating the random field, and to infer from that the three dimensional properties. The results have been verified numerically by Appel (1988).

The problem of the nesting of structures at different window radii has been investigated numerically by Appel (1988). It turns out that the nesting is very important for spectra $\langle|\delta^2|\rangle \sim k^n$ with $n = 1$. So much so in fact that the mass function is severely flattened at the low mass end. The mass spectrum taking account of nesting turns out to be almost independent of n.

Introducing some scaling hypotheses between the correlation functions of all orders brings in more information on the nature of the distribution than is contained in the simple two point functions. Schaeffer (1987) and Lucchin and Matarrese (1988) have calculated mass functions under specific hypotheses concerning how the correlation functions of high order are inter–related. Such hypotheses are clearly ad hoc, but it can be argued that they are at least consistent with the large scale distribution of galaxies.

7.2. What is the physics in windows?

Looking at a distribution smoothed to galaxy scales to study galaxies, or smoothed to cluster scales to study clusters is evident, given that we are not really interested in details of the substructures of these objects. The physics comes in when deciding what determines the scale of a galaxy or a cluster. In the former case it may be something to do with cooling and heating of gas (Couchman 1987b), while in the latter it may have something to do with dynamical timescales.

Windowing and then thresholding a distribution at a particular window scale produces a set of objects having a range of scales. Objects with their maxima just above the threshold give regions that are smaller than the window radius, for example. Changing the window radius results in a different set of objects, though there will be many points common to both windowed sets. It is therefore difficult to attach any precise physical significance to individual objects that are identified by windowing on a particular scale. This is particularly true of the smaller scales in the cold dark matter spectrum where the spectrum is like "$1/f$" noise.

7.3. Maxima or excursion sets?

The idea of introducing a window to smooth the density field is to identify contiguous regions that might collapse into a single object. The regions can be identified by the location of their density maxima, or by the boundary of the surface enclosing the region, the so-called "excursion set" (Adler, 1981). For high thresholds, there is almost a one to one correspondence between maxima and the excursion sets: each excursion set generally contains one maximum. The difference comes when defining the mass associated with an above threshold region. The mass contained within the excursion set is not distributed in the same way as the mass contained within a window radius of a maximum.

7.4. Where does gravity come in?

What ultimately determines which baryons form a coherent object is the balance between the force of gravity and the pressure forces that arise when the gas is heated. The idea is of course that these coherent regions may by approximated with a suitable window function. However, the accretion process is highly complex as can be seen from the study of Ryden and Gunn (1987) who look into the accretion of gas onto a high amplitude density fluctuation.

8. OBSERVATIONAL PROGRAMS

It is evident that our recent progress in understanding elliptical galaxies is due mainly to a set of decisive observations that have enabled us to discriminate different classes of ellipticals, and to systematize their properties. The question is "where do we go from here?". The following list presents some possible directions for data collecting.

1. Rotation curves of E's to great radii

2. Rotation (expansion?) curves of low surface brightness E's

3. Chemical abundance gradients in E's

4. Dynamics of halos of our own Galaxy and M31

5. Velocities in halo and bulge components of S galaxies

6. Stellar populations in S's and E's

7. Search for environment dependent properties

8. Relationship of dwarfs to normal galaxies

9. Search for Leo-type clouds of HI

10. Luminosity functions of S and E galaxies in H band

11. Galaxy evolution at $z = 1 - 2$

12. A search for "young" galaxies

13. Galaxy sub-clustering in clusters

14. Nature and distribution of gas in clusters

15. Detection of microwave background anisotropies

16. A study of the clustering of QSO absorption line clouds

One might protest that this amounts to just about everything that one could do with galaxies. That is indeed so.

9. MAGNETIC FIELDS

The galaxy formation process was inevitably a complex one and it is quite possible that we have been missing some essential ingredient. That ingredient could well be magnetic fields, especially if we follow the ideas on superconducting cosmic strings as put forward, for example, by Ostriker et al. (1986). Cosmic scale magnetic fields are a natural consequence of this kind of scenario.

Galaxies today have substantial magnetic fields, but we do not know how they were generated, nor even the origin of the seed field. The field was almost certainly amplified and organized during the galaxy formation process. How cosmic magnetic fields could influence galaxy formation is the subject of an article by Wassermann (1978). The broader issues of comsic fields, such as generation and amplification, are reviewed by Rees (1987).

We would have to rethink our ideas if, for example, the galaxies in clusters were systematically aligned in some way (see for example, Peebles and Struble, 1986). That might indicate that galaxies formed in the collapse of large scale structure, or it might support a theory based on cosmic strings.

10. WHAT DO WE MAKE OF THIS?

Ellipticals formed rather earlier than spirals, which simply means that they came from higher amplitude density fluctuations. There was dissipation involved in the formation process. The accretion of material onto the forming elliptical was, because of the richer environment, a somewhat irregular process. In contrast, the formation of a spiral may have been a continual but relatively quiescent process, there having been no neighboring fluctuations to disturb the process.

In this sense an elliptical is a galaxy that for a variety of reasons was unable to form a disk. The accumulated material simply went into the spheroidal system rather than into an organized disk.

342

Two effects come into play. There is firstly the higher density of the proto-galactic material. This predisposes to forming systems of shorter dynamical timescale in which stars could form relatively easily. (The gas would be cooled by the Compton process if the formation era were early enough). The more efficient star formation means that gas will be converted to stars before it can have collapsed to form a thin rotationally supported disk.

There is secondly the effect of the environment. A proto-galaxy sitting in glorious isolation can accumulate gas in a systematic way, facilitating the formation of a thin disk system. On the other hand, a proto-galaxy that is part of a sub-cluster that is undergoing continual merging events is strongly influenced by its ever changing environment. The gas will dissipate as differently moving streams collide, but if the result of the collisions is substantial star formation, the stellar system will have no organised sense of rotation.

If such a hypothesis were correct, then we should observe a K- or H- band luminosity function that is independent of both environment and of whether we are looking at disk galaxies or normal ellipticals.

There remains, however, the question of the origin of the dwarf ellipticals. Are they younger by-products of the galaxy-formation process, or are they the dying remnants of unused galactic building blocks?

11. ACKNOWLEDGMENTS

I would like to thank my students Lone Appel, Liu Xiang Dong, Jens Hesselbjerg an Henrik Vedel for continual arguments on the subject of galaxy formation in general, and the elliptical - disk question in particular.

Dave Burstein, Alan Dressler and Sandy Faber gave me the benefit of their views on Elliptical galaxies during the IAU Symposium No.130 at Balatonfured, Hungary. Those discussions contributed substantially to my understanding of the subject and I am grateful for their help.

References

[1] Adler, R.J. (1981) *The Geometry of Random Fields.* (Wiley).

[2] Appel, L. (1988) in preparation.

[3] Arimoto, N. and Yoshii, Y. (1987) Astron. Astrophys., **173**, 23.

[4] Bardeen, J.M., Bond, J.R. and Efstathiou, G. (1987) Astrophys. J., **321**, 28.

[5] Bardeen, J.M., Bond, J.R., Kaiser, N. and Szalay, A.S. (1986) Astrophys. J., **304**, 15.

[6] Barnes, J. and Efstathiou, G. (1987) Astrophys. J., **319**, 575.

[7] Bertschinger, E. (1987) Astrophys. J. (Lett), **323**, L103.

[8] Bingelli, B., Sandage, A. and Tammann, G. (1985) Astron. J., **90**, 1681.

[9] Bingelli, B. (1986) in *Nearly Normal galaxies: From the Planck Time to the Present,* ed S.M. Faber, (Springer-Verlag).

[10] Binney, J., Kormendy, J., and White, S.D.M. (1982) *Morphology and Dynamics of Galaxies.* (12th. SAAS-FEE lecture course).

[11] Blumenthal, G.R., Faber, S.M., Primak, J.R. and Rees, M.J. (1985) Nature, **311**, 517.

[12] Bond, J.R. (1986) in *Nearly Normal galaxies: From the Planck Time to the Present,* ed S.M. Faber, (Springer-Verlag).

[13] Bothun, G.D., Beers, T.C., Mould, J.R. and Huchra, J.P. (1986) Astrophys. J., **308**, 510.

[14] Brandenburger, R.H. (1986) in *Nearly Normal galaxies: From the Planck Time to the Present,* ed S.M. Faber, (Springer-Verlag).

[15] Burstein, D., Davies, R.L., Dressler, A., Faber, S.M., Stone, R.P.S., Lynden-Bell, D., Terlevich, R.J. and Wegner, G. (1987) Astrophys. J. Suppl, **64**, 601.

[16] Caldwell, N. (1987) Astron. J., **94**, 1116.

[17] Caldwell, N. and Bothun, G.D. (1987) Astron. J., **94**, 1126.

[18] Carlberg, R. (1988) Astrophys. J., **324**, 664.

[19] Couchman, H.M.P (1987a) Mon. Not. R. astr. Soc., **225**, 777.

[20] Couchman, H.M.P. (1987b) Mon. Not. R. astr. Soc., **225**, 795.

[21] Couchman, H.M.P. and Rees, M.J. (1986) Mon. Not. R. astr. Soc., **221**, 53.

[22] Davies, R.L., Burstein, D., Dressler, A., Faber, S.M., Lynden-Bell, D., Terlevich, R.J. and Wegner, G. (1987) Astrophys. J. Suppl, **64**, 581.

[23] Davis, M., Efstathiou, G., Frenk, C.S. and White, S.D.M. (1985) Astrophys. J., **292**, 371.

[24] Dekel, A. and Rees, M.J. (1987) Nature, **326**, 455.

[25] Dekel, A. and Silk, J. (1986) Astrophys. J., **303**, 39.

[26] Dressler, A., Faber, S.M., Burstein, D., Davies, R.L., Lynden-Bell, D., Terlevich, R.J. and Wegner, G. (1987) Astrophys. J. (Lett), **313**, L37.

[27] Dressler, A., Lynden-Bell, D., Burstein, D., Davies, R.L., Faber, S.M., Terlevich, R.J. and Wegner, G. (1987) Astrophys. J., **313**, 42.

[28] Djorgovski, S. and Davis, M. (1987) Astrophys. J., **313**, 59.

[29] Efstathiou, G. and Lahav, O. (1988) reported at IAU Symposium No.130.

[30] Evrard, A.E. (1987) preprint.

[31] Frenk, C.S., White, S.D.M., Efstathiou, G. and Davis, M. (1985) Nature, **317**, 595.

344

[32] Grinstein, B. and Wise, M.B. (1986) Astrophys. J., **310**, 19.

[33] Gunn, J.E. (1982) in *Astrophysical Cosmology,* ed Bruck, H.A., Coyne, V.G. and Longair, M.S., (Pontifical Acad. Sciences).

[34] Hunter, D. and Gallagher, J. (1985) Astrophys. J. Suppl, **58**, 533.

[35] Jedrzejewski, R.I., Davies, R.L. and Illingworth, G.D. (1987) Astron. J., **94**, 1508.

[36] Kaiser, N. (1984) Astrophys. J. (Lett), **284**, L9.

[37] Kormendy, J. (1985) Astrophys. J., **295**, 73.

[38] Kormendy, J. (1986) in *Nearly Normal galaxies: From the Planck Time to the Present,* ed S.M. Faber, (Springer-Verlag).

[39] Lucchin, F. and Matarrese, S. (1988) preprint.

[40] Madsen, J. (1987) Astrophys. J., **316**, 497.

[41] Martinez–Gonzalez, E. and Sanz, J.L. (1988) Astron. Astrophys., , in press.

[42] Matarrese, S., Lucchin, F. and Bonometto, S.A. (1986) Astrophys. J. (Lett), **310**, L21.

[43] McGlynn, T.A. (1984) Astrophys. J., **281**, 13.

[44] Ostriker, J.P., Thomson, C. and Witten, E. (1986) Phys. Lett., **180**, 231.

[45] Peacock, J.A. and Heavens, A.F. (1985) Mon. Not. R. astr. Soc., **217**, 805.

[46] Peebles, P.J.E. and Struble, M.F. (1986) Astron. J., **91**, 471.

[47] Politzer, H.D. and Wise, M.B. (1985) Astrophys. J. (Lett), **285**, L1.

[48] Quinn, P.J., Salmon, J.K. and Zurek, W.H. (1986) Nature, **322**, 329.

[49] Rees, M.J. (1985) Mon. Not. R. astr. Soc., **213**, 75P.

[50] Rees, M.J. (1987) Q. Jl. R. astr. Soc., **28**, 197.

[51] Ryden, B.S. and Gunn, J.E. (1987) Astrophys. J., **318**, 15.

[52] Schaeffer, R. (1987) Astron. Astrophys., **180**, L5.

[53] Silk, J. (1985) Astrophys. J., **297**, 1.

[54] Silk, J., Wyse, R.F.G. and Shields, G.A. (1987) Astrophys. J. (Lett), **322**, L59.

[55] Shu, F.H (1987) Astrophys. J., **316**, 502.

[56] Thuan, T.X., Gott III, J.R. and Schneider, S.E. (1987) Astrophys. J. (Lett), **315**, L93.

[57] Tremaine, S., Henon, M., and Lynden-Bell, D. (1987) Mon. Not. R. astr. Soc., **219**, 285.

[58] van Albada, T.S. (1982) Mon. Not. R. astr. Soc., **201**, 939.

[59] Vedel, H., Hesselberg, J.C. and Jones, B.J.T. (1987) in preparation.

[60] Wassermann, I. (1978) Astrophys. J., **224**, 337.

[61] White, S.D.M., Frenk, C.S., Davis, M. and Efstathiou, G. (1987) Astrophys. J., **313**, 505.

[62] White, S.D.M. and Narayan, R. (1987) Mon. Not. R. astr. Soc., **229,** 103.

[63] Wirth, A., and Gallagher, J. (1984) Astrophys. J., **282,** 85.

[64] Wyse, R.F.G. and Jones, B.J.T. (1984) Astrophys. J., **286,** 88.

[65] Yoshii,Y. and Arimoto, N. (1987) Astron. Astrophys., **188,** 13.

SELF-SIMILAR INHOMOGENEOUS COSMOLOGICAL MODELS

David Alexander and Robin M. Green
Department of Physics and Astronomy
University of Glasgow
Glasgow G12 8QT
Scotland U.K.

ABSTRACT. We consider inhomogeneous cosmological solutions which admit a self-similar symmetry of the second kind. These solutions have a non-zero cosmical constant which is freely interpreted in terms of a vacuum state. We discuss the evolution of such solutions and the possibility of patching from a similarity symmetry of the second kind to a similarity symmetry of the first kind. This patch must proceed via a phase change to effectively switch off the cosmical constant. We hope, by this method, to progress smoothly from an early universe model to a later universe model.

1. Similarity Solutions

The metric is taken to be spherically symmetric with r a comoving coordinate:

$$ds^2 = e^{\sigma(r,t)}dt^2 - e^{w(r,t)}dr^2 - R^2(r,t)d\Omega^2$$

The cosmical constant in this model introduces a fundamental scale which prevents a simple self-similar symmetry. However, Henriksen, Emslie and Wesson (1983) hereafter HEW, demonstrated the existence of a self-symmetry of the second kind where the similarity variable is given by

$$\xi = t'/r' = (3/\Lambda)^{\frac{1}{2}}\exp[(\Lambda/3)^{\frac{1}{2}}t]/r$$

This always proves to be possible if the equation of state $p_m(\rho_m)$ is allowed to be determined as part of the solution.

Following Podurets (1964) and Misner and Sharp (1964) we introduce the gravitational mass, which in dimensionless form is

$$M(\xi) = 2m/r$$

$$m = \int_0^R 4\pi\rho R^2 dR$$

N. Kaiser and A. N. Lasenby (eds.), The Post-Recombination Universe, 347–349.
© *1988 by Kluwer Academic Publishers.*

The self-similarity allows the field equations to be reduced to four ordinary differential equations, convenient variables being M, S, P, η - the last three being the dimensionless scale factor, pressure and density respectively. The other two metric tensors e^w, e^σ can be expressed algebraically in terms of these four variables, (HEW).

The system however admits two general integrals which may be taken as

$$C^2 = e^w/S^2$$

$$\Delta = M - \eta S^3/3 = 2m/r - (8\pi/3)\rho R^3/r$$

where C characterises the solution (see below) and Δ is a measure of the inhomogeneity.

These integrals allow the solution to be defined by the single equation

$$dM/dS = [M-3C(M-\Delta)(1-M/S)^{\frac{1}{2}}]/S[1-C(1-M/S)^{\frac{1}{2}}]$$

As the definition shows, $\Delta=0$ corresponds to solutions which are homogeneous in density. These were derived in analytic form by HEW, where the authors distinguished between open and closed solutions. In an open solution, $S \to \infty$ as $\xi \to \infty$ characterised by C<1, whereas $S \to 0$ as $\xi \to \infty$ in a closed solution (C>1). In fact, all solutions, homogeneous or not are spatially closed i.e. at a fixed time the universe has a finite volume.

For the inhomogeneous solutions ($\Delta \neq 0$) a more complicated behaviour is observed, see figure 1.

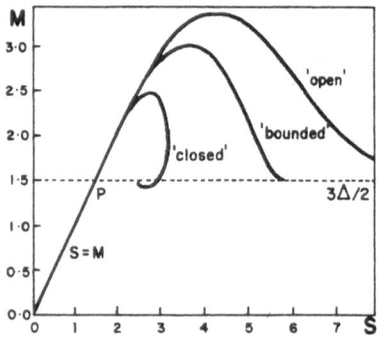

Figure 1. The figure shows the form of the inhomogeneous solutions for each of the regions discussed in the text. The dotted line is given by M=3Δ/2 and the point p is where these solutions are patched to the singular solution S=M and thus continued to the origin.

All solutions for which C<1 can be described as open in the above sense - the scale factor increases monotonically and the spacetime approaches a de-Sitter metric. By contrast, when C>1, the scale factor is bounded and tends to a finite non-zero value. In general, therefore, such solutions tend to an Einstein static model.

More importantly these solutions can be split into two distinct types: those which are well behaved with S tending to some S_{max} and those which develop shell-crossing singularities, characterised by the pressure and the 4-curvature going infinite while the 3-curvature remains finite.

We note that all ordinary inhomogeneous solutions fail to meet the dominant energy criterion $P+\eta>0$ at early times. Only the homogeneous solutions ($\Delta=0$) can be extended indefinitely back to $\xi=0$. There exists however a singular solution $M=S=\xi$ in which an equation of state can be freely imposed. This solution may be patched to an inhomogeneous solution on the hypersurface where $M=S$. The continuity conditions across the patch, namely $g_{\mu\nu}$ and $g_{\mu\nu,\lambda}$ continuous, make requirements of the inhomogeneous solution - if it is to be self-similar i.e. have to patch on the line $P+\eta=0$.

2. Patching Solutions.

We now consider the possibility of patching the solutions described above to solutions where the similarity is of the first kind. This necessarily involves some sort of phase change to wash out Λ since we would like $\Lambda=0$ in the later universe. Again across the patch we require the continuity of the metric and its derivative.

Wesson (1986) has suggested that such a patch can be obtained for the singular solution of HEW. However, in the more general case the severe restrictions imposed by symmetry and the continuity conditions make a patch possible only at a fixed time, $t=t_0$.

The lack of freedom in the choice of the equation of state caused by the symmetry being of the second kind, prevents us from obtaining a solution which is self-similar **and** satisfies the continuity conditions across the whole $t=t_0$ hypersurface.

We hope to look at a patched solution which does not have the restriction of self-similar symmetry and will indeed be a post-recombination solution. The evolution of such a model may lead to the formation of seeds for significant galaxy formation and thus give us a mechanism by which we can smoothly progress from the early universe to the late universe.

References
R. N. Henriksen, A. G. Emslie and P. S. Wesson, Phys. Rev. D 27,1219(1983).
M. A. Podurets, Astron. Zh. 41,28(1964) Sov. Astron. 8,19(1964)].
C. W. Misner and D. H. Sharp, Phys. Rev. 136, B571(1964).
P. S. Wesson, Phys. Rev. D 34, 3925(1986).
D. Alexander, R. M. Green and A. G. Emslie (1987) to be submitted to Astron. Astrophys.

COSMIC STRINGS: A PROBLEM OR A SOLUTION?

David P. Bennett
NASA/Fermilab Astrophysics Center
Astronomy and Astrophysics Center, University of Chicago

Francois R. Bouchet
Department of Astronomy, U.C. Berkeley
IGPP, Lawrence Livermore National Laboratory
On leave from Institut d'Astrophysique de Paris

ABSTRACT
We address, by means of Numerical Simulations, the most fundamental issue in
the theory of cosmic strings: the existence of a scaling solution. The resolution
of this question will determine whether cosmic strings can form the basis of an
attractive theory of galaxy formation or prove to be a cosmological disaster like
magnetic monopoles or domain walls. After a brief discussion of our numerical
technique, we present our results which, though still preliminary, offer the best
support to date of this scaling hypothesis.

INTRODUCTION
Although Cosmic strings are often discussed as possible sources of the primordial
density fluctuations, the fundamental question of how a string network evolves
in the early universe has not been resolved. Strings form when the universe is
at a GUT scale temperature, but they are expected to be relevant at much later
times because the majority of the string length at the time of formation is in the
form of infinitely long strings. These infinitely long strings cannot radiate away
into gravitational radiation, so they are expected to survive indefinitely. This is
the reason that cosmic strings might be relevant for galaxy formation, but it also
poses a potential problem for the cosmic string scenario. If the interactions between strings are neglected, it can be shown that the energy density of the infinite
strings scales roughly like non-relativistic matter. So if interactions were not important this would imply that cosmic strings (or their gravitational radiation)
would dominate the universe soon after the strings are formed. This "disaster" is
supposed to be avoided by the process by which the infinitely long strings cross
themselves and break off small loops which can decay into gravitational radiation.
This scenario has been studied analytically by Kibble (1985) and Bennett (1986).
Their work has shown that there are two possibilities: either the loop production is not sufficient to avoid a string-dominated universe, or the strings will settle
down to a scaling solution in which the number of strings crossing a given horizon
volume is fixed.

N. Kaiser and A. N. Lasenby (eds.), The Post-Recombination Universe, 351–354.
© 1988 by Kluwer Academic Publishers.

A great deal of work has already been done on the cosmic string theory of galaxy formation, *assuming* that a scaling solution does indeed exist, and there has been a great deal of speculation as to the *characteristics* of the assumed scaling solution. So far, however, all of this work is on uncertain ground because the basic details of string evolution are not understood. Albrecht and Turok (1985) have published preliminary results from their simulation two years ago. However, their program was fairly crude, and these results were criticized as inconsistent on the basis of analytical work (Bennett, 1986).

NUMERICAL TECHNIQUE

To Generate the initial conditions, we follow the general procedure introduced by Vachaspati and Vilenkin (1984). To minimize the problems caused by the degeneracies of these initial conditions, we have replaced the generated sharp corners by arcs of circles, and added small transverse velocities on the curved segments.

To evolve the generated configuration, we solve the partial differential equations which describe cosmic string motion in an expanding universe using a modified leapfrog scheme. A major difficulty is that the strings have physical discontinuities in $\dot{x}$ and x', which result from the reconnections that occur when strings cross. These "kinks" have a long lifetime, and may have important implications for the loop fragmentation pattern. To avoid the development of short wavelength instabilities near the kinks, we have introduced limited amount of numerical diffusion. This is accomplished by averaging the velocities over neighboring points *only* when instabilities start to develop. The onset of this instability is detected by comparing the values for the energy per unit proper length $\epsilon = \sqrt{x'^2/(1 - \dot{x}^2)}$ which is evolved with its own equation even though it is not independent of $\dot{x}$ and x'. When the evolved value for ϵ differs from the value calculated from $\dot{x}$ and x' by a few per cent, the velocity averaging is invoked. This procedure prevents the development of instabilities, and seems to preserve the kinks fairly well.

Most of the computer time is devoted to the detection of string crossings. To determine if two string segments crossed during the time step, we check the volume of the tetrahedron spanned by the four points on the two segments. If it changed sign during the step, the configuration is checked at the time the volume is zero, to see if a crossing did really occur (the positions of the points are extrapolated linearly between time steps). This procedure is *exact*.

Finally, when two segments have been determined to cross, we interchange partners, and average the positions and the velocities in the crossing region to minimize the amount of diffusion required in the subsequent evolution. At this stage, we also update a "genealogical tree", which records entire loop fragmentation and reconnection history. This enables us to get a posteriori a detailed picture of the string system evolution.

RESULTS

We have performed several runs in boxes of size $28\xi_0$, and one run on a $36\xi_0$ box (ξ_0 is the correlation length of the strings in the initial configuration). The simulation on the $36\xi_0$ box had $\sim 350,000$ points and 1000 strings initially. After 870 steps (21 cpu hours), the universe had expanded by a factor of 2.9 and 14000 new loops were produced. In an effort to "bracket" the scaling solution, we evolved several configurations with different initial horizon sizes, and thus different initial energy densities in long strings $\rho_L(t_0)$, to see if ρ_L scales as radiation as required for a scaling solution (long strings are defined to be of proper length $>$ ct). Fig. 2 shows the behavior of $\rho_L t^2/\mu$ as a function of time for several different runs. It is apparent that the different runs seem to be converging toward similar (constant) values with $\rho_L t^2/\mu \simeq 25$ or so. Thus, our preliminary conclusion is that we are seeing evidence for a scaling solution. It should be stressed, however, that our value for $\rho_L t^2/\mu$ is an *order of magnitude* larger than the value quoted by Albrecht and Turok, so our results cannot be considered to be consistent with theirs.

In fact, our results differ substancially from many of the generally accepted ideas about string evolution. The standard scenario for loop production holds that horizon sized "parent" loops break off the infinite string network and fragment into roughly 10 "child" loops of roughly equal sizes, but we find that in addition to the horizon sized parent loops, the infinite strings lose significant amounts of energy directly into small loops. The reason for this is presumably that the strings have a lot of short wavelength structure in the form of the kinks that are formed whenever strings cross. In addition, the large parent loops fragment much more efficiently than was previously thought, so that most (but not all) of the loops that are created are close enough to our lower cut-off on loop size (usually 10 points per loop) so that their chances of fragmenting further are significantly (or completely) suppressed.

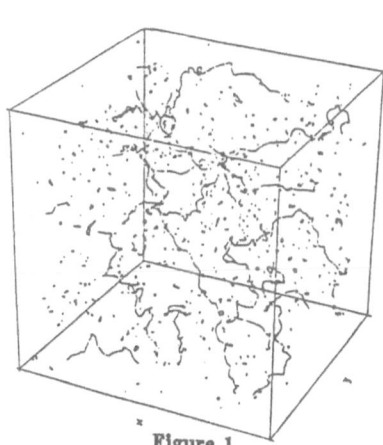

Figure 1

1/10th of a $28\xi_0$ configuration after expansion by a factor of 2.25. The cube has sides equal to $ct/2$.

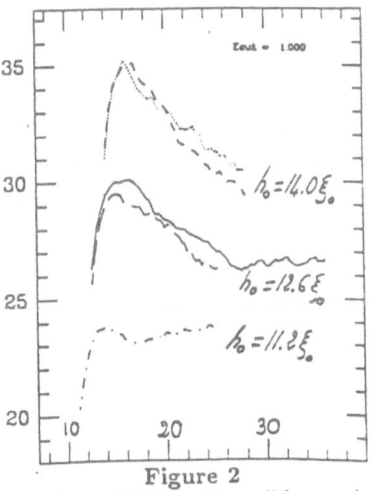

Figure 2

$\rho_L.t^2/\mu = f(h/\xi_0)$. The solid curve is for the run in a box of side $L=36\xi_0$, the others correspond to $L=28\xi_0$.

The fact that these very small loops play such an important role presents some difficult numerical problems. The most obvious problem is we cannot be sure of the correct loop distribution function because we don't know which of these small loops would fragment further if we had better resolution. Another difficulty is that our result for $\rho_L t^2/\mu$ has some dependence on our small loop cutoff. This is mainly because loop reconnection to the infinite strings is much less efficient for very small loops than for larger ones. Thus, reducing the lower cutoff increases the efficiency of loop production and decreases the scaling solution value of ρ_L/t^2. Thus, our determination of ρ_L/t^2 suffers from a systematic error due to our small loop cutoff. Only by fitting the free parameters of the analytic model of Albrecht and Turok (1985) and Bennett (1986) to these numerical results will we be able to resolve these questions.

CONCLUSION

Our results, though not definitive, lend support to the hypothesis of *existence* of a scaling solution although the string density at the scaling solution is an order of magnitude larger than that reported by Albrecht and Turok. Nevertheless more work is necessary in order to disentangle the precise characteristics of the scaling regime from numerical effects such as the cutoff on small loops.

REFERENCES

See the papers by R. Brandenberger, B. Rudak, A. Stebbins, and
S. Veeraraghavan in these proceedings.

A. Albrecht, and N. Turok, *Phys. Rev. Lett.* **54**, 1868 (1985).

D. P. Bennett, *Phys. Rev.* **D33**, 872, and **D34**, 3592 (1986).

T. W. B. Kibble, *Nucl. Phys.* **B562**, 227 (1985).

N. Turok, and P. Bhattacharjee, *Phys. Rev.* **D29**, 1557 (1984).

T. Vachaspati, and A. Vilenkin, *Phys. Rev.* **D30**, 2036 (1984).

Galaxy Formation with Neutrinos and Cosmic Strings

Robert H. Brandenberger[*]

Department of Physics
Brown University
Providence, R.I. 02912, USA

ABSTRACT : Galaxy and structure formation in a neutrino dominated universe with cosmic strings are investigated. Unlike in the usual adiabatic scenario strings survive neutrino free streaming to seed galaxies and clusters. The effective maximal Jeans mass is about $1.5 \times 10^{14} h_{50}^{-4} M_{\odot}$. Hence cluster formation is only marginally different than in the cold dark matter (CDM) and strings model. Galaxy masses are lower than with strings and CDM. The mass spectrum is flatter and the density profile about an individual loop is less steep, in better agreement with observations.

Hot dark matter particles have large thermal velocities at t_{eq} and hence cannot be gravitationally bound on small scales (free streaming). In models of formation of structure based on linear adiabatic perturbations all inhomogeneities on scales smaller than the maximal free streaming length λ_J are washed out. The mass M_J inside a ball of radius λ_J exceeds the galaxy mass. Hence in the above models galaxies can only form by fragmentation of larger-scale objects. This is a severe problem.

The **main result** of recent work[1,2] is that cosmic string loops which seed galaxies survive neutrino free streaming. By the time the loops decay the free streaming length has fallen below the galactic scale, enabling galaxies to form independently of clusters.

We have studied the dissipationless clustering[3] of neutrinos (with a mass chosen to produce $\Omega = 1$) about a seed loop. Starting from the Liouville equation for the phase space density we derived an integral equation (the Gilbert equation) for the neutrino energy density perturbation in Fourier space[1]. After numerically solving this equation for a spectrum of Fourier modes we obtain the following mass profile about a seed loop of mass M_1. $\delta M(r)$ is the neutrino mass accreted inside a ball of radius r.

$$\delta M(r) = \frac{3}{2} M_1 z_{eq} \left[1 - (1 + \frac{r}{L}) e^{-r/L} \right] \quad \text{with} \quad L \cong 8 h_{50}^{-2} \text{Mpc} \quad (1)$$

N. Kaiser and A. N. Lasenby (eds.), The Post-Recombination Universe, 355–357.

The second term in (1) represents the effect of neutrino free streaming. On scales larger than L there is essentially no growth loss compared to the cold dark matter (CDM) model. The mass M_J inside a ball of radius L is about $1.5 \times 10^{14} h_{50}^{-4} M_O$. We note that L is significantly smaller than the maximal Jeans length λ_J in an adiabatic model. For $r \ll L$ we get

$$\delta M(r) \cong \frac{3}{4} M_J z_{eq} \left[\frac{r}{L}\right]^2 \qquad (2)$$

The basic scenarios of structure formation are identical in cosmic string models with hot and cold dark matter. In both cases the largest loops at t_{eq} seed clusters, smaller ones galaxies. The radius of the seed of a cosmic structure is determined by requiring the correct number density. The number density of loops of radii R is given by a well determined distribution $n(R,t)$

$$n(R,t) = \nu R^{-5/2} t_{eq}^{1/2} t^{-2} \quad , \quad R < t_{eq}, \quad t > t_{eq}, \quad \nu \sim 10^{-2}. \quad (3)$$

There are important differences in the detailed predictions of cosmic string models with hot versus those with cold dark matter. We obtain the following results [2]:

1. Since cluster masses exceed M_J cluster formation is unchanged (hence the value for the mass per unit length μ is the same).

2. The density profile (2) leads to flat galaxy halo rotation curves. With CDM we get $v(r) \sim r^{-1/8}$.

3. The mass function of galaxies is less steep with HDM than with CDM and is in good agreement with the Schechter luminosity function for $M < 10^{14} M_O$. We get $n(M) \sim M^{-3/2}$ compared to $n(M) \sim M^{-5/2}$ for CDM, the reason being that the total nonlinear mass about small loops is proportional to R^3 and not to R as for CDM.

4. $n(M)$ asymptotically approaches $n(M) \sim M^{-1/2}$ for $M \ll M_{cu}$ where $M_{cu} \sim 10^7 M_O$. The corresponding cutoff mass for CDM is lower.

5. Galaxy masses are lower than with CDM. For objects with the mean separation of $10 h_{50}^{-1}$ Mpc we get a mass

$$M \simeq 5 . 10^{10} M_O h_{50}^6 \left[\frac{\sigma_c}{700}\right]^8 c \qquad (4)$$

where σ_c is the cluster velocity dispersion in kms^{-1} and c is a constant of order 1 which depends on ν and on the time of formation of cluster loops.

To derive the above galaxy mass it is important to take into account both baryon accretion and loop decay (which have been neglected in deriving (1) and (2)). Cosmic string loops decay by gravitational radiation. The decay time t_d of a loop of radius R is $t_d = (\gamma G\mu)^{-1}R$ with $\gamma \sim 5$. Hence small loops decay while the scale $x(R)$ (at which without loop decay $\delta\rho/\rho = 1$ today) is smaller than the free streaming length. This leads to a suppression of the neutrino perturbations on these scales which can be calculated by solving the Gilbert equation including loop decay.

However, baryon accretion between the time of recombination and the time $t_R(x)$ when $x(R)$ equals the free streaming length provides baryonic cores about which neutrinos can continue to cluster after t_d. We have calculated the growth of the perturbation $\delta = \delta\rho_B/\rho_B$ in baryons by solving the equation

$$\ddot{\delta} + \frac{4}{3t}\dot{\delta} - \frac{2\Omega_B}{3t^2}\delta = 4\pi G\delta\rho_s \qquad (5)$$

for the loop source perturbation $\delta\rho_s$ slowly decaying in time.

For a given R the suppression of the neutrino perturbations is more severe on smaller scales x, but the baryonic cores are more important. For galaxy loops the net suppression factor when combining the two effects is independent of x for x close to $x(R)$. This allows an easy determination of the galaxy mass (in deriving (4) we used $\Omega_B = 1/8$).

We conclude that a theory with hot dark matter and cosmic strings is a viable cosmological model. Similar results have also recently been obtained by Bertschinger and Watts [4].

* Also at DAMTP, University of Cambridge, Cambridge CB3 9EW, UK.

1) R. Brandenberger, N. Kaiser and N. Turok, 'Dissipationless cluster-ing of neutrino perturbations about a cosmic string loop', Phys. Rev. D , in press (1987).

2) R. Brandenberger, N. Kaiser, D. Schramm and N. Turok, 'Galaxy and structure formation with hot dark matter and cosmic strings', DAMTP preprint, July 1987.

3) J. Bond and A. Szalay, Ap. J. **274** , 443 (1984).

4) E. Bertschinger and P. Watts, MIT preprint (1987).

GAUGE INVARIANT COSMOLOGICAL PERTURBATION THEORY FOR COLLISIONLESS SCENARIOS

R.Durrer
Institut of Theoretical Physics
Schönberggasse 9
8000 Zürich
Switzerland

ABSTRACT.   Bardeen's gauge invariant cosmological perturbation theory is shortly reviewed in space-time splitting formalism. Then it is extended to the coupled Einstein-Liouville system. A gauge invariant version of the perturbation of the distribution function is derived and the perturbation of Liouville's equation is rewritten in this variable.

1.   INTRODUCTION

A gauge invariant formalism for a relativistic treatment of the perturbation theory of the coupled Einstein-Liouville equations on a Friedman universe is presented. This is useful for the treatment of dark matter in the early Universe. A numerical solution of the resulting system of linear differential equations (2,3,4,5) will be presented in the author's dissertation. The spectra obtained are comparable with those of refs. [3,4] .

2.   BARDEEN'S GAUGE INVARIANT FORMALISM

2.1 Gauge transformations

The notion of 'gauge invariance' in the context of cosmological perturbation theory means invariance under infinitesimal coordinate transformations,or changes in what is called background and what is called perturbation thereon. The transformation law is very simple:
     Be t a tensorfield , which is devided into a background contribution t_B and a small first order perturbation ϵt_1 .
     $$t = t_B + \epsilon t_1$$
An infinitesimal coordinate transformation can be given by the infinitesimal flow ϕ_ϵ of a vectorfield X . Then the transformation law of t_1 under so called gauge transformations is given by the Lie-derivative of the corresponding background variable : (See [5] .)
$$t_1 \rightarrow t_1 + L_X t_B \tag{1}$$

N. Kaiser and A. N. Lasenby (eds.), The Post-Recombination Universe, 359–362.

2.2 The perturbation variables

We will restrict ourselves to scalar perturbations with fixed wavenumber k. Of course everything can also be done for vector and tensor type perturbations. (See [1,2].)

We now adopt the conventional space-time splitting of a Friedman universe: $\mathcal{V} = \mathbf{R}_+ \times \mathcal{S}$, where in the unperturbed version $(\mathcal{S}, \gamma)$ is a 3-space of constant curvature $K = \pm 1, 0$. We are using conformal time, so the background metric is given by :

$$g_B = a^2(d\eta \otimes d\eta + \gamma_{ij} dx^i \otimes dx^j)$$

Let Y be the harmonic mode with wavenumber k . We denote the covariant derivative on $\mathcal{S}$ by $|$ and the Laplace-Beltrami operator on $\mathcal{S}$ by $\triangle$. So we have :

$$(\triangle + k^2)Y = 0 \text{ on } \mathcal{S}$$

From Y we can build the following scalartype vector and symmetric, traceless tensor:

$$Y_i \equiv -k^{-1} Y_{|i} \qquad\qquad Y_{ij} \equiv k^{-2}(Y_{|ij} + \frac{k^2}{3}\gamma_{ij}Y)$$

Then we can define the perturbations in the geometry with help of the following variables:

Lapse function:	$\alpha(\eta) = a(\eta)(1 + A(\eta)Y)$
Shift vector:	$\beta(\eta)^i = -B(\eta)Y^i$
3-metric on $\mathcal{S}$:	$\overline{g}(\eta) = a^2(\eta)[(1 + 2H_L)\gamma_{ij} + 2H_T Y_{ij}]dx^i \times dx^j$

Now one can calculate the perturbation of the scalar curvature on $\mathcal{S}$:

$$\delta \overline{R} = 4a^{-2}(k^2 - 3K)\mathcal{R} \quad , \quad \mathcal{R} = H_L + 1/3 H_T$$

The second fundamental form of the $\{\eta = \text{const.}\}$ hypersurfaces in $\mathcal{V}$ (the 'geometric slicing') , $K^{(g)}$, is expressed in a 'nearly orthonormal' frame (N^b, θ^i) which is determined by $g = -N^b \otimes N^b + \gamma_{ij}\theta^i \otimes \theta^j$, where N is the unit normal to the equal time slices of our space-time and N^b denotes the 1-form corresponding to N via $N^b(X) = g(N, X)$, for a vectorfield X .

One finds $\quad K^{(g)} = [-\frac{a'}{a^2}(1 + \theta_g Y)\gamma_{ij} - \frac{k}{a}\sigma_g Y_{ij}]\theta^i \otimes \theta^j$

with $\quad\quad\quad \theta_g = -A + (a'/a)^{-1}\frac{k}{3}B + (a'/a)^{-1}H_L'$ (perturbation of the expansion)

and $\quad\quad\quad \sigma_g = k^{-1}H_T' - B$ (the shear) .

If you now calculate the behaviour of this variables under gauge transformations you find, that the following combinations are gauge invariant:

$$\Phi \equiv \mathcal{R} - k^{-1}\frac{a'}{a}\sigma_g \qquad \Psi \equiv A - k^{-1}\frac{a'}{a}\sigma_g - k^{-1}\sigma_g'$$

We now proceed to the matter variables , which are given by the perturbed energy-momentum tensor T^μ_ν . Be u the normed timelike eigenvector of T^μ_ν and ρ its eigenvalue . We define the perturbations of this quantities as follows:

$$u = a^{-1}\begin{pmatrix} 1 - AY \\ vY^i \end{pmatrix} \qquad \rho = \rho_B(1 + \delta Y)$$

On the subspace of $T\mathcal{V}$ orthogonal to u we define π_L and π_T as the isotropic and anisotropic perturbation of the stress-tensor :

$$\Pi^i_{\ j} = p_B[(1 + \pi_L)\delta^i_{\ j} + \pi_T Y^i_{\ j}]$$

By looking at the transformation properties of the matter variables you find the following gauge invariant combinations :

$$\Pi = \pi_T \qquad \text{(anisotropic perturbation)}$$
$$\Gamma = \pi_L - \frac{c_s^2}{w}\delta \qquad \text{(amplitude of the entropy flux)}$$
$$V = v - k^{-1}H_T' \qquad \text{(velocity perturbation)}$$
$$\Delta_g = \delta + 3(1+w)\mathcal{R} \qquad \text{(one possible density perturbation variable)}$$

(Where $w = \rho_B/p_B$, $c_s^2 = p_B'/p_B' = $ (sound velocity)^2.)

The perturbed field equations are derived most easily by looking at the Gauss-Codazzi-Mainardi equations in the geometric slicing, which are the constraint equations and at the conservation equations in the matter slicing, which delivers the time evolution equation. The constraints are: ($\kappa = 8\pi G$)

$$\Phi = \frac{\kappa^2 a^2 \rho_B}{2k^2 + 3(1+w)\kappa^2 a^2 \rho_B}(\Delta_g + 3(1+w)(a/a)k^{-1}V) \qquad (2)$$

$$\Psi = \frac{\kappa^2 a^2 p_B}{k^2}\Pi - \Phi \qquad (3)$$

We are not going to write down the time evolution equations, which are identical to the Bianchi-identities. (See refs.[1,2].) They will of course also be a consequence of Liouville's equation , which is the time evolution equation for collisionless matter.

3.   THE DISTRIBUTION FUNCTION AND LIOUVILLE'S EQUATION

Now we are going to <u>expand</u> Bardeen's formalism to the distribution function on phase space. To avoid some complications we restrict ourselves to the case $K = 0$. The general relativistic phase space ,P_m, is the mass shell, a subbundle of the tangentbundle. We would like a division of the actual distribution function of the kind

$$f(x,p) = f_B(\eta,p) + \epsilon f_1(\eta,p)Y(x)$$

But of course the background distribution function is defined only on the unperturbed mass shell P_{Bm} . For the above division we therefore need an $\tilde{f}_B : P_m \to \mathbf{R}$.

A natural way to find such an $\tilde{f}_B$ is via an isomorphism $\iota : P_{Bm} \to P_m$.

It is clear that the splitting of f into background contribution and perturbation depends on ι . We choose the following ι: Be $(N = e_0, e_i)$ a tetrade of $\mathcal{V}$. For the unperturbed geometry we have $N_B = a^{-1}\partial_\eta$, $e_{Bi} = a^{-1}\partial_i$. We define

$$\iota : P_{Bm} \to P_m : (x, p^\mu e_{B\mu}) \mapsto (x, p^\mu e_\mu)$$

The distribution function is then splitted in the following way:

$$f(x,p) = (f_B \circ \iota^{-1})(\eta, p) + \epsilon f_1(\eta, p)Y(\bar{x})$$

If one now explores the gauge transformation properties of f_1 with respect to a vectorfield X, one has to take into account that in general the transformed tetrade, $\phi_\epsilon^*(e_\mu)$ is not the tetrade adapted to the <u>new</u> space-time splitting . Therefore the transformation law is not simply given by the Lie-derivative with respect to the tangent-vectorfield, TX on $T\mathcal{V}$ but with respect to the following vectorfield:

$$(TX)_{\parallel} \equiv X^{\mu}e_{\mu} + (p^{i}e_{i}(X^{0}) - \frac{a'}{a}X^{0})\partial_{p^{0}} + (p^{0}X_{0}{}^{,i} - \frac{a'}{a}X^{0})\partial_{p^{i}}$$

One easily shows that $(TX)_{\parallel}$ is tangent to P_{Bm} as it has to be for $L_{(TX)_{\parallel}}f_{B}$ to be defined. If one explicitely calculates the above transformation law for scalar typ gauge transformations,

$$X = tY\partial_{\eta} + lY^{i}\partial_{i}$$

one finds:

$$f_{1} \mapsto f_{1} - \frac{\partial f_{B}}{\partial p}\{p\frac{a'}{a}t + ip^{0}\mu kt\}$$

for $\mathbf{p} = p^{\mu}e_{\mu}$, $p = (\sum_{i}(p^{i})^{2})^{1/2}$, $\mu = \frac{\sum_{i}(p^{i}k^{i})}{pk}$. And therefore

$$\mathcal{F} = f_{1} - \frac{\partial f_{B}}{\partial p}\{p\mathcal{R} + ip^{0}\mu\sigma_{g}\} \quad \text{is gauge invariant.}$$

It is then easy to find the momentum integrated matter variables from $\mathcal{F}$, $(q = p^{0})$

$$
\begin{aligned}
\Delta_{g} &= \frac{1}{\rho_{B}}\int p^{2}q\mathcal{F}d\Omega dp \\
V &= \frac{i}{\rho_{B}+p_{B}}\int p^{3}\mu\mathcal{F}d\Omega dp \\
\Gamma &= \frac{1}{p_{B}}\int(p^{4}/3q - c_{s}^{2}p^{2}q)\mathcal{F}d\Omega dp \\
\Pi &= \frac{1}{2p_{B}}\int(p^{4}/q)(1 - 3\mu^{2})\mathcal{F}d\Omega dp
\end{aligned}
\tag{4}
$$

The perturbation of Liouville's equation in gauge invariant variables is then:

$$\partial_{\eta}\mathcal{F} + ik\mu\frac{p}{q}\mathcal{F} = ik\mu\frac{\partial f_{B}}{\partial p}[q\Psi - \frac{p^{2}}{q}\Phi]
\tag{5}$$

This Liouville equation together with 2,3 and 4 form a closed system of linear differential equation which we solved numerically for hot and cold dark matter .

After I had given this talk in Cambridge one of the participants, M.Panek, called my attention to similar work on this subject by M.Kasai and K.Tomita ([6]).

ACKNOWLEDGEMENT: I'm gratefull for many helpfull discussons I had with my supervisor Prof. N.Straumann.
This work was partially supported by the Swiss National Sience Foundation.

Bibliography

[1] J.M. Bardeen,*Phys. rev.* bf D22 , 1882 (1980)

[2] H. Kodama and M. Sasaki , *Prog of Theor. Phys. suppl.* **78** , (1984)

[3] J.R.Bond and A.S.Szalay, *Ap.J.* **274** , 443 (1983)

[4] P.J.E.Peebles,*Ap.J.* **259** , 442 (1982)

[5] N.Straumann,*General Relativity and Relativistic Astrophysics*,Springer 1985

[6] M.Kasai and K.Tomita *Phys.Rev.* **D33** , 1576 (1986)

EXTENDING N-BODY METHODS WITH SPH

August E. Evrard
Institute of Astronomy, Madingley Road, Cambridge CB3 0HA, UK

A numerical algorithm capable of following the self-consistent, three-dimensional hydrodynamic evolution of gas in a multi-component universe is briefly described. The method is the merged product of a particle-particle partice-mesh (P3M) N-body scheme with a particle-based hydrodynamic approach known as smoothed particle hydrodynamics (SPH).

I. Introduction

An ingredient currently lacking in simulations of cosmological structure is the ability to follow the hydrodynamics and thermal history of a gaseous component. Solving the general three dimensional problem poses a formidable challenge to any hydrodynamic method, as initial mass distributions sampled from Gaussian random fields will often lead to a high degree of substructure present at all stages of the evolution. A particle-based method known as SPH (Gingold and Monaghan (1977), Monaghan (1985)) appears best suited for cosmological application. Here, we briefly present a method designed to follow the hydrodynamics of a gas in a multi-component system consisting of gas and one or more collisionless components (e.g. dark matter).

Successful application of the code to problems of cosmological structure will require the ability to shock-heat initially cold gas and the ability for gas to dissipate through radiative cooling. In hierarchical clustering schemes, cold primordial gas will be shock-heated on collapse to an appropriate virial temperature. Radiative cooling will be important on galactic and smaller scales, while for groups and clusters of galaxies, the cooling timescales are long. Through the knowledge of local densities and temperatures, the algorithm described here is capable of modelling both these processes in a three dimensional and self-consistent fashion.

II. Method

The SPH method employs a set of particles to represent the distribution of gas. The particles are labelled with a local temperature or entropy, along with their mass and kinematic data. Local densities are determined by direct interpolation from the nearby particle distribution. For a set of particles each of mass m, the

N. Kaiser and A. N. Lasenby (eds.), The Post-Recombination Universe, 363–366.

density at particle i is

$$\rho_i = m \sum_j W(r_{ij}, h_i) \qquad (1)$$

The smoothing kernel $W(r_{ij}, h_i)$ is a compact function of pair separation r_{ij}, having an effective range given by the local smoothing scale h_i. This smoothing length determines the resolution of the code. It is allowed to vary both spatially (from particle to particle) and temporally to increase the accessible dynamic range. A typical choice of kernel is a Gaussian $W(u, h) = (\sqrt{\pi} h)^{-3} \exp(-u^2/h^2)$.

Given the local density ρ_i and the thermal energy ϵ_i (specified as input), an ideal gas equation of state $P_i = (\gamma - 1)\epsilon_i \rho_i$ specifies the pressure at the location of particle i. The gas forces acting on particle i can then be written as a sum of antisymmetric interactions with its neighbors j

$$\vec{F}_i^{gas} \equiv -\left(\frac{\vec{\nabla} P}{\rho}\right)_i = -m \sum_j \left(\frac{P_i}{\rho_i^2} + \frac{P_j}{\rho_j^2}\right) \vec{\nabla} W(r_{ij}, h_{ij}) \qquad (2)$$

where $h_{ij} = 0.5(h_i + h_j)$ is used to guarantee conservation of linear and angular momentum. The adiabatic change in internal energy is calculated from the energy exchange of each pairwise interaction. For high mach number collisions, thermal pressure is not sufficient to prevent free streaming of the gas particles. In real gases, molecular viscosity is effective in converting kinetic to thermal energy within roughly a molecular mean free path. To achieve the same result in the hydrodynamic simulation, an artificial viscosity is introduced. The purpose of the viscosity is to increase the pressure in regions of strongly converging flow from ρc^2 to ρv^2. Several forms of SPH viscosity have been considered in the literature, the forms I've adopted follow those used by Loewenstein and Mathews (1986) and Lattanzio et al. (1986).

The scheme has been implemented within the framework of the P3M N-body code described by Efstathiou et al. (1985). This code is optimal for large-scale structure simulations – long-range gravity is computed by FFT's and short-range gravity by a local direct summation. The local gas interactions fit nicely into the routine handling the short-range gravity, providing a natural mode of coupling between the two methods.

III. Example: Spherical Rich Cluster Evolution

As a relevant example, the turnaround and collapse of a spherical perturbation in an $\Omega = 1$ cosmology consisting of 90% dark matter and 10% gas is studied. The P3MSPH results are compared to those of the 1-d spherically symmetric code, modified to include collisionless shells of dark matter. The initial perturbation consists of a single radial cosine wave. Given a uniform sphere of radius x_t, one perturbs radii by $x_{new} = x_{old}(1 - \delta(x_{old})/3)$ and sets peculiar velocities according to growing mode linear theory $v = -(2/3)H x_{old}\delta(x_{old})$ where $\delta(x) = 0.5[1 + cos(\pi x/x_t)]$. The model is scaled to a size appropriate to a rich cluster of galaxies. It is begun at a redshift $z = 4$, has total mass $M_t = 3 \times 10^{15} M_\odot$

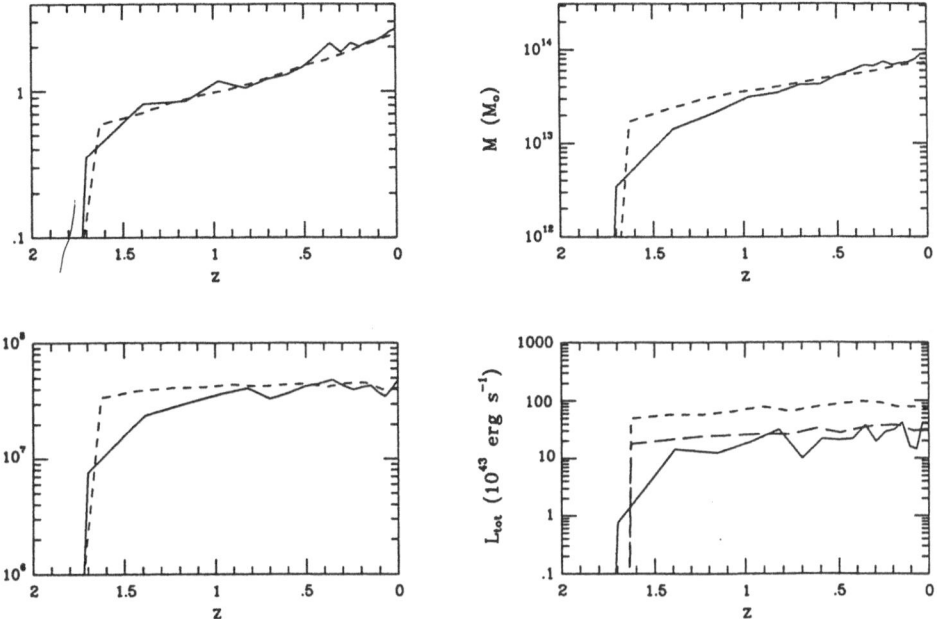

Figure 1 : Evolution of rich cluster core properties (see text for details).

and a comoving radius of $x_t \simeq 22$ Mpc for $H_o = 50$ km s^{-1} Mpc^{-1}. The initial gas temperature is $T = 10^5$ K.

A comparison of the properties of the cluster core is given in Figure 1. The core at some redshift z is defined by the radius at which the mean interior density equals 170 times the background density at that redshift. The core radius, interior gas mass, mean temperature, and total x-ray luminosity are shown as a function of redshift, with the P3MSPH results given by solid lines, the 1-d as dashed lines. The agreement is very good for all quantities save the luminosity, which is low in P3MSPH by a factor of ~ 3. The discrepancy results from the higher resolution of the 1-d calculation – since $L_x \propto \rho^2 T^{0.5}$, a significant contribution to the luminosity can be made by a relatively small amount of mass at high density. A high central density is achieved because the lack of angular momentum barriers in this problem leads to a nearly singular power-law density profile. When the resolution of the 1-d codes is reduced to that of the 3-d code (the inner ~ 200 kpc emission removed), the agreement is much improved, as shown by the long dashed line in Figure 1. The resolution limits of the 3-d P3MSPH should not pose a problem for determining x-ray luminosities in realistic collapses, where core-halo density structures typical of violent relaxation are expected, as long as the resolution is sufficient to define the core density value.

IV. Final Remarks

A more detailed explanation of the method can be found in Evrard (1988). Several applications are currently under way. The fate of cluster gas in a neutrino dominated universe has recently been investigated by Evrard and Davis (1988), who found that shock-heating in pancakes was ineffective in preventing the bulk of the gas from falling into the giant clusters forming at pancake intersections. The evolution and statistical properties of x-ray clusters in the cold dark matter model is also currently under study. Careful implementaton of cooling within the code should open up modelling of the galaxy formation era in the near future.

References

Efstathiou, G., Davis, M., Frenk, C.S. and White, S.D.M. 1985, *Ap. J. Suppl.*, **57**, 241.
Evrard, A.E. 1988, *M.N.R.A.S.*, in press.
Evrard, A.E. and Davis, M. 1988, Nature, in press.
Gingold, R.A. and Monaghan, J.J. 1977, *M.N.R.A.S.*, **181**, 375.
Lattanzio, J.C., Monaghan, J.J., Pongracic, H., and Schwarz, M.P. 1986, *SIAM J. Sci. Stat. Comp.*, **7**, 591.
Loewenstein, M. and Mathews, W.G. 1986, *J. Comp. Phys.*, **62**, 414.
Monaghan, J.J. 1985, *Comp. Phys. Rept.*, **3**, 51.

GALAXY FORMATION IN THE COLD DARK MATTER COSMOGONY

Carlos S. Frenk
Physics Department
University of Durham
England

ABSTRACT. The formation of dark galactic halos in the cold dark matter cosmogony can be studied using high resolution N-body simulations. Galactic halos are predicted to form at relatively recent epochs with properties and abundance similar to those inferred for the halos of real galaxies. Massive halos tend to form preferentially in high density regions and as result the galaxies that form within them are more strongly clustered than the underlying mass. This natural bias may be strong enough to reconcile the observed clustering of galaxies with the assumption that $\Omega = 1$.

1. INTRODUCTION

An understanding of galaxy formation requires viewing this process as an integral part of the growth of large scale structure in the Universe. In recent years several attempts to build such a comprehensive description have been made. The most thoroughly investigated and in many ways the most successful of the new cosmogonies is the cold dark matter (CDM) theory. In its standard form, it assumes a flat universe dominated by weakly interacting massive particles and primordial fluctuations with the scale-invariant form predicted by inflation (cf Blumenthal et. al. 1984). The resulting mass distribution is specified by only 2 "free" parameters, the initial amplitude of the fluctuation spectrum and a linear scale determined by the Hubble constant. In a universe with $\Omega = 1$, galaxies cannot be fair tracers of the mass; it has been assumed that in the CDM model galaxies form only near high peaks of the linear density field (Bardeen et. al. 1986, Davis et. al. 1985). The resulting galaxy distribution is specified by two further parameters which can be thought of as the abundance of galaxies and the relative strength of the galaxy and mass auto-correlation functions. (Bardeen 1986, Kaiser 1986). All four parameters can be simultaneously fixed by matching the model predictions to observed galaxy correlations. However, it will be argued below that in the CDM cosmogony galaxies are naturally biased relative to the mass and that the 2 parameters introduced by the high peak model are in fact redundant.

The CDM model reproduces the morphology of the nearby galaxy distribution including the presence of large filaments and voids and its quantitative correlation properties. It predicts the observed luminosity function, mass and mass-to-light ratio of Abell clusters and is consistent with our inferred infall velocity towards Virgo and with our observed motion through the microwave background (Davis et. al. 1985, White et. al. 1987a). The predicted level of fluctuations in the microwave background are well below current upper limits (Bond and Efstathiou 1984, Vittorio and Silk 1984). Significant discrepancies with present data are found only on the largest scales where the model may not produce sufficient superclustering (Bahcall and Soneira 1983). On galactic scales, a preliminary study indicated that the model produces galactic halos with about the observed

N. Kaiser and A. N. Lasenby (eds.), The Post-Recombination Universe, 367–370.
© 1988 by Kluwer Academic Publishers.

abundance and potential depth and that these halos have flat rotation curves at radial distances beyond a few tens of kiloparsecs (Frenk *et. al.* 1985, Frenk 1987).

Many of the predictions of the CDM model have been obtained from numerical simulations using high-resolution N-body techniques (Efstathiou *et. al.* 1985). These can follow accurately the late non-linear phases of structure growth. In this paper I will discuss new N-body results concerning galaxy formation in the CDM cosmogony, obtained in collaboration with Marc Davis, George Efstathiou and Simon White. To study the formation of galactic halos, we followed the evolution of 9 cubic patches of a CDM universe, of initial physical size 2 Mpc and varying initial density, from a redshift of 6 to the present day. (Here and below $H_0 = 50$ km/s/Mpc, the value required by our standard CDM model.) The simulations have 32768 particles of mass $\sim 6 \times 10^9 M_\odot$ and a spatial resolution of ~ 2 kpc at the start and ~ 14 kpc at the end. To study the clustering of galaxies we carried out 3 simulations of larger cubes, 50 Mpc on a side, using 262144 particles of mass $3.5 \times 10^{10} M_\odot$ and a force resolution of $50/(1+z)$ kpc. These simulations will be described in detail in two forthcoming papers (Frenk *et. al.* 1987 and White *et. al.* 1987b).

2. THE FORMATION OF DARK HALOS

Galaxy formation in the CDM cosmogony is a recent and protracted process which continues until the present day. Dark matter halos of galactic size are assembled in abundance only after $z \sim 3$. Prior to $z \sim 1$, the evolution is dominated by mergers, often involving two or more lumps of comparable mass. By $z \sim 1$ the merger rate has dropped significantly and by $z \sim 0.5$ most halos have acquired most of their final mass. These evolve to their final state in a fairly quiescent fashion, others accrete small satellites and about 30% undergo major merger events before the present. Thus the formation of dark halos involves susbtantial violent relaxation at relatively recent times. Mergers are extremely efficient at destroying substructure and even halos with the characteristic potential well depth of groups and clusters are very smooth at the present. A simple prescription for modelling the locations of galaxies within these halos suggests that the galaxies may often be able to survive the merging of their halos.

The most important properties of dark matter halos at $z = 0$ are:

• *Rotation curves.* The potential wells of dark halos can be characterised by a "rotation curve", $V(r) = [GM(r)/r]^{1/2}$, where $M(r)$ is the mass within a sphere of radius r centred on the halo. For most halos the rotation curves turn out to be flat at radii larger than the resolution limit of the calculations (*cf* Fig. 1 of Frenk *et. al.* 1985 and Fig. 3 of Frenk 1987); they resemble the measured rotation curves in the outer parts of spiral galaxies (Rubin *et. al.* 1985). Only for the most massive halos do we find rotation curves that rise significantly at $50 - 100$ kpc. These are the objects which have undergone most merging, and in some cases their rotation curves reach values well above those typical of spiral galaxies. We conjecture that such objects correspond to the halos of ellipticals or of small groups of galaxies. The generation of flat rotation curves seems to be closely related to the shape of the CDM fluctuation spectrum (Quinn *et. al.* 1986, Davis *et. al.* 1987).

• *Angular momentum.* Most halos tend to be slowly rotating with typical values of $v/\sigma \sim 0.1$, where v is the bulk rotation speed and σ the *rms* velocity dispersion. The low rotation speeds are partly due to the outward transfer of angular momentum which occurs as substructure is erased. This purely gravitational process can be very efficient at spinning down the inner regions of dense clumps (Frenk *et. al.* 1985, Barnes and Efstathiou 1986).

• *Shapes.* Dark halos are generally triaxial with, perhaps, a slight preference for near prolate configurations. The axial ratios can be quite extreme: in a few systems the ratio of the two longest axes approaches 3, and a ratio of 2 is quite common. There is a weak tendency for the centres of halos to be rounder than their outer regions. There is no correlation of shape with rotation speed.

• *Abundance.* The potential well of a dark halo can be described by a characteristic velocity V_c which we define as the value of the rotation curve in the region where it is nearly independent of radius. The predicted abundance of halos can then be compared with the observed abundance of galaxies in the following way. With each halo of characterisitic velocity V_c we associate a luminosity L by equating the cumulative distributions of V_c for the halos and of L for the galaxies. If the predicted abundance of halos is correct, then the inferred $L - V_c$ relation should agree both in shape and amplitude with the observed relations, Tully-Fisher for spirals and Faber-Jackson for ellipticals. The result of this comparison -in which there are *no* adjustable parameters- is quite remarkable: the model predictions lie in between the two observed relations (*cf* Fig. 12 of Frenk *et. al.* 1987).

3. NATURAL BIAS

The growth rate of gravitational fluctuations depends on density. Clumps in a high density region collapse earlier and accrete faster than similar objects in a low density region. As a result the typical mass and velocity dispersion of clumps is greater in protoclusters than in protovoids. Clumps are more clustered than the underlying mass distribution and this bias increases with potential well depth. To assess the importance of this effect on the *galaxy* distribution we adopted simple algorithms to model galaxy mergers and to avoid multiple galaxy formation within the same halo. The results are not especially sensitive to these procedures which are described in detail elsewhere (White *et. al.* 1987b).

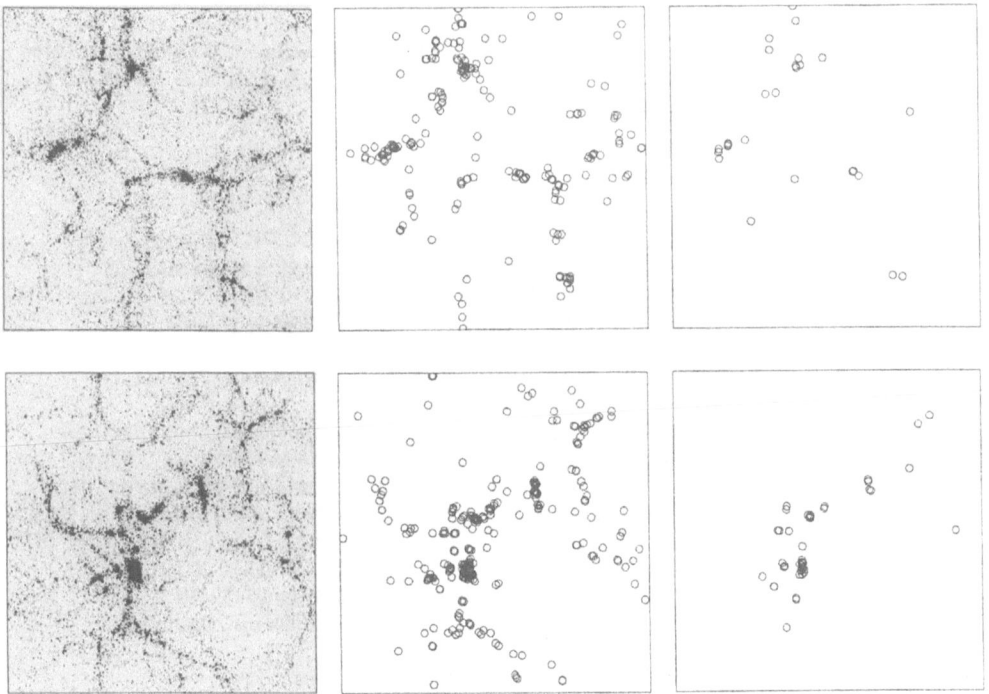

Figure 1. Distribution of mass and "galaxies" in two separate regions of a simulated cube of side 50 Mpc. Each row shows projections of a slice of width 1/8 of the cube. The left-hand column corresponds to the mass, and the middle and right-hand columns to "galaxies" with $V_c > 100$ and 200 km/s respectively.

Figure 1 shows the distributions of mass, of "galaxies" with $V_c > 100$ km/s and "galaxies" with $V_c > 200$ km/s in one of our 50 Mpc box simulations. The nature of the bias is quite clear in this picture. The same structures are delineated by each component, but the contrast is enhanced in the galaxy distributions. In regions devoid of bright galaxies, the mass density is also low and, as a result, these regions tend to be empty of fainter objects or, at most, to contain very few of them. The spatial auto-correlation functions of the galaxies have a higher amplitude than that of the mass. The mean correlation enhancement is 1.8 for "galaxies" with $V_c > 100$ km/s and 5 for "galaxies" with $V_c > 250$ km/s. For the latter the correlation function agrees well with that of observed bright galaxies. Thus, "natural biasing" appears to reconcile the dynamics of galaxy clustering with a flat universe.

4. CONCLUSIONS

It is clear from the models discussed here that the phase of galaxy formation in the CDM cosmogony may well be observable. Various indirect lines of evidence already seem to indicate a recent epoch of galaxy formation. However, both the observational data and the predictions of the model need to be elaborated further. Another aspect which is amenable to observational testing is the mechanism of natural biasing. This predicts that the strength of galaxy clustering should depend on luminosity. A sample of widely different intrinsic luminosities would be required for this test.

Acknowledgements. I thank Marc Davis, George Efstathiou and Simon White for permission to present our results prior to joint publication. I acknowledge a travel grant from NATO and grants from the SERC and the Nuffield Foundation.

REFERENCES

Bahcall, N. and Soneira, R., 1983. *Astrophys. J.*, **270**,20.

Bardeen, J.M., 1986. In *Inner Space/Outer Space*, (ed. E.W.Kolb, M.S.Turner, D.Lindley, K.Olive and D.Seckel, University of Chicago Press), 212.

Bardeen, J.M., Bond, J.R., Kaiser, N. and Szalay, A.S., 1986. *Astrophys. J.*, **304**,15.

Barnes, J. and Efstathiou, G. 1986. *Preprint.*

Blumenthal, G.R., Faber, S.M., Primack, J.R. and Rees, M.J., 1984.*Nature*, **311**,517.

Bond, J.R. and Efstathiou, G. 1984. *Astrophys. J.*, **285**,L45.

Davis, M., Efstathiou, G., Frenk, C.S. and White, S.D.M., 1985.*Astrophys. J.*, **292**, 371.

Davis, M., Frenk, C.S., White, S.D.M. and Efstathiou, G., 1987. *Preprint.*

Efstathiou, G., Davis, M. Frenk, C.S. and White, S.D.M., 1985.*Astrophys. J. Supp.*, **57**, 241.

Frenk, C.S., 1987. In *Nearly Normal Galaxies: from the Planck Time to the Present,* (ed S.M. Faber, Springer-Verlag), 421.

Frenk, C.S., White, S.D.M., Davis, M. and Efstathiou, G., 1987. *Astrophys. J.*,in press.

Frenk, C.S., White, S.D.M., Efstathiou, G. and Davis, M., 1985.*Nature*,**317**, 595.

Kaiser, N., 1986. In *Inner Space/Outer Space*, (ed. E.W.Kolb, M.S.Turner, D.Lindley, K.Olive and D.Seckel, University of Chicago Press), 258.

Quinn, P.J., Salmon, J.K. and Zurek, W.H., 1986. *Nature*, **322**, 392.

Rubin, V.C., Burstein, D., Ford, W.K. and Thonnard,N., 1985. *Astrophys. J.*, **289**, 81.

Vittorio, N. and Silk, J. 1984. *Astrophys. J.*, **285**, L39.

White, S.D.M., Frenk, C.S., Davis, M. and Efstathiou, G., 1987a.*Astrophys. J.*, **313**, 505.

White, S.D.M., Davis, M., Efstathiou, G. and Frenk, C.S. 1987b. *Nature*, in press.

The Growth of Angular Momentum Near Density Peaks

A.F. Heavens
Department of Astronomy, University of Edinburgh, Blackford Hill,
Edinburgh, EH9 3HJ, U.K.

We present the results of calculations of the tidal torques acting on the material in the neighbourhood of local density maxima in the linear regime, using the techniques pioneered by Doroshkevich (1970) and extended by Peacock & Heavens (1985) and Bardeen et al. (1986). The principal assumption involved is that the density field at early times is Gaussian noise. We are interested in the effect of the overdensity (or "height") of the peak on the angular momentum of the collapsing object. It has been argued (Blumenthal et al. 1984, Hoffman 1986) that the height of a peak is *anticorrelated with its angular momentum*, on the basis that very overdense peaks collapse early, so there is not much time for the tidal torques to act. However, this argument ignores any systematic dependence of the *magnitude* of the tidal torques on the height of the peak. This paper demonstrates that there is such a systematic effect; high peaks do experience greater tidal torques, counteracting somewhat the shorter time during which the torques effectively act. For realistic power spectra, the two effects very nearly cancel. We do, however, predict an *anticorrelation of* $J/M^{5/3}$, a quantity which is related to the *spin parameter*, λ. Unfortunately, the distributions of angular momentum are very broad, and could easily mask such an effect.

Method

We expand the density and velocity fields in Taylor expansions abound the local maximum in the density, and calculate the angular momentum of the material within constant overdensity surfaces. This amounts to calculating the tidal torques on the material. We find that high peaks experience somewhat stronger torques, on average, than the low peaks. The stronger torques on high peaks are counteracted by the fact that high peaks collapse earlier than low peaks.

If we assume that angular momentum growth effectively ceases when the overdensity $\delta\rho/\rho$ reaches a certain value (of order unity), then the time for the torques to act, will be roughly proportional to $\nu^{-3/2}$ (where ν is the overdensity of the peak in units of the r.m.s.), provided collapse occurs when the density parameter is not much less than unity. The distributions of final acquired angular momentum are shown in Fig 1. The balance between higher torques and shorter collapse times depends on the power spectrum. The adiabatic cold dark matter spectrum, when smoothed on galactic scales, and a spectrum characteristic of adiabatic neutrino fluctuations both lead to a rough cancellation of the two effects.

N. Kaiser and A. N. Lasenby (eds.), The Post-Recombination Universe, 371–373.
© 1988 by Kluwer Academic Publishers.

We have also considered the distribution of $j_{5/3} \propto J/M^{5/3}$, and find that this should be rather insensitive to the smoothing length imposed. The universality of the $j_{5/3}$ distribution then means that a correlation $J \propto M^{5/3}$ is predicted in this model. A very similar correlation is indeed seen in the simulations of Barnes & Efstathiou (1987).

The expected values of $J/M^{5/3}$ are quite useful, as they depend only weakly on H_0 and Ω_0. The main factor determining the angular momentum of a system is the overall amplitude of the fluctuation spectrum, characterised by the redshift at which a one-sigma perturbation collapses, z_1. With $H_0 \equiv 100$ h kms^{-1}Mpc^{-1} and $M \equiv 10^{12}$ M_{12} $M_\odot$, we have, for $\Omega_0 = 1$

$$J = 1.5 \times 10^{70} \; h^{-1/3} \; (1 + z_1)^{-0.5} \; j_{5/3} \; M_{12}^{5/3} \; \text{kgm}^2\text{s}^{-1}$$

Since the mass of a peak does generally depend on height (Peacock & Heavens 1985), there is a systematic shift of the mean value of $j_{5/3}$ to lower values for higher peaks. This may be some use as a discriminant of peak height. It does mean, however, that the range of expected values is very large. Characteristic values are in the range 10^{-2} to 10^{-1}. Hence for z_1 not very much larger than unity, we may expect angular momenta in the range 10^{68} – 10^{69} $M_{12}^{5/3}$ kgm^{2}s^{-1}.

The *spin parameter* λ is approximately $6 \; j_{5/3} \; \nu^{1/2}$, so a modest anticorrelation of λ with height *is* predicted. In this sense, Blumenthal et al. and Hoffman are correct in their conclusions. Barnes & Efstathiou only sample effectively a small range in ν, so the width of the distributions can easily mask the anticorrelation (a systematic shift of only a factor two is predicted for the range they cover).

Discussion

For power spectra with large fluctuations on scales much larger than the small-scale smoothing length, peaks above a certain threshold are strongly clustered, as first pointed out by Kaiser (1984) and Politzer & Wise (1984). It is not surprising, therefore, that for such power spectra, the high peaks experience greater tidal forces. For steep power spectra, high peaks are not much more strongly clustered than the low peaks, so the tidal torques vary much less between peaks of different height, and the collapse time is the principal determinant of final angular momentum. Most currently fashionable power spectra occur near the transition between torque-dominated and collapse-time-dominated behaviour.

Our analysis allows us to refine the argument of Blumenthal et al. (1984) and Hoffman (1986), which attempts to explain morphological trends with environment on the basis of angular momentum. They argued that high peaks, which, at least for some power spectra, are strongly clustered (see above) should acquire less angular momentum and become elliptical galaxies. This argument, although including the collapse time effect, ignores the systematic variations in tidal torques, and is only valid for very steep power spectra. However, the analysis here suggests that if the angular momentum is weighted by a power of the mass, it may act as a discriminant of peak height. The measure $J/M^{5/3}$ is useful in this regard, as its predicted values are very insensitive to cosmological parameters and the smoothing length, and it is related to the spin parameter. There is no anticorrelation of height with *angular momentum*, but there is a modest anticorrelation with *spin parameter*, for all power spectra. However, given the width of the distributions, one should probably look elsewhere for an explanation of the preponderance of ellipticals in clusters.

Bardeen, J.M., Bond, J.R., Kaiser, N. & Szalay, A.S., 1986. *Astrophys. J.*, **304**, 15.
Barnes, J. & Efstathiou, G., 1987. *Astrophys. J.*, **319**, 575.
Blumenthal, G.R., Faber, S.M., Primack, J.R. & Rees, M.J., 1984. *Nature*, **311**, 517.
Doroshkevich, A.G., 1970. *Astrofizika*, **6**, 581.
Hoffman, Y., 1986. *Astrophys. J.*, **301**, 65.
Kaiser, N., 1984. *Astrophys. J.*, **284**, L9.
Peacock, J.A. & Heavens, A.F., 1985. *Mon. Not. R. astr. Soc.*, **217**, 805.
Politzer, H.D. & Wise, M.B., 1984. *Astrophys. J.*, **285**, L1.

Fig. 1. The final angular momentum acquired by peaks of different height, taking into account the shorter collapse times of higher peaks. The figure shows probability distributions for j given the peak height, for a power spectrum not far removed from Cold Dark Matter or Massive Neutrino models with n=1. (γ measures the steepness of the smoothed spectrum, and is defined by Bardeen *et al.*).

Fig. 2. The distribution of $J/M^{5/3}$. This quantity is useful for comparison with observation, because the normalisation of the ordinate is independent of almost everything except one characteristic collapse epoch, and is related to the dimensionless spin parameter λ. There is a systematic shift to lower values for higher peaks (for all power spectra).

Fig. 1

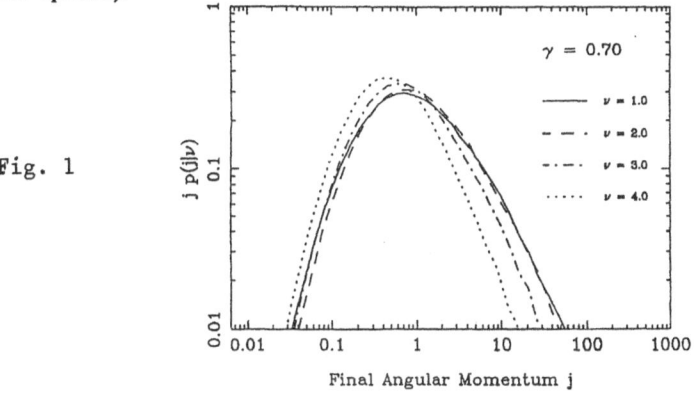

Final Angular Momentum j

Fig. 2

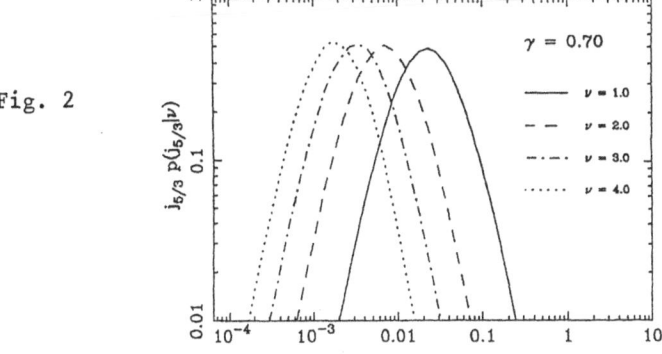

Final Angular Momentum Measure $j_{5/3}$

THE MORPHOLOGICAL LUMINOSITY FUNCTION

J.L. Sanz and E. Martínez-González
Departamento de Física Moderna
Universidad de Cantabria
39005-Santander, Spain.

ABSTRACT. We obtain the morphological luminosity function for galaxies assuming that they form at high density regions of a filtered Gaussian random field. We compare our results with some observations of the LF for E and E+S0 galaxies. Using scale-free spectra, we obtain bell-shaped curves that follow the observations.

1. INTRODUCTION

The luminosity function LF, for all the galaxies is well represented by the Schechter function $\phi(L) \propto L^{\alpha} e^{-L/L_*}$, with the parameters $\alpha \simeq -1$ and $L_* \simeq 10^{10} L_\odot$. On the other hand, observations of the LF for different Hubble types indicate bell-like shape distributions once the dwarf galaxies are eliminated from the corresponding sample. Recently Sandage et al. (1985) have studied the different morphological LF for galaxies in the Virgo cluster, fitting the observations to Gaussian distributions in several cases. Our aim is to obtain the LF within the scheme of biased scenarios, with different filtering scales and thresholds for the different morphological types, following the results of Martínez – – González and J.L. Sanz (1988),

2. THE DISTRIBUTION OF INTERVALS ALONG A LINE

Assuming that the initial density fluctuations can be represented by a Gaussian random field and defining the "seed" regions that are able of collapsing to form galaxies in terms of a global bias ν and a filtering scale R(following Kaiser 1984, structures form only at high density regions), we are able to calculate the distribution of intervals between sucessive ν-values along any arbitrary line in the 3D-case (Martínez – – González and Sanz 1987). Rice (1954) and Longuet-Higgins (1957) calculate this distribution for the 1D and 2D-cases, respectively, and the level $\nu = 0$.

N. Kaiser and A. N. Lasenby (eds.), The Post-Recombination Universe, 375–377.

3. THE MASS DISTRIBUTION

The thresholds that will be used here are high enough to consider the
isodensity regions (i.e. intersections of the random field with the
level ν) to be approximately spherical. This is also supported by 2D
numerical simulations of Gaussian random fields (Appel 1987, Coles
1988). using the spherical contours we are able to calculate the probability density of their radii P(R) from the probability density of
intervals F(l) calculated in the previous section. The result is

$$P(R) = -\frac{1}{2} R^2 \frac{d}{dR}\left[R^{-1} F(R)\right] \quad .$$

Now, the mass probability density can be easily calculated
using the relation $m = \frac{4}{3}\pi R^3 \rho_b$, where ρ_b is the background density. If
we assume a constant m/L ratio, where L is the luminosity, we can
obtain the probability density for the luminosity and therefore for the
absolute magnitude M.

4. RESULTS

Two free parameters appear in our approach that we choose to be the
level ν and the mean luminosity $\langle L \rangle$. We have considered scale-free
spectra $P(k) \propto k^n$, n=0,-2, which correspond to white noise an to the
effective index associated to standard CDM models at the scale of galaxies. The comparison with the observations by Sandage et al. (1985) for
E and E+SO galaxies is shown in the figures below. The curves a refer
to an index n=-2 whereas b is related to n=0. In the four cases we
have considered a threshold ν =2.5. It is clear that the case n=-2 is a
better fit for both observations, where we find a wider curve as compared to the n=0 case. The upper arrow indicates where our approach
break down because the distribution of intervals is valid only until
$l \simeq 6 \langle l \rangle$. More detailed calculations and comparison with other

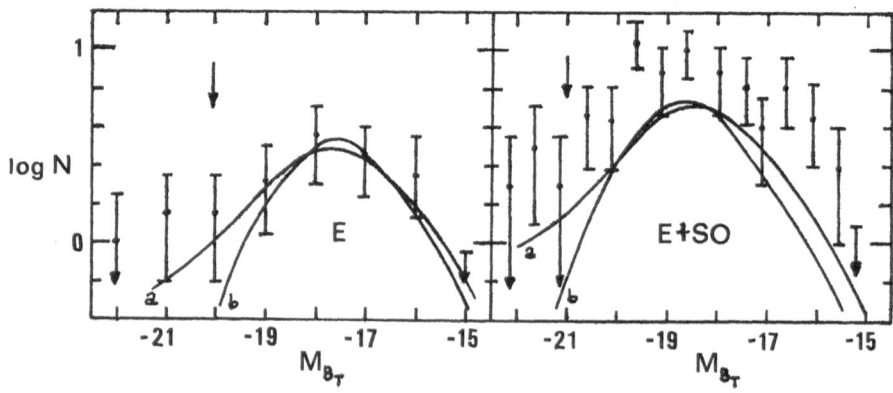

observations will be given elsewhere (E. Martínez-González and J.L. Sanz 1987 b).

REFERENCES

Apple, L. (1987), private communication.
Coles, P. in Proceedings of NATO Advanced Study Institute, "The post-
    -recombination Universe", Kaiser, N. and Lasemby, A.N. eds.,
    Reidel (1988).
Kaiser, N. (1984), Ap. J. (Letters) **284**, L9.
Longuet-Higgins, M.S. (1957), Phil. Trans. Roy. Soc. London **A 249**, 321.
Martínez-González, E. and Sanz, J.L. (1987 a) in Proceedings of I.A.U.
    Symposium no. 130, "The Structure of the Universe", Balatonfured,
    Hungria, Reidel 1988, in press.
Martínez-Gonźalez, E. and Sanz, J.L. (1987 b), Ap. J. submitted.
Rice, S.O. (1954), in "Selected papers on noise and stochastic proce-
    sses", ed. N. Wax, Dover Publ., New York, p. 133.
Sandage, A., Binggeli, B. and Tammann, G.A. (1985) , Ap. J. **90**, 1759.

Index